EARTH SYSTEM SCIENCES

EARTH SYSTEM SCIENCES

Editor

Anđela Vasilescu

scitus
academics

Earth System Sciences
Edited by **Anđela Vasilescu**

This Book is Printed on Acid-Free Paper. ♾

ISBN: 9781681178943
Library of Congress Control Number: 2017948922

SCITUS Academics LLC,
616, Corporate Way, Suite 2, 4766,
Valley Cottage, NY 10989

www.scitusacademics.com

Notice

Printed in United States of America.

Preface

Earth system science (ESS) is the application of systems science to the Earth sciences. In particular, it considers interactions between the Earth's "spheres"—atmosphere, hydrosphere, cryosphere, geosphere, pedosphere, biosphere, and, even, the magnetosphere—as well as the impact of human societies on these components. Earth system science assumes a holistic view of the dynamic interaction between the Earth's spheres and their many constituent subsystems, the resulting organization and time evolution of these systems, and their stability or instability. Earth changes its shape without a break. Earthquakes occur while this is happening. Earthquakes can hurt people badly. In order to prevent their undesire effects geophysicists, seismologists to be more precise, try to understand dynamic system of Earth. They locate faults that may produce earthquakes. Earth science affects our everyday lives. For example, meteorologists study the weather and watch for dangerous storms. Hydrologists study water and warn of floods. Seismologists study earthquakes and try to predict where they will strike. Geologists study rocks and help to locate useful minerals. Geophysicists and geologists has also vital role to determine near surface structures. The studies of Earth's history and of the physical and chemical properties of the substances that make up our planet, are of great significance to our understanding both of its past and its future. The geological and other environmental processes on Earth and the composition of the planet are of vital importance in locating and harnessing its resources.

This book is intended for research scholars, geologists, civil engineers, mining engineers, and environmentalists. It includes the coverage on the solid earth (geosphere), the atmosphere, the hydrosphere, and the biosphere; it also addresses related aspects of planetary and space sciences. The text also will be used by students, and it will continue to be of value to them throughout their subsequent professional and research careers. Hopefully, this work will provide the reader with a useful foundation for discussing and evaluating specific environmental issues, as well as for developing ideas for problem solving.

Table of Contents

CHAPTER 1

Recognition of geochemical footprints of mineral systems in the regolith at regional to continental scales

P. de Caritat[a,b], P. T. Main[a], E. C. Grunsky[c] & A. W. Mann[d]

[a]*Geoscience Australia, GPO Box 378, Canberra, ACT 2601, Australia*

[b]*Research School of Earth Sciences, Australian National University, Canberra, ACT 2601, Australia*

[c]*Department of Earth and Environmental Sciences, University of Waterloo, Waterloo, Canada N2L 3G1*

[d]*PO Box 778, South Fremantle, WA 6162, Australia*

ABSTRACT

Understanding the character of Australia's extensive regolith cover is crucial to the continuing success of mineral exploration. We hypothesise that the regolith contains geochemical fingerprints of processes related to the development and preservation of mineral systems at a range of scales. We test this hypothesis by analysing the composition of surface sediments within greenfield regional-scale (southern Thomson Orogen) and continental-scale (Australia) study areas. In the southern Thomson Orogen area, the first principal component (PC1) derived in our study [Ca, Sr, Cu, Mg, Au and Mo at one end; rare earth elements (REEs) and Th at the other] is very similar to the empirical vector used by a local company (enrichment in Sr, Ca and Au concomitant with depletion in REEs) to successfully site exploration drill holes for Cu–Au mineralisation. Mapping of the spatial distribution of PC1 in the region reveals several areas of elevated values and possible mineralisation potential. One of the strongest targets in the

PC1 map is located between Brewarrina and Bourke in northern New South Wales. Here, exploration drilling has intersected porphyry Cu–Au mineralisation with up to 1 wt% Cu, 0.1 g/t Au, and 717 ppm Zn. The analysis of a comparable geochemical dataset at the continental scale yields a compositionally similar PC1 (Ca, Sr, Mg, Cu, Au and Mo at one end; REEs and Th at the other) to that of the regional study. Mapping PC1 at the continental scale shows patterns that (1) are spatially compatible with the regional study and (2) reveal several geological regions of elevated values, possibly suggesting an enhanced potential for porphyry Cu–Au mineralisation. These include well-endowed mineral provinces such as the Curnamona and Capricorn regions, but also some greenfield regions such as the Albany-Fraser/western Eucla, western Murray and Eromanga geological regions. We conclude that the geochemical composition of Australia's regolith may hold critical information pertaining to mineralisation within/beneath it.

Keywords: Australia, mineral systems, regolith, geochemistry, National Geochemical Survey of Australia, multivariate statistics, compositional data analysis

1. INTRODUCTION

Two of the great challenges to mineral exploration, particularly in Australia, are (1) gaining a better understanding of the almost ubiquitous regolith cover and (2) making better use of this knowledge to unearth new mineral wealth within or beneath this cover. These challenges arise from the fact that the efficiency of most traditional geophysical and geochemical exploration methods, initially developed for areas of bedrock outcrop and shallow subcrop, is blunted by the thickness of the intervening weathered and/or transported material.

We hypothesise here that one way of characterising the cover and studying distal footprints of mineral systems, both national priorities of the UNCOVER Initiative (http://www.uncoverminerals.org.au/), is through surface (regolith) geochemical surveys. We test the above hypothesis by analysing the composition of surface sediments within greenfield regional-scale (southern Thomson Orogen) and continental-scale (National Geochemical Survey of Australia or NGSA) study areas. The present study is to our knowledge the first attempt at investigating the upscaling of a regional Australian geochemical survey to the continental scale, and drawing implications from this for mineral exploration.

2. BACKGROUND

2.1 CRC LEME Thomson region geochemical survey

During 2005–2006, 'the Cooperative Research Centre for Landscape Environments and Mineral Exploration (CRC LEME)' carried out surface regolith sampling over about ten 1:250 000 scale map sheets in northwestern New South Wales, in collaboration with Geoscience Australia (GA) and the Geological Survey of New South Wales (GSNSW) (Caritat & Lech, 2007; Lech & Caritat, 2007). This survey was one of a number of pilot projects aimed at defining fit-for-purpose methodologies for a potential future national survey. A few sampling media were targeted; most pertinent here are the shallow catchment outlet sediment samples (10–25 cm depth), which were collected at 79 sites for Mobile Metal Ion® (MMI) analysis. Quality control data are reported in Caritat and Lech (2007).

2.2 Southern Thomson Orogen regional geochemical survey

The Southern Thomson Project (http://www.ga.gov.au/scientific-topics/minerals/unlocking-resource-potential/southernthomson) is a collaboration between GA, GSNSW and the Geological Survey of Queensland (GSQ). It is aimed at better understanding the mineral systems potential of this region within the southern Thomson Orogen straddling the New South Wales–Queensland border. As part of this project, a surface regolith geochemical survey was conducted over the Eulo and Toompine 1:250 000 scale map sheets in Queensland, with the aims of (1) characterising the geochemical composition of the regolith over the area, and (2) identifying any surface expression of basement lithologies and/or mineral systems detectable from the surface (Main & Caritat, 2016). Fieldwork for this phase of the project took place during October–November 2013 and included the collection of 41 Top Outlet Sediment (TOS; 0–10 cm depth) samples for MMI analysis. Quality control data are reported in Main and Caritat (2016).

2.3 Combined southern Thomson regional geochemical survey

In the following, the combined CRC LEME and GA-GSNSW-GSQ surface geochemistry projects are collectively referred to as the southern Thomson (geochemical) survey for brevity. In total 120 (79 CRC LEME + 41 GA-GSNSW-GSQ) near-surface samples were taken over the combined area of ~185 000 km^2, yielding an average sample density of ~1/1540 km^2. A list of

30 elements common to both datasets and with comparable analytical ranges and distributions was collated (see Methods). A simplified geological map of the area also showing relevant mines and mineral occurrences is available in the Supplementary Papers (Figure SP1).

2.4 National Geochemical Survey of Australia

The NGSA was an ultra-low-density geochemical survey of the surface regolith of Australia, aimed at bridging a gap in the documentation and understanding of the multi-element composition of Australia's surface regolith to match its well-known and utilised continental-scale geophysical datasets (e.g. magnetic, radiometric and gravity datasets). Samples of transported sediment were collected from near the outlets of »1200 large river catchments at depths of 0–10 cm (TOS) and »60–80 cm (Bottom Outlet Sediments; BOS). The NGSA covered over 80% of Australia (>6 million km2) at an average sample density of »1/5200 km2 (Caritat & Cooper, 2011a, 2016; Lech, Caritat, & McPherson, 2007). The TOS samples analysed by MMI at 1191 sites are discussed herein. The same 30 elements common to the southern Thomson regional dataset were selected (see Methods). A simplified geological map of the area also showing relevant mines, mineral deposits and mineral occurrences is available in the Supplementary Papers (Figure SP2).

2.5 Mineral systems concept

Although ore bodies themselves can be less than a few metres or tens of metres in width, the processes leading to their formation and preservation, summarised by the 'mineral systems' concept of Wyborn, Heinrich, and Jaques (1994) (Figure 1a), commonly affect a much larger area. Thus, relatively small, high-value end products, the economically defined ore zones, can derive from processes operating at large scales potentially leaving extensive footprints, for instance when mineralising fluids permeate a suitable lithology leaving behind a tell-tale legacy of hydrothermally altered rock indicating the mineralisation corridor. In a more recent version of the mineral systems model, the temporal dimension and the critical spatio-temporal window for ore deposition are introduced (Huston et al., 2016; Figure 1b). When sampling surface sediments at low density, such as outlet sediments in large catchments, the objective is more to test if large-scale elements of the mineral system were intersected rather than to detect the precise location of ore bodies. The latter must be the objective of follow-up work. It is worth noting that parts of a mineral system may be under-sampled by geochemical surveys, especially those applying an ultra-lowdensity approach.

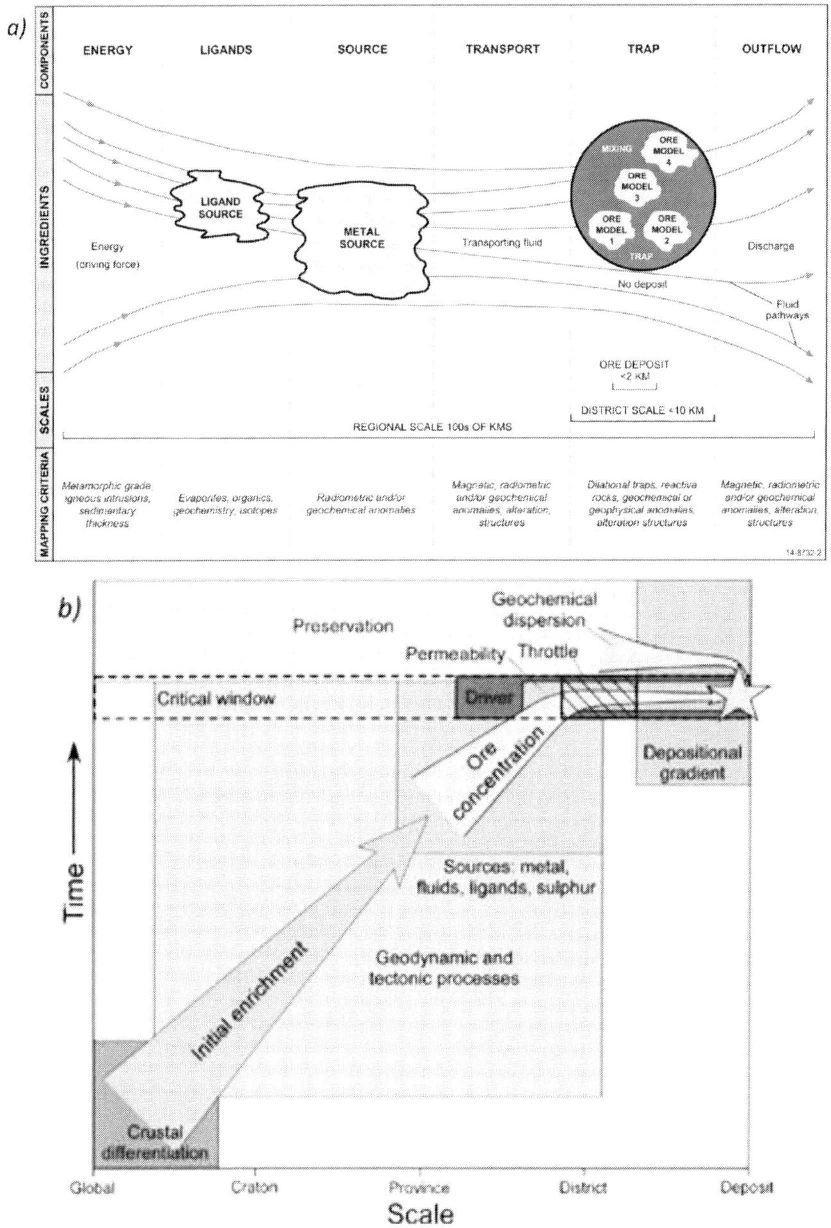

Figure 1. (a) Schematic representation of the original mineral systems concept showing the range of spatial scales involved and mapping criteria (modified after Knox-Robinson and Wyborn, 1997); and (b) casting of the conceptual model in a time–space context (Huston et al., 2016).

The mineral systems approach is gaining in popularity as a useful concept (e.g. Blewett, 2014; Huston et al., 2016; Jaques, Jaireth, & Walshe, 2002; McCuaig, Beresford, & Hronsky, 2010; Murphy et al., 2011) and is a practical link for the application of (low-density) surface geochemical mapping to mineral exploration (Reimann, Ladenberger, Birke, & Caritat, 2016).

2.6 Geochemical footprints of mineral systems in the regolith

Mineral systems processes such as mineralisation and alteration modify the mineralogical and geochemical composition of rock. Through weathering over large timespans, distinct geochemical characteristics can be imparted to the regolith either directly by in situ weathering of modified (e.g. mineralised/ altered) rock, or via transport of this modified geochemical signature in the bedrock through the overlying regolith (e.g. Anand, Aspandiar, & Noble, 2016). The existence of such geochemical haloes at or near the surface is well established in a variety of environments, although the exact formation mechanisms are still debated and most likely specific to individual conditions (e.g. Cameron, Hamilton, Leybourne, Hall, & McClenaghan, 2004; Kelley, Kelley, Coker, Caughlin, & Doherty, 2006). It is clear that geochemical anomalies can propagate through (even transported) regolith cover via pore water (e.g. seismic pumping), bubbles, plant roots, capillary pressure, animals (e.g. termitaria), nanoparticles, or electro-chemical cells (op. cit.; Anand et al., 2016; Hamilton, 1998). Once at or near the surface, lateral mechanical transport, including colluvial, alluvial and eolian, can further disseminate the geochemical signature. Tailored geochemical protocols and methods, including hydrogeochemistry, biogeochemistry, soil gas and soil hydrocarbon chemistry, and weak and/or selective geochemical extractions, such as MMI (Mann, 2010), have been developed over recent decades to enhance the detection of these subtle elemental or isotopic fingerprints, which have to be detected against the background of the natural geochemical signature of the regional weathering rocks at the site or within the catchment.

3. METHODS

In this contribution, we focus on the surface sediment samples and the MMI method of extraction. The MMI method was designed to extract only weakly bound elements adsorbed on the surfaces of soil particles (e.g. Fe-oxyhydroxides, clay minerals, organic matter), based on the idea that these elements could have migrated up through a regolith profile (including potentially sedimentary basin sequences) that developed post-mineralisation (e.g. Mann, 2010; Mann, Birrell, Mann, Humphreys, & Perdrix, 1998). In both the regional (southern Thomson) and national (NGSA) studies, comparable sample collection, preparation and analysis methods were applied, allowing for a comparison of results at the regional and continental scales, a first for Australia. Briefly, topsoil samples were collected at the outlets of large catchments targeting floodplain or overbank landforms (with an inevitable and variable contribution of eolian material being recognised locally). The samples were air-dried, homogenised and sieved to a coarse (<2 mm) fraction (approximating 'whole rock' composition) before being digested with the multi-element MMI reagent and analysed by inductively coupled plasma-mass spectrometry at the ALS/SGS Laboratories in Perth, Western Australia. A comprehensive quality control and assessment program was integral to each study from inception to reporting. Data and analysis quality between the two datasets were deemed comparable; for more detail, please refer to the original reports from the CRC LEME (Caritat & Lech, 2007), southern Thomson project (Main & Caritat, 2016) and NGSA (Caritat & Cooper, 2011b; Caritat, Cooper, Lech, McPherson, & Thun, 2009; Caritat, Cooper, Pappas, Thun, & Webber, 2010; Lech, Caritat, & McPherson, 2007) projects.

The resulting MMI compositional data from the regional southern Thomson and continental NGSA geochemical surveys were analysed statistically and geospatially to produce maps of element associations, some of which we interpret as being indicative of components of mineral systems (Wyborn et al., 1994). The following data analysis methodology was applied:

1) A list of elements common to both the regional and national datasets was established, from which elements with an excessive proportion (>»30%) of censored values (i.e. below the lower level of detection, LLD), or those deemed not fit for purpose (i.e. with poor precision or accuracy), were removed. This resulted in 30 elements (Ag, Au, Ba, Ca, Cd, Ce, Co, Cu, Er, Gd, La, Li, Mg, Mn, Mo, Nd, Ni, Pb, Pr, Rb, Sc, Sm, Sr, Tb, Th, Ti, U, Y, Yb, Zn) being retained in both datasets.

2) A statistical approximation of the censored or missing values was performed in a step called imputation. This was achieved using the nearest-neighbour methodology of the 'impRZilr' function of the 'robCompositions' package (Hron, Templ, & Filzmoser, 2010) in the R statistical language and environment (CRAN, 2016). The method replaced censored values with robust estimates of concentrations, generally below the LLD for the element, but occasionally by values slightly above the LLD. This was considered acceptable given the lower precision of element determination at concentrations near the LLD.

3) In accordance with the contemporary practice known as compositional data analysis (CoDA), the raw concentration values (in parts per

million, or ppm) were transformed in order to circumvent compositional data limitations of closure, relativity and non-normality (Aitchison, 1999; Reyment, 1989); here the centred logratio (clr) method was applied. Without such transformation, classic statistical methods may not be applicable to compositional data without a risk of distortion and misinterpretation of results (McKinley et al., 2016).

4) The multivariate statistics method of principal component analysis (PCA; e.g. Jolliffe, 1986) was applied to the clr-transformed data in order to (1) reduce dimensionality from 30 non-independent variables (elements) to a more manageable set of orthogonal (independent) principal components (PCs) that are linear combinations of the elements and account for the bulk of the data variability, and (2) reveal the element associations that are statistically the most significant. The dominant PCs generally represent 'structure' in the data, which is controlled by mineral stoichiometry and reflects dominant geological processes. Thus, PCA has been deemed a useful 'process discovery' tool (Caritat & Grunsky, 2013).

5) The PCs accounting for the most variability in the data, generally the first four to six PCs, were mapped and geospatially interpolated using ordinary kriging (spherical semi-variogram models, variable searching radius, 12 nearest input sample points, using ArcGIS v.10.2) to produce continuous, coloured raster maps of PC values. These maps were then interpreted in the context of potential mineralisation footprints. Here we focus on the results for the first principal component (PC1), while other PCs, some of which pertain to lithology, weathering, mineralisation, etc., are discussed elsewhere (e.g. Caritat & Grunsky, 2013; Main & Caritat, 2016).

4. RESULTS AND DISCUSSION

4.1 Multivariate statistical analysis

Table 1 presents the summary statistics for the above 30 elements analysed after MMI extraction in both the regional (southern Thomson) and continental (NGSA) studies. As expected the range (max¡min) of most elements is larger (commonly by a factor of two or more) for the NGSA than for the southern Thomson datasets. This occurs because the larger an area of investigation, the more likely one is to sample significantly different lithologies, regolith materials and environmental conditions in general.

Table 1. Summary statistics for 30 elements analysed after MMI extraction in the regional (southern Thomson) and continental (NGSA) studies.

	Southern Thomson ($N = 120$)					NGSA ($N = 1191$)				
Element	LLD	Min	Median	Mean	Max	LLD	Min	Median	Mean	Max
Ag	0.001	0.0004	0.004	0.007	0.070	0.001	0.0001	0.005	0.011	0.302
Au	0.0001	0.00004	0.0004	0.0004	0.0014	0.0001	0.00001	0.0002	0.0004	0.020
Ba	0.01	0.19	2.51	3.14	12.5	0.01	0.01	0.87	1.41	15.3
Ca	10	176	638	634	1470	10	7.7	390	423	3680
Cd	0.001	0.0006	0.004	0.005	0.023	0.001	0.0002	0.005	0.01	1.75
Ce	0.005	0.002	0.091	0.162	1.23	0.005	0.001	0.101	0.39	12.1
Co	0.005	0.004	0.232	0.289	1.14	0.005	0.002	0.178	0.26	2.58
Cu	0.01	0.29	1.03	1.18	4.29	0.01	0.01	0.96	1.16	30.5
Er	0.0005	0.0006	0.021	0.028	0.131	0.0005	0.0004	0.0112	0.038	2.23
Gd	0.001	0.002	0.060	0.073	0.364	0.0005	0.0006	0.029	0.07	1.61
La	0.001	0.0001	0.015	0.039	0.273	0.001	0.0002	0.027	0.12	3.31
Li	0.005	0.002	0.010	0.014	0.118	0.005	0.0003	0.008	0.019	0.788
Mg	1	45	240	260	2000	1	2	140	193	2920
Mn	0.01	0.16	6.8	9.2	72.5	0.005	0.079	3.99	6.83	59.7
Mo	0.005	0.002	0.009	0.011	0.036	0.005	0.0003	0.007	0.011	0.276
Nd	0.001	0.0004	0.085	0.143	0.961	0.001	0.0008	0.092	0.27	6.87
Ni	0.005	0.071	0.450	0.543	2.70	0.005	0.006	0.411	0.723	16.1
Pb	0.01	0.002	0.05	0.07	0.28	0.01	0.002	0.04	0.19	5.9
Pr	0.001	0.0001	0.012	0.021	0.158	0.001	0.0002	0.016	0.054	1.25
Rb	0.005	0.022	0.071	0.082	0.259	0.00025	0.003	0.107	0.146	2.63
Sc	0.005	0.002	0.009	0.014	0.136	0.005	0.0007	0.008	0.051	6.84
Sm	0.001	0.0002	0.030	0.045	0.288	0.001	0.0004	0.025	0.063	1.34
Sr	0.01	0.95	6.24	6.82	23.8	0.01	0.05	2.40	3.52	35.6
Tb	0.001	0.0002	0.008	0.010	0.051	0.001	0.0002	0.004	0.011	0.247
Th	0.0005	0.0005	0.015	0.026	0.166	0.0005	0.0002	0.021	0.101	3.52
Ti	0.003	0.002	0.030	0.088	2.16	0.003	0.002	0.019	0.09	26.5
U	0.001	0.001	0.036	0.045	0.315	0.001	0.001	0.063	0.121	3.49
Y	0.005	0.004	0.172	0.219	1.00	0.005	0.003	0.098	0.29	4.16
Yb	0.001	0.0006	0.014	0.018	0.081	0.001	0.0003	0.007	0.027	1.56
Zn	0.02	0.02	0.07	0.13	0.80	0.02	0.01	0.12	0.48	19.5

LLD: lower level of detection; Min: minimum (after imputation; see text); Max: maximum.

The only exception noted to this trend is for Mn, which has a »20% larger range, and indeed a maximum value also, in the southern Thomson than in the NGSA datasets. The medians (robust estimators of the central values) from both studies are within a factor of two of each other for most elements, with greater discrepancies noted only for Ba and Sr (median southern Thomson/median NGSA of 2.6 and 2.9, respectively). This characteristic is believed to reflect the fact that the majority of the southern Thomson samples are from the Eromanga geological region, which has been noted previously to be highly enriched in MMI Ba and Sr (Caritat, Cooper, Mann, & Prince, 2011), consistent with a dominantly sedimentary environment.

The results of the PCA conducted (separately) on both datasets are presented in Table 2 for PC1, with the scaled and ordered loadings (e.g. Joliffe, 1986) also illustrated in Figure 2. For the southern Thomson dataset (Figure 2a), PC1 has the most negative scores for Sm (lowest), Nd, Pr, Th, La and Ce and the most positive scores for Mo, Au, Mg, Cu, Sr and Ca (highest). The total variance explained by PC1 here is 49.3%. For the NGSA dataset (Figure 2b), PC1 has the most negative scores for Sm (lowest), Pr, Nd, Gd, Tb, Ce, La and Th and the most positive scores for Mo, Au, Cu, Mg, Sr and Ca (highest). The total variance explained by PC1 here is 53.3%.

The associations found at the negative ends of the PC1 scores of both datasets are mainly light lanthanides or rare earth elements (REEs), with the exceptions of Tb, which is a heavy REE, and Th, which is an actinide. REEs commonly behave in a very similar way in geological environments due to their comparable physico-chemical characteristics (Henderson, 1984), and thus this clustering of REEs is not surprising. Light REEs typically are more abundant in granitoid rocks and shales than in other rock types; they are typically hosted in heavy minerals/resistates (e.g. monazite, xenotime, and allanite) or in lesser concentrations but in more ubiquitous minerals such as feldspars, biotite and pyroxenes (e.g. Reimann & Caritat, 1998). During weathering, the relative abundance of REEs in the parent rock can be preserved in the soil profile, or alternatively a slight enrichment in the less mobile light REEs can be observed, and the mobile fraction of REEs (i.e. that not locked into primary silicates or resistates) can be washed down the profile by (acidic) precipitation (e.g. Ling et al., 2015; Mina_rik, _Zigov_a, Bendl, Sk_rivan, & _St'astn_y, 1998; Nesbitt, 1979). REEs have been found to adsorb readily on clay minerals, organic matter and metal oxyhydroxides in soil profiles (e.g. €Ohlander, Land, Ingri, & Widerlund, 1996; Prud^encio, Figueiredo, & Cabral, 1989).

Table 2. Loadings resulting from the PCA conducted for 30 elements analysed after MMI extraction in the regional (southern Thomson) and continental (NGSA) studies

Element	Southern Thomson	NGSA
Ag	0.199	0.151
Au	0.209	0.193
Ba	0.181	0.160
Ca	0.244	0.227
Cd	0.144	0.143
Ce	−0.213	−0.229
Co	0.177	0.157
Cu	0.233	0.208
Er	−0.123	−0.185
Gd	−0.164	−0.230
La	−0.225	−0.226
Li	0.194	0.183
Mg	0.231	0.216
Mn	0.046	0.081
Mo	0.200	0.184
Nd	−0.248	−0.232
Ni	0.197	0.175
Pb	0.148	−0.027
Pr	−0.239	−0.234
Rb	0.040	0.093
Sc	−0.080	−0.171
Sm	−0.249	−0.237
Sr	0.240	0.227
Tb	−0.202	−0.229
Th	−0.229	−0.220
Ti	−0.068	−0.015
U	0.088	0.019
Y	−0.163	−0.217
Yb	−0.082	−0.171
Zn	0.010	0.031

The associations found at the positive ends of the PC1 scores of both datasets are mainly reminiscent of carbonaterelated alkaline earth metals (Ca, Sr and Mg) and ore-related\ precious/transition metals (Au, Cu and Mo). The alkaline earth metals association is consistent with the geochemical signature expected from sedimentary rocks (Caritat et al., 2011) and/or soil carbonate/calcrete. Note that the formation of carbonate in the regolith occurs under oxidising and alkaline conditions and can be biomediated (McQueen, Hill, & Foster, 1999; Reith et al., 2011; Schmidt Mumm, & Reith, 2004). The precious/transition metals association is likely a signature of ore-related elements having moved through the alluvial/colluvial sediments post-deposition (Anand et al., 2016; Mann,

Birrell, Fedikow, & Souza, 2005) and is generally consistent with the regolith signature of mineralisation including porphyry- associated deposits such as Cadia-Ridgeway and Northparkes (McQueen, 2005). In conclusion, it appears that PC1 dominantly reflects weathering (both dissolution/bleaching as revealed by the REEs, and neoformation of regolith carbonate), with an intriguing addition of metals possibly reflecting mineralisation.

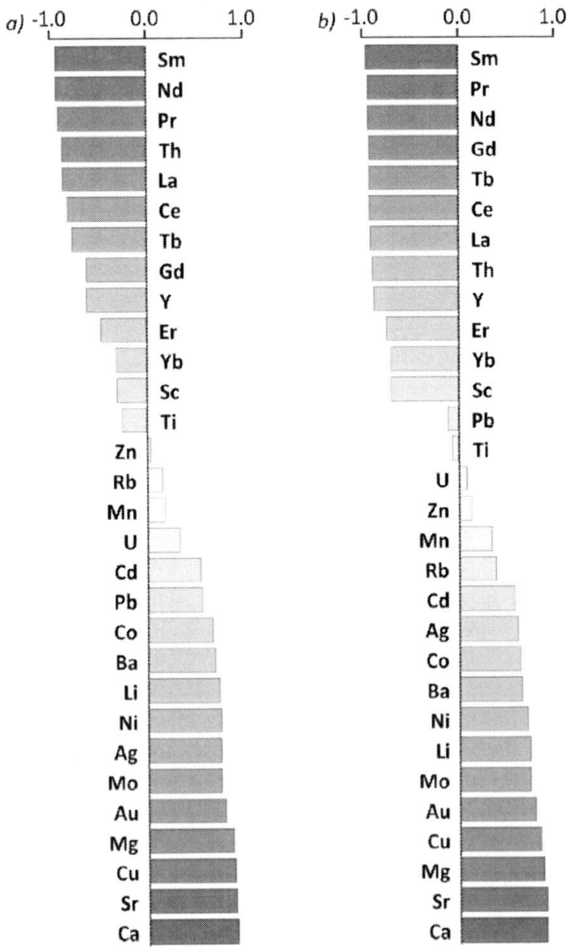

Figure 2. Loadings resulting from the PCA conducted for 30 elements analysed after MMI extraction.

4.2 Recognition of mineral systems footprints in the southern Thomson region

A local mineral exploration company also using MMI geochemistry in the southern Thomson region (at higher spatial resolution) has developed an empirical geochemical vector of enrichment in Sr, Ca and Au concomitant with depletion in REEs (without multivariate compositional statistical analysis) to successfully site drill holes targeting Cu–Au mineralisation in the region (J. Macauley, personal communication, 2013). This empirical vector coincidentally is very similar to PC1 developed above using multivariate compositional statistics. Based on this, the map of the spatial distribution of PC1 across the southern Thomson study area (Figure 3a) is hypothesised to have the potential to show enhanced prospectivity for Cu–Au mineralisation (warm colours). In a mineral systems context, this map potentially shows areas of metal deposition or enrichment, with a particular emphasis on the potential for porphyry Cu and/or Au mineralisation.

The above hypothesis is further supported by the location of the Warraweena mineral occurrence (Figure 3b, tenements EL7252/7253) reported by another local mineral exploration company between Brewarrina and Bourke (http://www.thom sonresources.com.au/ projects/ porphyry-copper-gold), which falls within a region of high positive PC1 values as calculated above.

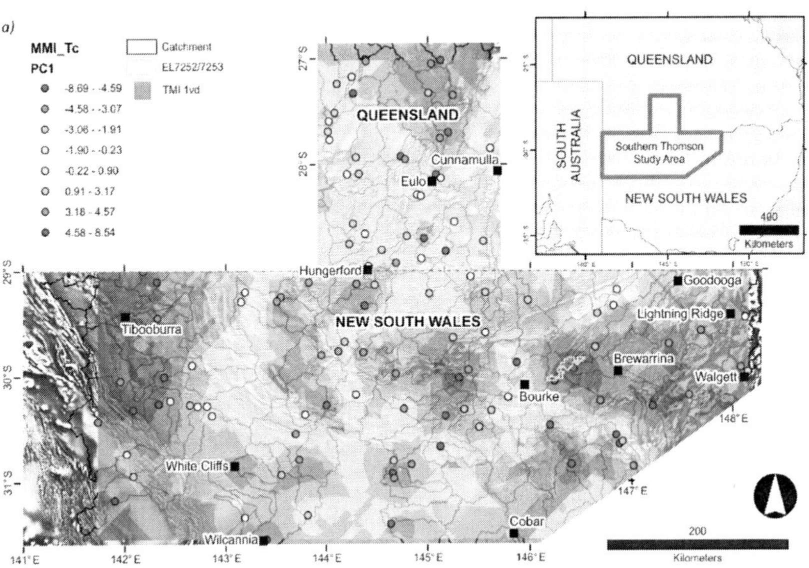

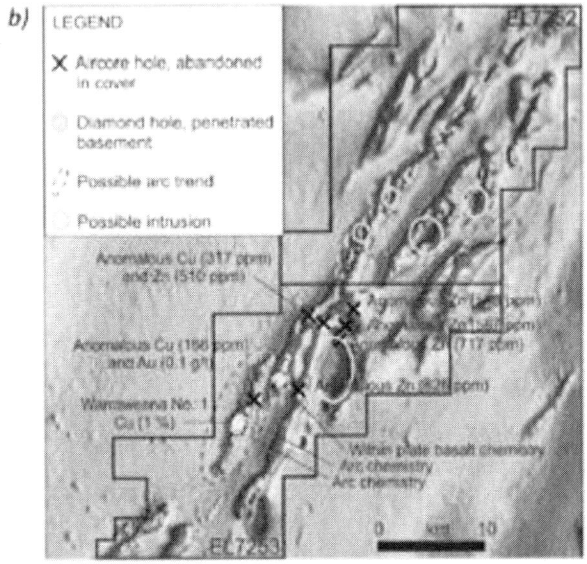

Figure 3. (a) Map of PC1 of the MMI extraction of coarse (<2 mm) surficial sediments from the southern Thomson survey study area. Values at sampling sites are shown by coloured circles; kriging interpolation is shown as a semi-transparent continuous raster based on the same colour scale (eight percentile-based classes). First vertical derivative of the total magnetic intensity (TMI 1vd) surface (Nakamura, 2015) is underlain (as shaded relief), as are catchment boundaries (as black/grey polygons). The green polygon shows the location of Thomson Resources Ltd tenements EL7252/7253, enlarged in (b), with the location of exploration drill holes over a TMI image (http://www.thomsonresources.com.au/ projects/ porphyry-copper-gold). Map (b) reproduced with permission from the Chief Executive Officer, Thomson Resources Ltd.

Here a diamond drill hole spudded by Mid Eastern Oil Ltd in 1965 (Warraweena no.1; Figure 3b) intersected mineralisation with up to 1 wt% Cu within the basement; three follow- up diamond drill holes drilled by Newcrest Mining Ltd failed to produce economic intersections (op. cit.). The basement sequence, including the informally named Warraweena volcanics of unknown age, contains very fine-grained lavas and volcaniclastic units of calc-alkaline basaltic andesite composition with island arc affinities similar to Ordovician rocks of the Macquarie Arc (Burton, Dadd, & Vickery, 2008; Glen et al., 2010), which host world-class porphyry Cu–Au deposits such as Cadia/Ridgeway and Northparkes in the adjacent Lachlan Fold Belt (http://www. thomsonresources.com.au/projects/porphyry-copper-gold).

Burton et al. (2008) concluded that if the interpretation of a volcanic arc present in the Bourke area under approximately 100 m of cover were correct, the potential of the region for hosting porphyry, epithermal to mesothermal vein and skarn style Cu–Au § Mo § Ag mineralisation could be enhanced. Thomson Resources Ltd followed this up with a detailed aeromagnetic survey leading to an aircore drilling campaign at Warraweena (Figure 3b), which resulted in further encouraging results, including intersections of up to 317 ppm Cu, 717 ppm Zn and 0.1 g/t Au (E. Rothery, personal communication, 2014; http://www.thom sonresources. com.au/projects/ porphyry-copper-gold).

If the patterns demonstrated by the MMI PC1 map of the regolith were indicative of areas with porphyry Cu–Au mineralisation potential in the basement (e.g. via hydromorphic dispersion and capillary rise during regolith carbonate formation), then several further regions of interest are revealed by the Thomson regional geochemical survey. These include the eastward continuation of the Warraweena area towards Walgett and spreading north and south; four regions in the centre of the study area (one »150 km northwest of Cobar, one »50–150 km west of Bourke, one »40 km northeast of Hungerford, and one »40 km northwest of Eulo); a region »80 km east of White Cliffs; around White Cliffs itself; around Wilcannia; and in the northwestern part of the area stretching from the Koonenberry Belt (»80 km west of White Cliffs and Wilcannia) north to Tibooburra. Historic mining and/or known mineral occurrences have been reported at some of these sites, e.g. the Albert Goldfield, and the Mount Brown and Warratta inliers in the Tibooburra–Milparinka region (McQueen, 2007). Interestingly, the well-endowed Cobar area is not highlighted by positive PC1 values; we believe this is the result of (1) the relatively elevated concentrations of REEs (e.g. Sm, Nd, La, Ce) and Th in this part of the map driving PC1 to negative values, and (2) the different style of mineralisation at Cobar to that inferred at Warraweena. Please also note the particularly low sampling density in this part of the map, implying a greater degree of uncertainty about interpretation here compared with other parts of the regional map.

4.3 Upscaling to the continental scale

Switching to the national scale, PC1 from the NGSA's MMI dataset, described above, was mapped in the same way as the regional southern Thomson PC1 map (Figure 4). Spatially, the 'opposing' associations of carbonate C mineralisation, as one end-member, and REEs C Th, as the other end-member, define patterns of warm and cool colours, respectively, across the continent (Figure 4). These continental-scale geochemical patterns indicate where each association can be expected to predominate in surficial (transported) regolith. The Ca– Sr–Mg–Cu–Au–Mo association, which is very similar to that found in the Thomson regional survey described above, occurs in the following geological regions, as defined by Blake & Kilgour (1998) (from west to east): Pilbara– Capricorn–northern Yilgarn, Albany-Fraser, western Eucla, eastern Eucla, western Amadeus, Victoria, Adelaide, Eromanga, southern Georgina–Isa, Curnamona–western Murray and southern Surat. Given the vast differences in scale and sampling density between the two surveys compared here, it is surprising how closely the regional and continental patterns (Figures 3a and 4) are consistent with one another (e.g. the national-scale map 'predicts' that the Thomson area will have 'highs' to the east, southwest and west and a northwesttrending 'low trough' coming up from the south through the middle of the study area; compare Figure 3a and Figure SP3 in Supplementary Papers). If this association were indeed indicative of Cu–Au mineral potential at the national scale, by extension of the regional case study from the southern Thomson region described above, then we suggest that some of these regions could be worthy of more detailed exploration effort and investment.

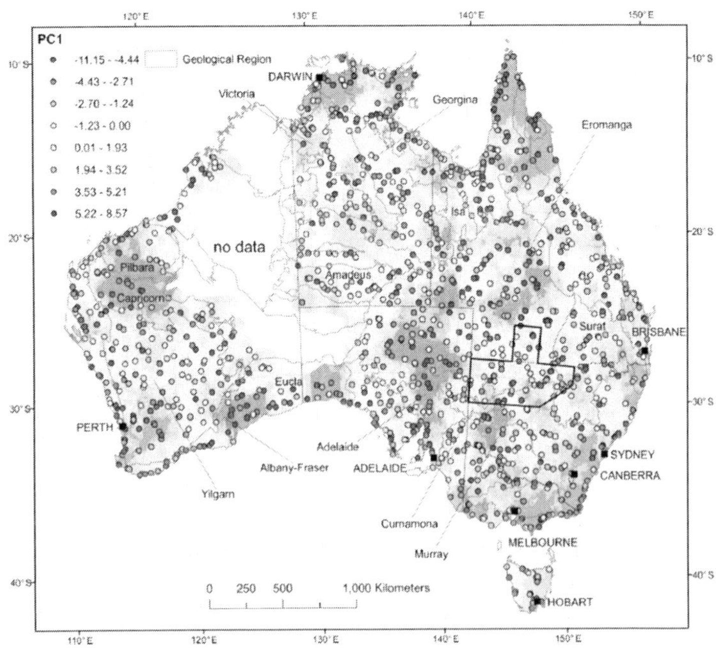

Figure 4. Map of PC1 of the MMI extraction of coarse (<2 mm) surficial sediments from the NGSA project study area. Values at sampling sites are shown by coloured circles; kriging interpolation is shown as a continuous raster based on the same colour scale (eight percentile-based classes). Geological regions (Blake & Kilgour, 1998) are overlain (as grey polygons); those discussed in the text are labelled. The southern Thomson project area (Figure 3) is shown by the black polygon.

It is noted that several of the areas highlighted in Figure 4 by warm colours are in regions of thick sedimentary basin cover (e.g. Eromanga Basin, Eucla Basin). Thus, confirmation of mineral potential would need to be established by followup investigations, including where possible using drill holes, in order to screen out any false positive anomalies. Similarly, it appears that the MMI PC1 indicator developed here is quite reliant on weathering processes, including the formation or presence of carbonate in the (near-)surface regolith. There is, however, more to it than just the presence of calcrete, as indicated by the spatial patterns being in places quite different from published regolith carbonate distribution maps (e.g. Wilford, Caritat, & Bui, 2015, and references therein). Regions known to have Cu–Au mineralisation (cfr. Figure SP2 in the Supplementary Papers) are not all highlighted particularly strongly by the method (e.g. Macquarie Arc in New South Wales), and this should also be the subject of further work.

5. CONCLUSIONS

Multivariate statistical analysis of (appropriately transformed) compositional data obtained on surface regolith in the southern Thomson region (northern New South Wales/southern Queensland) and across the continent using the same chemical extraction method (Mobile Metal Ion®) has shown some interesting results.

The composition of the first principal component is almost identical for both datasets, from the regional to the continental scales (Ca–Sr–Mg–Cu–Au–Mo association as one end-member; light REEs–Th as the other). This principal component is interpreted to reflect weathering processes with potentially an overprint from mineralisation as shown by the presence of the precious/transition metals.

The spatial patterns of both studies are surprisingly comparable over the southern Thomson region, given the considerable differences in scale and data density.

Further, exploration results (by a junior exploration company) and rigorous statistical analysis (this work) have independently yielded a remarkably similar geochemical indicator, one that is a proven empirical vector to mineralisation for that explorer on the ground and, additionally, that statistically accounts for the most variability in the data as shown by this study. Another mineral exploration company has reported encouraging drilling results from an area that this study shows to have high first principal component values.

If the Ca–Sr–Mg–Cu–Au–Mo association were indicative of the formation of soil carbonate with an overprint of elements dispersed from porphyry Cu–Au mineralisation in the Thomson region, as suggested by local exploration results, then a number of areas with this characteristic are highlighted and are potentially worthy of further investigations:

- At the regional scale: eastward continuation of the Warraweena area towards Walgett and spreading north and south; four regions in the centre of the study area (one ~150 km northwest of Cobar, one ~50–150 km west of Bourke, one ~40 km northeast of Hungerford, and one ~40 km northwest of Eulo); a region ~80 km east of White Cliffs; around White Cliffs itself; around Wilcannia; and in the northwestern part of the area stretching from the Koonenberry Belt to Tibooburra;
- At the national scale: Pilbara–Capricorn–northern Yilgarn, Albany-Fraser, western Eucla, eastern Eucla, western Amadeus, Victoria, Adelaide, Eromanga, southern Georgina–Isa, Curnamona–western Murray and southern Surat geological regions.

In terms of national exploration priorities, this study shows that potentially valuable information is held within the geochemical composition of surficial (transported) regolith, allowing discrimination between more and less prospective regions. This suggests that the geochemical characterisation of the cover in Australia (and the understanding of its evolution through time) is worthy of investment by industry, academia and government. Regolith geochemistry brings a value-adding complementary toolkit to consider in conjunction with geological, geophysical and geochronological datasets. The present study demonstrates a viable workflow to upscale prospectivity analysis from regional studies to a continental perspective for porphyry Cu–Au mineralisation types; a similar methodology could be applicable to other mineralisation types from other regional greenfield case studies with (sub-) economic drilling intersections.

ACKNOWLEDGEMENTS

The studies reported here would not have been possible without Commonwealth funding through the Cooperative Research Centre Program, the Onshore Energy Security Program, and Geoscience Australia appropriation. We thank all the landowners for granting access to their properties for the purposes of sampling, and the laboratory staff for assistance with preparing and analysing the samples. ALS/SGS Mineral Services, Perth, are acknowledged for providing the MMI analyses for the NGSA. We thank Ian Roach, Matilda Thomas, David Champion, Roger Skirrow, Karol Czarnota and Richard Blewett (Geoscience Australia) for their constructive reviews of the original manuscript, which was significantly improved as a result. David Cohen and two other anonymous journal referees and the Special Issue editors are acknowledged for their efforts in critically reviewing this paper. Figure 3b is reproduced with permission from the Chief Executive Officer, Thomson Resources Ltd. PdC and PTM publish with permission from the Chief Executive Officer, Geoscience Australia.

REFERENCES

1. Aitchison, J. (1999). Logratios and natural laws in compositional data analysis. Mathematical Geology, 31, 563–580. doi: 10.1023/A:1007568008032
2. Anand, R. R., Aspandiar, M. F., & Noble, R. R. P. (2016). A review of metal transfer mechanisms through transported cover with emphasis on the vadose zone within the Australian regolith. Ore Geology Reviews, 73, 394–416. doi: 10.1016/j.oregeorev.2015.06.018
3. Blake, D., & Kilgour, B. (1998). Geological Regions of Australia 1:5,000,000 scale [Data Set]. Geoscience Australia, Canberra. Retrieved from http://www.ga.gov.au/metadata-gateway/metadata/record/gcat_a05f7892-b237-7506-e044-00144fdd4fa6/GeologicalCRegionsCofCAustralia%2CC1%3A5C000C000Cscale
4. Blewett, R., (2014). UNCOVER: unlocking Australia's hidden mineral potential. Geoscience Australia Insights, 7 July 2014. Retrieved from http://www.ga.gov.au/news-events/news/latest-news/uncover-unlockingaustralias-hidden-mineral-potential
5. Burton, G. R., Dadd, K. A., & Vickery, N. M. (2008). Volcanic arc-type rocks beneath cover 35 km to the northeast of Bourke. Quarterly Notes of the Geological Survey of New South Wales, 127, 1–23. Retrieved from http://digsopen.minerals.nsw.gov.au/
6. Cameron, E. M., Hamilton, S. M., Leybourne, M. I., Hall, G. E. M., & McClenaghan, M. B. (2004). Finding deeply buried deposits using geochemistry. Geochemistry: Exploration, Environment, Analysis, 4, 7–32. doi: 10.1144/1467-7873/03-019
7. Caritat, P. de, & Cooper, M. (2011a). National Geochemical Survey of Australia: The Geochemical Atlas of Australia. Geoscience Australia Record, 2011/20, 557 pp (2 Volumes). Retrieved from http://www.ga.gov.au/metadata-gateway/metadata/ record/gcat_71973
8. Caritat, P. de, & Cooper, M. (2011b). National Geochemical Survey of Australia: Data Quality Assessment. Geoscience Australia Record, 2011/21, 478 pp (2 Volumes). Retrieved from http://www.ga.gov.au/metadatagateway/metadata/record/gcat_71971
9. Caritat, P. de, & Cooper, M. (2016). A continental-scale geochemical atlas for resource exploration and environmental management: the National Geochemical Survey of Australia. Geochemistry: Exploration, Environment, Analysis, 16, 3–13. doi: 10.1144/geochem2014-322
10. Caritat, P. de, & Grunsky, E. C. (2013). Defining element associations and inferring geological processes from total element concentrations in Australian catchment outlet sediments: multivariate analysis of continental- scale geochemical data. Applied Geochemistry, 33, 104–126. doi: 10.1016/j.apgeochem.2013.02.005
11. Caritat, P. de, & Lech, M. E. (2007). Thomson Region Geochemical Survey, Northwestern New South Wales. Cooperative Research Centre for Landscape Environments and Mineral Exploration Open File (Report, 145), 1021 pp C CD-ROM. Retrieved from http://crcleme.org.au/Pubs/OFRSindex.html
12. Caritat, P. de, Cooper, M., Lech, M., McPherson, A., & Thun, C. (2009). National Geochemical Survey of Australia: Sample Preparation Manual. Geoscience Australia

Record, 2009/08, 28 pp. Retrieved from http://www.ga.gov.au/metadata-gateway/metadata/record/gcat_68657

13. Caritat, P. de, Cooper, M., Mann, A., & Prince, P., (2011). Lithological signatures and exploration implications from overbank sediment sampling and MMI Analysis: preliminary findings from the National Geochemical Survey of Australia. 25th International Applied Geochemistry Symposium, Rovaniemi, Finland. 22–26 August 2011, Final Programme & Abstracts, 159. Retrieved from http://www.iags2011. fi/ technical_programme.htm
14. Caritat, P. de, Cooper, M., Pappas, W., Thun, C., & Webber, E., (2010). National Geochemical Survey of Australia: Analytical Methods Manual. Geoscience Australia Record, 2010/15, 22 pp. Retrieved from http://www.ga.gov.au/metadata-gateway/metadata/record/gcat_70369
15. CRAN (The Comprehensive R Archive Network) (2016). The Comprehensive R Archive Network. Retrieved from http://cran.r-project.org/
16. Glen, R. A., Saeed, A., Hegarty, R., Percival, I. G., Bodorkos, S., & Griffin, W. L. (2010). Preliminary Zircon Data and Tectonic Framework for the Thomson Orogen, Northwestern NSW. Geological Survey of New South Wales Report, GS2010/0379, 27 pp. Retrieved from http://digsopen.minerals.nsw.gov.au/
17. Hamilton, S. M. (1998). Electrochemical mass-transport in overburden: a new model to account for the formation of selective leach geochemical anomalies in glacial terrain. Journal of Geochemical Exploration, 63, 155–172. doi: 10.1016/S0375-6742(98)00052-1
18. Henderson, P. (Ed.) (1984). Rare Earth Elements Geochemistry. Developments in Geochemistry, vol 2. Amsterdam: Elsevier. ISBN 9780444421487
19. Hron, K., Templ, M., & Filzmoser, P. (2010). Imputation of missing values for compositional data using classical and robust methods. Computational Statistics & Data Analysis, 54, 3095–3107. doi: 10.1016/j. csda.2009.11.023
20. Huston, D. L., Mernagh, T. P., Hagemann, S. G., Doublier, M. P., Fiorentini, M., Champion, D. C., ...Bastrakov, E. (2016). Tectono-metallogenic systems – The place of mineral systems within tectonic evolution, with an emphasis on Australian examples. Ore Geology Reviews, 76, 168–210. doi: 10.1016/j.oregeorev.2015.09.005
21. Jaques, A. L., Jaireth, S., & Walshe, J. L. (2002). Mineral systems of Australia: an overview of resources, settings and processes. Australian Journal of Earth Sciences, 49, 623–660. doi: 10.1046/j.1440-0952.2002.00946.x
22. Joliffe, I. T. (1986). Principal component analysis. Berlin: Springer-Verlag. doi: 10.1007/b98835
23. Kelley, D. L., Kelley, K. D., Coker, W. B., Caughlin, B., & Doherty, M. E. (2006). Beyond the obvious limits of ore deposits: the use of mineralogical, geochemical, and biological features for the remote detection of mineralization. Economic Geology, 101, 729–752. doi: 10.2113/gsecongeo.101.4.729
24. Knox-Robinson, C. M., & Wyborn, L. A. I. (1997). Towards a holistic exploration strategy: using geographic information systems as a tool to enhance exploration. Australian Journal of Earth Sciences, 44, 453–463. doi: 10.1080/08120099708728326
25. Lech, M. E., & Caritat, P. de (2007). Recent results from a geochemical survey in the New South Wales part of the Thomson Orogen: Implications for mineral exploration. 23rd International Applied Geochemistry Symposium,(Oviedo, Spain. 14–19 June 2007, Program & Abstracts, 49. Retrieved from https://www.appliedgeochemists.org/ images/stories/IAGS_2007/Extended_Abstracts.pdf
26. Lech, M. E., Caritat, P. de, & McPherson, A. A. (2007). National Geochemical Survey of Australia: Field Manual. Geoscience Australia Record, 2007/08, 53 pp. Retrieved from http://www.ga.gov.au/metadata-gateway/metadata/record/gcat_65234
27. Ling, S., Wu, X., Ren, Y., Sun, C., Liao, X., Li, X., & Zhu, B. (2015). Geochemistry of trace and rare earth elements during weathering of black shale profiles in Northeast Chongqing, Southwestern China: their mobilization, redistribution, and fractionation. Chemie der Erde-Geochemistry, 75, 403–417. doi: 10.1016/j.chemer.2015.07.004
28. Main, P., & Caritat, P. de (2016). Geochemical Survey of the Southern Thomson Orogen, Southwestern Queensland and Northwestern New South Wales – The Chemical

Composition of Surface and Near-Surface Catchment Outlet Sediments. Geoscience Australia Record, in press2016/11, 136 pp. Retrieved from http://www.ga.gov.au/metadata-gateway/metadata/record/83230

29. Mann, A. W. (2010). Strong versus weak digestions: ligand-based soil extraction geochemistry. Geochemistry: Exploration, Environment, Analysis, 10, 17–26. doi: 10.1144/1467-7873/09-216
30. Mann, A. W., Birrell, R. D., Fedikow, M. A. F., & Souza, H. A. F. de (2005). Vertical ionic migration: mechanisms, soil anomalies, and sampling depth for mineral exploration. Geochemistry: Exploration, Environment, Analysis, 5, 201–210. doi: 10.1144/1467-7873/03-045
31. Mann, A. W., Birrell, R. D., Mann, A. T., Humphreys, D. B., & Perdrix, J. L. (1998). Application of the mobile metal ion technique to routine geochemical exploration. Journal of Geochemical Exploration, 61, 87–102. doi: 10.1016/S0375-6742(97)00037-X
32. McCuaig, T. C., Beresford, S., & Hronsky, J. (2010). Translating the mineral systems approach into an effective exploration targeting system. Ore Geology Reviews, 38, 128–138. doi: 10.1016/j.oregeorev.2010.05.008
33. McKinley, J. M., Hron, K., Grunsky, E. C., Reimann, C., Caritat, P. de, Filzmoser, P., Boogaart, K. G. Van Den, & Tolosana-Delgado, R. (2016). The single component geochemical map: fact or fiction? Journal of Geochemical Exploration, 162, 16–28. doi: 10.1016/j.gexplo.2015.12.005
34. McQueen, K. G. (2005). Ore deposit types and their primary expressions. In: Butt, C. R. M., Cornelius, M., Scott, K. M., & Robertson, I. D. M. (Compilers and Editors), Regolith expression of Australian Ore Systems – a compilation of geochemical case histories and conceptual models. Cooperative Research Centre for Landscape Environments and Mineral Exploration (CRC LEME), Monograph, 1–14. Retrieved from http://crcleme.org.au/Pubs/Monographs/RegExpOre.html
35. McQueen, K. G. (2007). A thirsty and confusing diggings: the Albert Goldfield, Milparinka-Tibooburra, north-western NSW. Journal of Australasian Mining History, 5, 67–96. Retrieved from http://www.mininghistory.asn.au/journal/
36. McQueen, K. G., Hill, S. M., & Foster, K. A. (1999). The nature and distribution of regolith carbonate accumulations in southeastern Australia and their potential as a sampling medium in geochemical exploration. Journal of Geochemical Exploration, 67, 67–82. doi: 10.1016/S0375-6742(99)00071-0
37. Mina_rík, L., _Zigov_a, A., Bendl, J., Sk_rivan, P., & _St'astn_y, M., 1998. The behaviour of rare-earth elements and Y during the rock weathering and soil formation in the _Rí_cany granite massif, Central Bohemia. The Science of the Total Environment, 215, 101–111. doi: 10.1016/S0048-9697(98)00113-2
38. Murphy, F. C., Hutton, L. J., Walshe, J. L., Cleverley, J. S., Kendrick, M. A., McLellan, J., ... Nortje, G. S. (2011). Mineral system analysis of the Mt Isa-McArthur River region, Northern Australia. Australian Journal of Earth Sciences, 58, 849–873. doi: 10.1080/08120099.2011.606333
39. Nakamura, A. (2015). Magnetic anomaly map of Australia [Data Set]. Geoscience Australia, Canberra. Retrieved from http://www.geoscience.gov.au/gadds
40. Nesbitt, H. W. (1979). Mobility and fractionation of rare earth elements during weathering of a granodiorite. Nature, 279, 206–210. doi:10.1038/279206a0
41. Ohlander, B., Land, M., Ingri, J., & Widerlund, A. (1996). Mobility of rare earth elements during weathering of till in northern Sweden. Applied Geochemistry, 11, 93–99. doi: 10.1016/0883-2927(95)00044-5
42. Prud^encio, M. I., Figueiredo, M. O., & Cabral, J. M. P. (1989). Rare earth distribution and its correlation with clay mineralogy in the clay-sized fraction of Cretaceous and Pliocene sediments (central Portugal). Clay Minerals, 24, 67–74. doi: 10.1180/claymin.1989.024.1.06
43. Reimann, C., & Caritat, P. de (1998). Chemical elements in the environment – factsheets for the geochemist and environmental scientist. ISBN 3-540-63670-6. Springer-Verlag, Berlin, Germany, 398 pp.

44. Reimann, C., Ladenberger, A., Birke, M., & Caritat, P. de (2016). Low density geochemical mapping and mineral exploration: application of the mineral system concept. Geochemistry: Exploration, Environment, Analysis, 16, 48–61. doi: 10.1144/geochem2014-327
45. Reith, F., Etschmann, B., Dart, R. C., Brewe, D. L., Vogt, S., Schmidt Mumm, A., & Brugger, J. (2011). Distribution and speciation of gold in biogenic and abiogenic calcium carbonates – Implications for the formation of gold anomalous calcrete. Geochimica et Cosmochimica Acta, 75, 1942–1956. doi: 10.1016/j.gca.2011.01.014
46. Reyment, R. A. (1989). Compositional data analysis. Terra Nova, 1, 29–34. doi: 10.1111/j.1365-3121.1989.tb00322.x
47. Schmidt Mumm, A., & Reith, F. (2004). The biogeochemistry of calcrete forming processes. In: Roach, I. C. (Editor), Regolith 2004, Proceedings of the CRC LEME Regional Regolith Symposia, November 2004, Cooperative Research Centre for Landscape Environments and Mineral Exploration (CRC LEME), Monograph. 316–320. Retrieved from http://www.crcleme.org.au/Pubs/Monographs/Regolith2004.html
48. Wilford, J., Caritat, P. de, & Bui, E. (2015). Modelling the abundance of soil calcium carbonate across Australia using geochemical survey data and environmental predictors. Geoderma, 259–260, 81–92. doi:10.1016/j.geoderma.2015.05.003
49. Wyborn, L. A. I., Heinrich, C. A., & Jaques, A. L. (1994). Australian proterozoic mineral systems: Essential ingredients and mappable criteria. The Australian Institute of Mining and Metallurgy (AusIMM) Annual Conference, Darwin, Northern Territory. 5–9 August 1994, 109–115. Retrieved from http://www.ausimm.com.au/publications/ publication.aspx?IDD6547

CHAPTER 2

Framework for the optimization of operation and design of systems with different alternative water sources

Lisa J. Blinco[1,2], Martin F. Lambert[1,2], Angus R. Simpson[1,2] and Angela Marchi[1,2]

[1]*School of Civil, Environmental and Mining Engineering, The University ofAdelaide, Engineering North N136, North Terrace Campus, Adelaide, South Australia 5005, Australia*

[2] *Cooperative Research Centre for Water Sensitive Cities, Monash University LPO, PO BOX 8000, Clayton, Victoria 3800, Australia*

AUTHORS' CONTRIBUTIONS

LJB was responsible for the development of the conceptual framework, review of literature and drafting the manuscript. MFL provided feedback and recommendations in the development of the framework and helped to draft the manuscript. ARS provided feedback and recommendations in the development of the framework and helped to draft the manuscript. AM provided feedback and recommendations in the development of the framework and helped to draft the manuscript. All authors read and approved the final manuscript

ABSTRACT

Water security has become an increasing concern for many water system managers due to climate change and increased population. In order to improve the security of supply, alternative sources such as harvested stormwater, recycled wastewater and desalination are becoming more commonly used. This brings about the need for tools to analyze and optimize systems that use such sources, which are generally more complex than traditional water systems. Previous methodologies have been limited in their scope and cannot be applied to all types of water sources and systems. The framework presented in this paper has been developed for holistic analysis and optimization of water supply and distribution systems that use alternative water sources. It includes both design and operational decision variables, water and energy infrastructure, simulation of systems, analysis of constraints and objectives, as well as policies and regulations which may affect any of these factors. This framework will allow users to develop a comprehensive analysis and/or optimization of their water supply system, taking into account multiple types of water sources and consumers, the effect of their own design and operational decisions, and the impact of government policies and different energy supply options. Two case study systems illustrate the application of the framework; the first case study is a harvested stormwater system that is used to demonstrate the importance of simulation and analysis prior to optimization, the second utilizes four different water sources to increase security of supply and was optimized to reduce pump energy use.

Keywords: Water distribution systems, Integrated water resources management, Decision-making, Conceptual framework, Alternative water sources Optimization

1. INTRODUCTION

A changing climate and increasing population have put a strain on traditional water resources, which typically rely on natural catchment water. This has made water security an increasing concern for many water system managers, who have investigated options for reducing demand and supplementing supply. Alternative water sources, such as harvested stormwater, recycled wastewater and desalination, are increasingly being used to improve water security of cities and towns. Methods for simulation, analysis and optimization of traditional potable water distribution systems (WDSs) cannot necessarily be directly transferred to systems that use alternative water sources. Therefore there is a need to develop a methodology specifically for alternative water source systems, which includes both hydraulic and hydrologic considerations, as well as the many additional parameters and variables associated with alternative water sources. There are many modelling tools used in current practice for integrated water management, such as eWater Source, WEAP (Water Evaluation and Planning System) and Mike Basin. These modelling tools do not include hydraulic simulation, and therefore may not accurately represent performance of urban water networks. Moreover, this framework is not software, rather its purpose is to guide water system managers in how to best simulate and optimize their systems, particularly those that integrate multiple water sources, and natural and human-made systems. The framework should be used to determine which system components need to be modelled, which type of modelling tools are most appropriate, what regulations and policies need to be taken into account and how to evaluate the performance of the system.

The framework introduced in this paper can be applied to the optimization of the design and operation of water supply and distribution systems from source to consumer, considering multiple traditional and alternative sources, multiple uses and multiple objectives. Electrical energy sources and their effect on electricity prices and greenhouse gas (GHG) emissions are included, as are several types of government policies that may affect the design, operation, data and evaluation of the system. The objectives of this paper are to (1) develop a generalized framework that could be applied to any water supply and/or distribution system optimization problem and (2) outline the application of this framework to two case study systems with a focus on optimizing their operation.

2. LITERATURE REVIEW

Since 2000, there has been significant consideration of the concept of water security (Cook and Bakker 2012) as water is increasingly seen as a fundamental and finite resource (Bogardi et al. 2012). Consequently, the use of alternative sources, such as harvested stormwater, desalination, recycled wastewater and rainwater, has gained traction (Fielding et al. 2015). Harvested stormwater schemes are often decentralized and used for non-potable supply such as household gardening and irrigation of public reserves (Naylor et al. 2012), however, in some cases are also used for potable supply (McArdle et al. 2011). While desalination is a climate independent (and therefore more reliable) source, is often not the most cost effective or environmentally sensitive option (Becker et al. 2010; Miller et al. 2015). Recycled wastewater is also climate independent, and generally used for large scale non-potable applications (Muga and Mihelcic 2008; Oron et al. 2014), however, it can also be used for indirect or direct potable supply (Rodriguez et al. 2009; Nagal 2015). Domestic rainwater tanks are increasing in popularity and have benefits of reducing water usage from utilities and reducing stormwater runoff from houses (Campisano and Modic 2012; Umapathi et al. 2013). Demand management strategies have also been used to reduce per capita consumption and therefore reduce the pressure on limited water supplies (Dawadi and Ahmad 2013; Friedman et al. 2014).

Some alternative sources, such as harvested stormwater, introduce additional complexity to the problem of modeling and optimization than has been previously considered for traditional water systems (Marchi et al. 2016). There is, for example, the need to consider the supply and distribution systems together, rather than separately, as it is less likely that there will be large storages isolating the supply side from the distribution side. When including the supply side, longer simulation times often need to be used, requiring rainfall and evaporation scenarios to be taken into account. The security of supply with regard to climate change needs to be considered (Paton et al. 2014; Cai et al. 2015), as some sources are climate dependent and some are climate independent. The social acceptability of using particular sources for particular applications and the willingness of consumers to pay more for alternative source systems to be constructed and maintained may need to be incorporated (Hwang et al. 2006; Londoño Cadavid and Ando 2013; Fielding et al. 2015). The perception of risks associated with alternative water source systems by water system managers may also present a barrier to the implementation and success of such systems (Dobbie and Brown 2012; West et al. 2016). Many alternative sources also have associated externalities that result in either cost or benefit to the user, such as reduced effluent flow to the ocean or receiving water body by reusing wastewater and reduced urban stream flows by harvesting stormwater (Marchi et al. 2016).

The increased use of alternative water sources then raises the question of how such systems should by analyzed and optimized to ensure they are implemented as effectively as possible. Stokes et al. (2014) developed a framework for optimizing the cost and GHG emissions of WDSs, taking into account both the design and operation of the system, energy sources and GHG emission factors. This study, however, was applicable only to traditional WDSs, with no consideration of the supply side and alternative water sources. Chung et al. (2008) developed a mathematical model for evaluating integrated water supply systems with decentralized treatments. Multiple sources, uses, transportation and treatment systems can be considered, however only surface water, groundwater and recycled wastewater sources are included. This model does not incorporate any optimization procedure, only analysis of different options developed by the user. Makropoulos et al. (2008), with further developments in Rozos and Makropoulos (2013), produced a decision-support tool for modeling the urban water system from source to tap. The software can be used to select combinations of water saving strategies and technologies, including how much water from each type of demand (for example domestic, commercial) is obtained from each source and how the system is operated. It uses a demand-oriented, water balance approach and does not include capability for other types of simulation models such as hydraulic and hydrologic modeling.

Uncertainty, particularly with regard to climate change, is an important consideration that has been taken into account in several methodologies. Paton et al. (2014) developed a framework for water supply system planning with alternative sources and climate change considerations, while Beh et al. (2014, 2015) developed two methods for optimal sequencing of urban water supply augmentation options under deep uncertainty regarding demands and climate. The research by both Paton et al. (2014) and Beh et al. (2014, 2015) considered only the planning of water supply projects, and did not optimize the specific design or operation of the systems. Sequencing is also considered in Cai et al. (2015), however, in this case it is applied to planning of drought mitigation strategies in agricultural systems. They consider multiple decision stages in which options such as infiltration ponds, parallel terraces, irrigation triggering threshold and irrigation water sources can be implemented. Marchi et al. (2016) developed a methodology for optimizing the design of harvested stormwater systems taking into account future climate scenarios; however, it does not apply to other types of alternative sources or optimization of system operation. It does include a detailed analysis of the associated externalities, such as reduced peak flows and improved economic value of properties near stormwater schemes. Ashbolt et al. (2014) introduced a framework for planning of short-term operations for water systems using surface water, groundwater, desalination, and recycled wastewater with multiple objectives and multiple inflow replicates to account for uncertainty. Long-term operating strategies and the design of the system were not included and the operating strategies considered were limited to bulk water transfers and not the operation of pumps and smaller storages.

3. FRAMEWORK FOR THE OPTIMIZATION OF ALTERNATIVE WATER SOURCE SYSTEMS

The framework presented in the current paper was developed to guide the modeling and optimization of water supply and distribution systems that use alternative water sources. It is comprised of several components and sub-components that fit within an optimization structure, for example, a multi-objective evolutionary algorithm (Fig. 1). The options component [OPT] describes the potential 'decision variables' that are available in an optimization problem, that is, the factors that can be changed in order to produce a different outcome. This includes both the initial design of the water supply and distribution infrastructure and the long- and short-term rules that govern the operation of the system once it has been commissioned. The infrastructure component [INF] describes the physical components of the system that need to be modeled and the data associated with each, including both water infrastructure and energy infrastructure, which may affect the evaluation of electrical energy cost and life-cycle GHG emissions. There is also a government policy component [G] that covers the policies from regulating bodies that may affect other aspects of the framework. The analysis component [ANL] describes the simulation of each potential system configuration and evaluation against objectives and constraints. The optimization algorithm [OA] investigates different possible combinations of decision variables from the options component, models the system according to the infrastructure component and evaluates it using the analysis component to find the optimal solution(s).

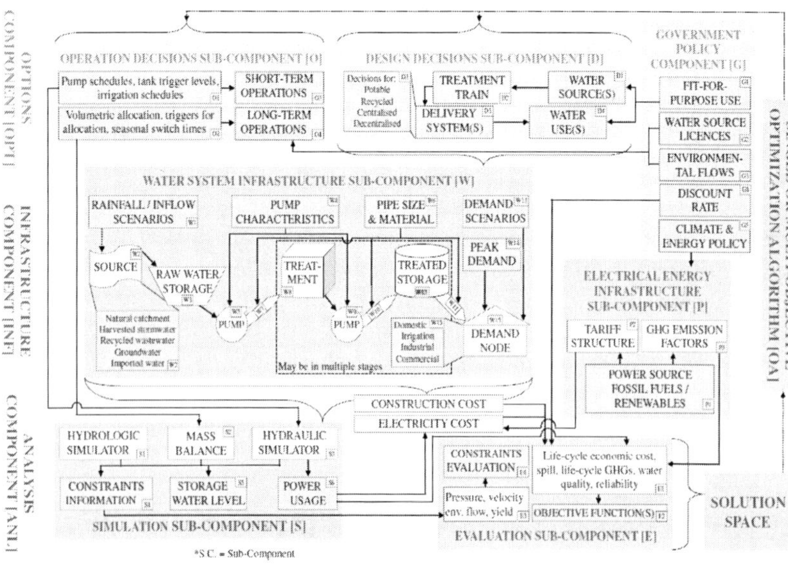

Figure. 1 Schematic of the framework for optimizing the design and operation of alternative water source systems

Details of the components and sub-components are shown in Fig. 1 and described in Sections 'Options component [OPT]', 'Infrastructure component [INF]', 'Government policy component [G]' and 'Analysis component [ANL]'. Table 1 summarizes the parameters that need to be considered in the optimization and simulation of alternative water source systems with respect to the different items that are presented in Fig. 1 and in the following sections. There are three (non-exclusive) categories that each parameter may be placed in – decision variables, parameters that are set, and uncertain parameters. Decision variables are parameters that the user may be able to examine using optimization. It is important to note that in most optimization problems, not all of these parameters will be available as decision variables at once, and it is likely that only a small number will be considered. For example, when optimizing pump operations for an irrigation system, only the first three 'decision variables' shown in Table 1 (pump schedules, tank trigger levels, and demand scheduling) may be considered. The remaining parameters that are designated as decision variables in Table 1, particularly those relating to the design of the system (for example, delivery system layout and pump sizing) would already be set and not able to be optimized if the existing infrastructure cannot be modified. The parameters that are set are those that very rarely, if ever, are able to be optimized by the user. These include parameters that would be controlled by external sources, for example consumers of domestic or commercial demands, pipe manufacturers and higher level government and regulatory bodies; and also parameters that need to be predefined to a known or assumed value before optimization or simulation can be performed, for example, fire demand/reserve, hydrologic/hydraulic variables and objective and constraint selection and definition. The final category, uncertainty, designates those externally set or predefined variables that are not well known or may be subject to change in the future and therefore may need to be considered in a sensitivity analysis. While the selected values of decision variables have an impact on the performance of a system, they are generally within the control of the decision maker, and therefore are not classed as 'uncertain'. It is important to note that the categorization in this table is indented as an indication of how each parameter is typically treated. There are, of course, exceptions to this, as almost all of the parameters could be considered as decision variables if desired and have some associated uncertainty. For example, environmental flows have been designated as an externally set parameter, as it is likely that the operator of a system will have to meet requirements set by an external organization such as the Environmental Protection Agency. They may, however, want to investigate providing greater environmental flows, or show the benefits of reducing their environmental flow requirements and being able to supply more water elsewhere.

Table 1 Summary of parameters for the design and operation of alternative water source systems

Parameter	Decision variable[a]	Parameter that is set	Uncertain parameter	Relevant items in Fig. 1
Operational inputs [O]				
Pump schedule	X			O1
Tank trigger levels	X			O1
Tank/storage maximum and minimum allowable levels		X		O1, W3, W11
Demand pattern (irrigation, agriculture)	X			O1, D4, W13
Demand pattern (domestic, commercial, industrial)		X	X	O1, D4, W13
Demand flow rate (peak, average, peak day)		X	X	O1, D4, W14
Valve settings or operating rules	X			O1
Pump pressure settings	X			O1
Volumetric allocation of water	X			O2
Reservoir trigger levels	X			O2
Switch time between operating regimes	X			O2
Priority ranking of operating rules		X		O3, O4
Design inputs [D] and water infrastructure [W]				
Water source selection	X			D1, W2
Water source infrastructure (layout, capacity)	X			D1, W2
Treatment type selection	X			D2, W8
Treatment infrastructure (layout, capacity,	X			D2, W8

treatment rate/level)				
Delivery system type selection	X			D3
Delivery system layout (lengths, elevations, junctions, tank locations)	X			D3, W7, W10, W12, W15
Pipe material and diameters	X			D3, W7, W10, W12
Pipe parameters (unit cost, pipe wall roughness (ε), wall thickness, embodied energy)		X	X (ε)	D3, W6, W7, W10, W12
Pump sizing	X			D3, W5, W9
Pump performance characteristics and cost		X		D3, W4
Tank sizing (capacity, height, diameter)	X			D3, W3, W11
Fire demand/reserve		X		D3, W11
Water user type selection	X			D4
Rainfall/streamflow series		X	X	W1
Reservoir capacity and volume curve	X			W3
Pond (e.g. wetland) capacity and volume curve	X			W3
Prioritization rules for demands types		X		W15
Other inputs [P], [G] and [S]				
Power source selection		X	X	P1, P3, G5
Electricity tariff structure and cost		X	X	P2
GHG emission factors		X	X	P3, G5
Fit-for-purpose requirements		X		G1
Water license amounts		X		G2
Environmental flow amounts		X		G3

Discount rate		X	X	G4
Hydrologic variables (e.g. permeability)		X		S1
Hydraulic variables (e.g. water temperature)		X		S3
Optimization problem formulation [E]				
Objective selection		X		E1
Objective function(s)		X		E2
Constraint selection		X		E3
Constraint limits (maximum and minimum)		X		E4
Penalty costs		X		E4

[a]Note: Parameters specified as decision variables are shown in bold throughout Sections 'Options component [OPT]' and 'Infrastructure component [INF]'

Options component [OPT]

The options component covers the potential decision variables (and the range of possible choices for the decision variables) for an optimization problem. This component is split into two sub-components; the operational decisions sub-component [O] and the design decisions sub-component [D]. Design decisions include elements that can be changed before a system is constructed, such as the layout and capacities, materials and other properties of the various infrastructure components. Operational decisions include elements that can be changed after construction during the daily management of the system, such as the operating rules for pumps and valves and allocation of water from different sources.

Operational decisions Sub-component [O]

Both short- and long-term operations are considered in the operational decisions sub-component. The critical aspects of this sub-component (items in **BOLD** can be optimized), as shown in Fig. 1 and Table 1 are:

[O1]the specific short term operating strategies including **PUMP SCHEDULES** (when pumps are turned on or off based on time), **TRIGGER LEVELS** (water levels in tanks or other storages that determine when pumps or valves turn on or off), **IRRIGATION OR DEMAND SCHEDULES** (for systems where they can be pre-determined), **VALVE SETTINGS AND OPERATING RULES**, and **PRESSURE SETTINGS FOR PUMPS** (to maintain the set pressure at a particular point).

[O2]the specific long term operating strategies including **VOLUMETRIC ALLOCATION OF WATER** from different alternative sources, **TRIGGER LEVELS** (for example in reservoirs) that determine allocations from different sources or water demand restriction levels, **SWITCH TIMES BETWEEN DIFFERENT OPERATING REGIMES** (for example between different trigger level sets for different seasons) and power source selection.

[O3]the overall short-term operating strategy, including operating rules that are optimized in [O1] and operating rules that are pre-set and are not to be optimized (acting as constraints). Where there are multiple operating rules, the priority of each rule and order they are enforced in is important to consider.

[O4]the overall long-term operating strategy, including operating rules that are optimized in [O2] and operating rules that are pre-set and are not to be optimized. Again, the priority and order of the rules is important to consider.

Most systems have multiple operating conditions to meet and therefore multiple operating rules will be in place. Prioritization of the different operating rules is important, and this may be set by the operator or be chosen by the optimization tool. This component requires information from the government policy sub-component ([G] in Fig. 1), specifically in terms of water source licensing and environmental flow regulations. These policies would typically be regulated by local or state government departments or the environmental protection authority. Operational rules set in this sub-component will inform the simulation sub-component [S] as they will need to be represented in any simulation model(s) of the system.

Design decisions Sub-component [D]

This sub-component incorporates all of the design decisions that are available to the designer for the entire water supply and distribution system, from source to user. The critical aspects of this sub-component (items in **BOLD** can be optimized), as shown in Fig. 1 and Table 1 are:

[D1]the **WATER SOURCES SELECTED** to be used including natural catchments, harvested stormwater, recycled wastewater, groundwater, imported water, domestic rainwater, desalination, domestic greywater and sewer mining; and the **LAYOUT AND CAPACITY OF SOURCE INFRASTRUCTURE**.

[D2]the **TYPES OF TREATMENT SELECTED** including centralized treatment at plants such as mechanical filtration, chemical dosing, ultraviolet treatment and ozonation, and decentralized in situ treatments such as gross pollutant traps, wetlands and biofilters; and the **LAYOUT,**

CAPACITY, DOSING RATES AND RETENTION TIMES FOR TREATMENT FACILITIES.

[D3]the **TYPE AND CONFIGURATION OF THE DELIVERY SYSTEM** used including potable, non-potable (for example dual reticulation systems to deliver recycled water), centralized and decentralized, and the infrastructure design variables such as **SYSTEM LAYOUT, PIPE SIZES, LENGTHS AND MATERIALS, PUMP SIZING, VALVE SIZING, AND TANK SIZING.**

[D4]the **TYPES OF WATER USERS** that are supplied by the system including potable, irrigation, agriculture, industrial, non-potable domestic/commercial and firefighting, and the demand rate and pattern for water use (for example, scheduling of irrigation demands).

Regulations on fit-for-purpose water use from the government policy component [G] in Fig. 1 inform what water sources can be used for particular applications and these are likely to be specified by state or federal government departments or health authorities. Generally, sources such as harvested stormwater and recycled wastewater cannot be used for potable supply and rather serve non-potable demands in dual-reticulation systems or are supplied to irrigation, agricultural and industrial users. There may be some systems, however, in which necessary approvals have been obtained to use these sources for potable supply. The design decisions are inputs to the water system infrastructure sub-component [W] which describes the system elements and data to be modeled.

Infrastructure component [INF]

The purpose of this component is to describe the infrastructure that needs to be modeled in order to evaluate the objectives and constraints of the problem. There are two sub-components; the water system infrastructure sub-component [W] and the electrical energy infrastructure sub-component [P]. Water system infrastructure includes the specific aspects of the water supply and distribution system and the data required, including construction and maintenance costs. Electrical energy infrastructure includes the power source (fossil fuel types and renewable types) and the electricity price and GHG emission factor data needed.

Water system infrastructure Sub-component [W]

This sub-component includes the specific infrastructure aspects of the water system design and the relevant data that is needed in order to simulate it. Most systems and optimization problems will not require all of these factors to be considered or modeled; however, the purpose of this framework is to cover a large range of the possible requirements for an optimization and hence the scope is intentionally broad.

The water system infrastructure sub-component [W] as shown in Fig. 1 represents a system with one water source, one treatment plant, one storage tank and one demand node. In reality, many systems will have more than one of each of these components, particularly the treated storage [W11] and demand node [W15]. Pumping of water between storages may occur in multiple stages, particularly when there is a large difference in elevation. For typical centralized potable WDSs, all treatment will occur at one water treatment plant. In decentralized systems such as for harvested stormwater schemes, however, treatment may occur in multiple stages. For example, a gross pollutant trap may be located on an urban creek before the water is collected in a harvest pond, then the water may be pumped to be treated through a wetland, and then treated again in a treatment plant.

The critical aspects of this sub-component (items in **BOLD** can be optimized) as shown in Fig. 1 and Table 1 are:

[W1] the rainfall or inflow scenarios for the water source; for example rainfall or streamflow scenarios for natural catchments and stormwater sources, or a sewer system flow pattern for recycled wastewater. Sources such as desalination and, depending on the temporal scale of the optimization, groundwater, do not usually require an inflow scenario. Rainfall and streamflow scenarios may be a data series obtained from measurements at gauging stations or modeled in a hydrologic simulation program [S1]. Multiple inflow scenarios may be used, particularly for systems with highly variable inflows. Losses such as evaporation and infiltration may also need to be taken into account for sources with large open storages such as reservoirs and natural water ways.

[W2] the **SOURCE TYPE** as described in [D1] with input from [W1].

[W3] the raw water storage; this may be a reservoir (typical for a natural catchment), a harvest pond for a stormwater system, a tank (for example for a recycled wastewater system) or an aquifer for groundwater. Associated data including **CAPACITY, A VOLUME CURVE, ELEVATION, HEIGHT AND DIAMETER** is required.

[W4] characteristics of available pumps such as performance curves (head, efficiency, and power against flow), cost, rated speed and variable speed pump (VSP) information where applicable

[W5] the **PUMP** transferring water from the raw water storage to a treatment facility, requiring data from [W4].

[W6] pipe size and material information such as available diameters, unit costs, pipe wall roughness, wall thickness and embodied energy. For new pipes, this information will be easily obtained from the pipe manufacturer. For existing systems, however, there may be some uncertainty in these parameters if detailed records of the 'as constructed' system and any pipe replacements have not been kept. In addition to this, the pipe wall roughness of existing pipes will generally be uncertain. Pipe wall roughness increase as pipes age, and pipe condition assessment may be needed to provide an estimate.

[W7] the **PIPE SYSTEM** transferring water from the raw water storage to the treatment facility, **PIPE LENGTHS AND LAYOUTS** need to be known as well as information from [W6].

[W8] the **TREATMENT FACILITY** that will produce water of the required quality based on the source type and demand type. Characteristics of the individual treatment methods as described in [D2] need to be known.

[W9] the **PUMP** transferring water to a treated storage, requiring the same data as [W5].

[W10] the **PIPE SYSTEM** transferring water to a treated storage, with the same information as [W7] required.

[W11] the **TREATED STORAGE**, for example, a tank or multiple tanks that are typically at a high elevation point of the network in order to supply demands by gravity. Data required includes the **ELEVATION, HEIGHT, DIAMETER** and maximum and minimum allowable water levels.

[W12] the **PIPE SYSTEM** transferring water from the treated storage to consumers, which again requires information as in [W7]. This pipe system is likely to be more complex than those in [W7] and [W10], particularly for systems with many different demand nodes. For systems with only one source of water, [W7] and [W10] are likely to be single pipelines. For decentralized systems with only one specific consumer, [W12] will also most likely be a single pipeline. Most systems, however, have much more than one demand point and as such distribution systems are often looped or branched systems that require more complex analysis than single pipelines.

[W13] demand scenarios that will be applied to the demand nodes, consisting of a pattern of demand multipliers over a day, week or year. There may be multiple demand scenarios required for a system, for example, if there are different types of demand nodes (such as domestic, commercial, industrial) or different seasonal demands.

[W14] the peak demand is the demand rate that is typically used to size the system components and so will affect the design of the system. The demand scenarios [W13] are more likely to affect the operation of the system as the demand varies over the simulation time. The peak day demand (average demand over the peak day), the peak hour demand (the average demand over the hour with maximum consumption in the peak day) and average demand rates may also be required. Fire loading demands and other emergency conditions will affect the design of the system, for example storage tanks should be sized to be able to provide demand in the case of fires, other emergencies and system failures (e.g. if the supply to the tank is cut off).

[W15] the demand nodes for the consumers, these may be different types of users as described in [D4] and require information from [W13] and [W14]. Different types of users will have different demand rates [W14] and demand patterns [W13]. When simulating the system, an average demand rate will often be used with the demand pattern, rather than the peak demand. Systems with multiple demand nodes may prioritize different types of demands over other, for example, irrigation systems using non-potable water may prioritize high profile sport fields over reserves with no formal usage.

Choices made in the optimization of the design decisions sub-component [D] in Fig. 1 will be inputs to the water system infrastructure sub-component. There may be other parameters that are not decision variables in the optimization (as differentiated in Table 1) though are still required by this sub-component in order to simulate the system. The construction and maintenance costs of each of the infrastructure components needs to be known in order to calculate the initial construction cost and ongoing costs as part of life-cycle economic costing. Information collected through this sub-component will be input to the simulation sub-component [S] depending on the types of simulation models used and to the evaluation sub-component [E] through the construction cost or other factors calculated for the specific objectives of a problem.

Electrical energy infrastructure Sub-component [P]

The electrical energy infrastructure sub-component includes any power infrastructure that affects the electricity prices and GHG emission factors. The critical aspects of this sub-component as shown in Fig. 1 and Table 1 are:

[P1] the breakdown of power sources including fossil fuel sources such as coal and oil, and renewable sources such as solar, wind and hydrothermal.

[P2] the electricity price tariff structure, which may be a peak and off-peak structure, or multi-part (more than two price levels) and could include a peak demand charge which applies to the peak electricity power usage in each month.

[P3] the GHG emission factor, which is based on the power source breakdown [P1] and may vary with time, either in the short-term (with sources that do not have storage such as solar panels and wind turbines) or the long-term (as fossil fuel sources tend to be phased out and renewable sources become more popular).

Climate and energy policy [G5] in the government policy component in Fig. 1 will affect the power source breakdown and electrical energy pricing now and into the future. This is likely to be regulated by a federal government department or body. Information from this sub-component is used to calculate electrical energy costs in order to evaluate life-cycle economic costs and also to calculate life-cycle GHG emissions in the evaluation sub-component [E].

Government policy component [G]

The government policy component covers policies by regulating bodies at any level (local, state, federal) that may affect other aspects of the framework. These policies need to be considered over the operational life-span of the system, for example, climate and energy policy may affect future energy sources and therefore affect future GHG emissions. The critical aspects of this component as shown in Fig. 1 and Table 1 are:

[G1] fit-for-purpose water use, which may be regulated by state or federal governments or health agencies and affects which water sources [D1] and water uses [D4] can be combined in the design decisions sub-component. It may also guide which design decisions (for example, treatment) are appropriate.

[G2] water source licenses, which may be regulated by local or state governments or the environmental protection agency, depending on the catchment size, and will affect the amount of water available from particular sources for allocation in long-term operations [O4].

[G3] environmental flows, which similarly to water source licenses may be regulated by local or state bodies depending on the size of the catchment and affect the amount of water available for allocations [O4].

[G4] the discount rate applied to operational costs and GHG emissions in life-cycle analysis [E1]. This is unlikely to be set by a government body and rather will be informed from outside the decision making team, generally by recommendations from economists.

[G5] climate and energy policy set by state and federal governments will affect the energy sources available now and in the future, therefore affecting GHG emission factors and any GHG objectives [P].

Analysis component [ANL]

The analysis component uses information from the options, infrastructure and government policy components to simulate the system and evaluate how it performs relative to the objectives and constraints. Within an optimization algorithm, the analysis component is used to assess multiple combinations of decision variables from the options component to determine how they perform. There are two sub-components within the analysis component; the simulation sub-component [S] and the evaluation sub-component [E]. The simulation sub-component includes the modeling aspects of the problem and the key variables that are required to be output from the models in order to evaluate the system. Optimization objectives and constraints are defined in the evaluation sub-component, which also provides information to the optimization algorithm as to which of the potential solutions perform best.

Simulation Sub-component [S]

The simulation sub-component considers the type of simulation model that is most applicable to the particular system and problem, and specifies the key variables that need to be calculated in the model(s). The critical aspects of this sub-component as shown in Fig. 1 and Table 1 are:

[S1] the hydrologic simulator, which is required if rainfall scenarios need to be transformed to streamflow, typically for systems using natural catchment water or harvested stormwater.

[S2] the mass balance model, which may be required for systems that have multiple water sources with long-term allocation decisions, particularly if there are different rainfall and evaporation scenarios to be considered for the storages.

[S3] the WDS hydraulic simulator, which is required to analyze pump and pipe systems that transfer water between different storages and treatments and to consumers.

[S4] information on constraints, such as yield from a hydrologic model, environmental releases and system reliability from a mass balance model, and nodal pressures, pipe velocities, pump switches and tank levels from a hydraulic model.

[S5] the water levels in storages, which are important particularly when considering operational decisions, such as trigger levels, and for constraints, such as system reliability.

[S6] the power usage from any pumps or treatment facilities, which are important in informing the ongoing electrical energy costs as part of life-cycle economic costing. Generally a WDS hydraulic simulator is required to model detailed pump operations and therefore accurately estimate the pump power usage.

Each of the three types of models will require different simplifications or assumptions depending on the particular system. For example, mass balance modeling will generally only consider one pump operating point so may not accurately calculate the pump power usage. When deciding which type of model to use for a particular problem, the user will need to consider the different simplifications, assumptions, advantages and disadvantages of each model. Trade-offs between accuracy of outputs and simulation run times need to be considered. For example, when optimizing both short- and long-term operations of a system, there is likely to be a trade-off between using a hydraulic simulator for detailed hydraulic information and using a mass balance model for shorter run times. Most problems may ideally use elements from more than one type of model; however, using multiple models will increase computational complexity and run times. Wherever possible, the most applicable type of model should be selected and augmented with the required elements from other types of models. Depending on the particular system and optimization problem, there may be other key variables that need to be calculated in the simulation models. For optimization of pumping operations, which is the focus of the case studies in this paper, storage water levels and pump power usage are the most important. Existing hydrologic, mass balance and hydraulic simulators, for example, MUSIC, WATHNET and EPANET, have often been used in conjunction with optimization algorithms and should be taken advantage of where possible rather than creating individual simulators for different problems.

Information from the operation decisions sub-component [O] will be input to the simulation sub-component as the overall operating strategy for the system ([O3] and [O4]) will need to be modeled. Short-term operations are likely to be considered in a hydraulic simulator and long-term operations, including allocations, in a mass balance model. Parameter data on the physical components of the system from the water system infrastructure sub-component [W] are also required as inputs for this sub-component. Constraint information is provided to the evaluation sub-component to compare the systems performance against specified requirements. Energy usage is used to calculate objective functions such as life-cycle economic costs and life-cycle GHG emissions. Simulating systems prior to optimization is an important step to help inform the formulation of the optimization problem and provide a check that results from the optimization are reasonable.

Evaluation Sub-component [E]

The purpose of the evaluation sub-component is to compare the performance of each of the potential solutions to the objectives and constraints of the problem. The critical aspects of this sub-component as shown in Fig. 1 and Table 1 are:

[E1] the specific objective(s) to be considered in the optimization; typically, minimizing life-cycle economic cost is a primary objective (or a component of that such as construction cost or operational cost individually). Other possible objectives include minimizing spills from reservoirs and other storages, minimizing life-cycle GHG emissions (or a component of that such as embodied energy from construction or operational emissions), minimizing supplemental potable water supply (in systems using non-potable sources), maximizing water quality, maximizing reliability and minimizing environmental impact.

[E2] the objective function(s) to be optimized; multiple objectives may be evaluated as individual functions in a multi-objective optimization algorithm or combined into a single function for use in a single objective optimization algorithm. It is important to consider how each objective should be formulated, for example, when optimizing short-term pump operations to minimize ongoing costs, the objective function may be evaluated in terms of cost per volume of water pumped, as this will take into account the amount of water delivered to consumers. Reliability of a system may be formulated in different ways, for example minimizing the time spent with water restrictions applied or minimizing the time spent below a certain storage level. Some objectives may be more difficult to quantify, such as minimizing environmental impact, so more specific objectives may be required, for example, maximizing environmental flow or minimizing the change in a water body's natural hydrological regime. Simplifications and assumptions may be required to formulate some objectives as mathematical functions. When performing multi-objective optimization, trade-offs between the different objectivesshould be considered by the development of Pareto fronts, allowing the decision maker to determine which Pareto optimal solution best fits their needs (see examples in Wu et al. 2010a, b, 2012a, b, 2013).

[E3] the specific constraints to be considered as described in [S4]

[E4] the evaluation of the constraints compared to the limits set by the user; maximum and/or minimum values for each constraint need to be specified. Some constraints may be flexible, for example, if an environmental flow is set by a regulator, the operator could consider increasing the set environmental flow as a decision variable in the optimization. There are several different ways constraints can be incorporated into the optimization algorithm. Penalty functions are often used for single-objective problems. They add value (often a monetary amount) to the objective function in a minimization problem and remove value from the objective function in a maximization problem based on the magnitude of the constraint violation, therefore making solutions that violate constraints less desirable (Nicklow et al. 2010). Care must be taken when formulating penalty functions to keep solutions that only slightly violate constraints in consideration during the optimization process, while ensuring the feasibility of the final optimal solutions. For multi-objective problems, a constraint-handling technique that will ensure feasible solutions are retained in preference to infeasible solution is often employed. An example of this is the constraint tournament selection procedure introduced by Deb et al. (2002).

Information about the objectives is received from the simulation sub-component [S] and from the calculation of construction, maintenance and electrical energy costs based on the water system infrastructure sub-component [W] and simulation sub-component. A discount rate for costs or GHG emissions may be set in the government policy sub-component [G] which will impact the ongoing costs and emissions in a life-cycle analysis. The discount rate may be informed by economists, such as the Stern review which recommends low discount rates for projects that lead to the production of GHG emissions (Stern 2006). Information about the performance of each potential solution in relation to the objectives and constraints is provided to the optimization algorithm in order to find the best solutions.

4. OPTIMIZATION ALGORITHM [OA]

The optimization algorithm is used to determine which solution(s), out of many potential solutions to the problem, performs best in relation to the objective function(s). The procedure used to set up the optimization will depend on the type of algorithm chosen; however, the first step is generally to define the decision variables, objectives and constraints of the problem. This will then guide what aspects of the system need to be modeled and what data is required in order to take into account all of the decision variables and that will provide information for all of the objectives and constraints. Multiple potential solutions to the problem form the 'solution space' and the optimization algorithm guides the search of this solution space towards the global optimum. The size of the solution space depends on the number of decision variables and number of choices available for those decision variables. More complex problems are often described as having a more 'rugged' solution space, meaning the optimization algorithm is more likely to get trapped in local optima and will have more difficulty finding the global optimum. When a single objective optimization algorithm is used, one optimal solution will be found, while in multi-objective optimization, a Pareto front will be developed with multiple solutions representing different trade-offs between the objectives.

Most optimization algorithms have parameters that need to be defined by the user, such as the number of generations or iterations and the population size in evolutionary algorithms. Although the choice of these parameters does not influence the components shown in Fig. 1, they have an effect on the optimal solutions found by the algorithm. In general, the most effective set of parameter values to use will vary between different optimization problems and the size of the solution space can only give some indication of what parameter values to use. In fact, multiple parameter sets should be tested in order to find the most appropriate values for the specific problem. Ideally, the chosen parameter set should find the same optimal solution regardless of the starting point or initial solution(s) for the optimization. Dandy et al. (1996) presented an improved genetic algorithm formulation for optimization of WDS design. Five different parameter sets were trialed on both their improved genetic algorithm and a comparatively simple genetic algorithm. They acknowledged that parameter selection does require some judgement on the part of the user, however, they found their optimization results to be relatively insensitive to the parameter choice, particularly for the improved genetic algorithm. As well as the effect of various parameter values, different optimization algorithms will be more suited to different problems. This issue has been addressed by the development of hybrid algorithms, such as AMALGAM (a multi-algorithm, genetically adaptive multiobjective approach proposed by Vrugt and Robinson (2007)), which combines several different optimization algorithms to improve search efficiency. These hybrid algorithms also have the benefit of requiring little to no parameter specification by the user.

5. SENSITIVITY ANALYSIS

As identified in Table 1, values of some input parameters (for example, describing the network or water demand loadings) are uncertain or subject to change in the future. Sensitivity analysis can be performed to account for a wide range of possible future conditions when optimizing and simulating systems. Variation of a particular parameter may result in different Pareto fronts (in multi-objective optimization) or different optimal solutions (in single objective optimization), as seen in Wu et al. (2010b) when they considered variations in discount rates. These various Pareto fronts or optimal solutions along with the various parameter values that produced them can then be provided to the decision maker. Sensitivity analysis will also help to identify if there are any uncertain parameters that do not affect the optimal results. Robustness of the optimized solutions can also be explored *a-posteriori*: in general, solutions that perform well for many different possible conditions are more desirable from the decision makers' point of view. Climate change provides an additional source of uncertainty for the parameters identified in Table 1 – detailed discussion of this is omitted from Sections 'Demand', 'Rainfall and streamflow', 'Electricity and GHG emissions' and 'Discount rate' as it is covered in Section 'Climate change'.

5.1 Demand

In some applications, such as irrigation and agriculture, the demand rate and pattern may be deterministic [O1], either the water supplier has control over the consumption, or may be able to work with those who do to determine an optimal demand schedule. For other applications, such as domestic, commercial and industrial, the demand rate and pattern depends on the consumption of water by multiple individual users [D4, W13, W14, W15], and therefore has greater uncertainty. Historical consumption can provide some level of assurance as to how water may be used in the future, at least on an aggregated scale. Diurnal, weekly and seasonal demand variations need to be considered. In the future, factors such as climate change, population growth and water saving initiatives will affect how water is consumed and therefore impact demand rates and patterns. Emergency conditions and system failure are by their nature unpredictable and this should be taken into account when designing and operating WDSs.

An example of how demand uncertainty can be considered in the optimization of WDS design is the study by Basupi and Kapelan (2015). The demand at each time step was based on a normal distribution with a gradually increasing mean (based on deterministic demand forecasts) and an increasing standard deviation. Monte Carlo or Latin Hypercube simulation was included in their methodology to consider multiple demand scenarios. Each solution in the Pareto front was also further analyzed against three demand projections – average, optimistic (low overall demand) and pessimistic (high overall demand). Their results demonstrated the value of flexible WDS design over deterministic approaches when considering uncertainty.

5.1.1 Rainfall and streamflow

Rainfall and streamflow inputs [W1] may be required for systems using natural catchment water, harvested stormwater or imported water, and they are often treated with higher uncertainty than demands (Seifi and Hipel 2001; Reis et al. 2005). Within the current climate, there may be multiple realizations of possible rainfall and streamflow series (for example dry or wet years). Beh et al. (2015) considered rainfall, as well as population and temperature, as uncertain variables in their optimal sequencing methodology for water supply system augmentation. They considered both climate and hydrologic variability: seven possible future climate scenarios provided different forecasted rainfall reductions, and within each of these seven scenarios, 20 stochastic replicates of the rainfall sequence were produced. Different Pareto fronts were produced for each climate scenario, with the more severe scenarios finding solutions that required greater system augmentation and therefore had higher costs and

GHG emissions. The robustness of each Pareto solution was calculated based on the average reliability and vulnerability of the solution over the 20 rainfall sequences for the particular climate scenario.

5.1.2 Electricity and GHG emissions

Power source(s) [P1], electricity tariffs and costs [P2] and GHG emission factors [P3] will generally be known for the present time, however, it may not be clear how they will change in the future. The mix of power sources changes naturally over time, as different power plants are built or decommissioned. This change in power source types over time, as well as technical advancements will affect the cost and GHG emissions associated with electrical energy generation. The electricity market and economic factors will also affect the cost of electrical energy over time. Changes in electricity and GHG emissions can be an important factor to consider during an optimization problem, as shown in the following examples. Blinco et al. (2014) studied the optimization of pump operations in WDSs in relation to the minimization of GHG emissions and the use of different power source scenarios, showing that optimal tank trigger levels can be influenced by the variation in emission factors. Wu et al. (2012a) considered three different electricity tariff scenarios, which increased over time, and three different GHG emission factor scenarios, which decreased over time, in the optimization of WDS design. The different electricity tariff and emission factor scenarios affected the solutions found in the Pareto front and their overall costs and GHG emissions.

5.1.3 Discount rate

A discount rate [G4] may be used in life-cycle analysis for both ongoing economic costs and ongoing GHG emissions. In practice, discount rates on economic costs vary significantly between different organizations, generally from 2 to 10% (Rambaud and Torrecillas 2005), while many water utilities use discount rates in the range of 6 to 8% (Wu et al. 2010a). When selecting discount rates, consideration should be given to whether both economic costs and GHG emissions are discounted, if they have the same discount rate, and if intergenerational equity is taken into account using social discount rates. Various social discount rates have been proposed for discounting ongoing costs; the Intergovernmental Panel on Climate Change (IPCC) adopted a zero discount rate over a period of 100 years, after which no consideration for future costs or benefits is given (Fearnside 2002), other suggestions include 1.4% (Stern 2006) for projects that are impacted by climate change, 2–4% (Weitzman 2007) and a time declining rate (Gollier and Weitzman 2010). Wu et al. (2010b) gave an example of a sensitivity analysis of discount rates in the optimization of WDS design for minimization of costs and GHG emissions. Discount rates of 0, 1.4, 2, 4, 6, 8% and a time declining rate were applied to the economic costs, with GHG emissions either not discounted at all, or discounted at the same rate as costs. They found that the different discount rate scenarios produced different Pareto fronts; when GHG emissions were discounted, the solutions tended to have lower capital costs and higher operating emissions.

5.1.4 Climate change

Management of water resources in the developed world has been based on an assumption of stationarity – that is, 'that natural systems fluctuate within an unchanging envelope of variability' (Milly et al. 2008). The effects of human-induced climate change make this assumption no longer valid (Milly et al. 2008), and introduce additional sources of uncertainty for many parameters. Uncertainty introduced by climate change is twofold – firstly, the impacts of climate change increase the uncertainty of future weather conditions; and secondly, our response to the threat of climate change and the types of adaption methods that will be utilized in the future are uncertain. Climate change affects the magnitude and temporal and spatial distribution of rainfall, temperature and other environmental factors, thus the possible rainfall and streamflow series to consider for the future will likely be different to the present. Changes to temperature and other environmental factors will also affect the hydrology of natural and urban catchments and therefore change how rainfall will transform to runoff or streamflow. Climate change impacts will also affect how people consume water, for example, higher temperatures and lower rainfall may drive people to water their gardens more. In order to simulate future climate conditions, general circulation models (GCMs) are often used in conjunction with future emissions scenarios. According to Mpelasoka and Chiew (2009), 'GCMs are the best tools available for simulating global and regional climate systems', however, the information provided is generally too coarse for applications to catchment runoff, and therefore some kind of downscaling is required. The modeling uncertainty of both the GCMs and downscaling methods increases the uncertainty of future climate scenarios (Paton et al. 2013). In 2000, the IPCC introduced several emissions scenarios (termed SRES scenarios) projecting future global GHG emissions (IPCC 2000). The various scenarios are based on different assumptions of the mix of energy generating technologies (fossil fuel or non-fossil fuel dominant) and population, economic and technological growth (IPCC 2007).

The extent to which we can reduce our GHG emissions will affect the magnitude of climate change impacts on rainfall and temperature. With the growing concerns of climate change and sustainability, renewable sources such as solar and wind will become more prevalent and replace fossil fuel sources such as coal and gas. This may affect electricity pricing and GHG emissions from power generation. Multiple future power source scenarios assuming different levels of climate change mitigation may need to be considered. Other climate change adaption strategies include economic incentives such as carbon taxes and cap and trade systems, which may affect economic analysis of WDSs. As discussed in Section 'Discount rate', when climate change and intergenerational equity are considered, social discount rates of 0, 1.4, 2–4% and time declining rates have been proposed.

Paton et al. (2013) analyzed the sources of uncertainty relating to climate change and their impact on water supply security. They considered 19 different scenarios with different combinations of six SRES scenarios, seven GCMs and six demand projections, as well as 1000 stochastic rainfall replicates. They found that the impact of the different sources of uncertainty on the optimal solutions varied over the 40-year planning period, with some having a greater effect in the short-term and others a greater effect in the long-term. Roshani and Filion (2014) investigated the impact that different climate change abatement strategies have on water main rehabilitation. They consider six carbon abatement strategies with different combinations of two discount rates (1.4 and 8%) and three carbon tax scenarios (no tax, 'fast and deep', and 'slow and shallow'). Using a low discount rate and implementing a carbon tax encouraged the optimization algorithm to find solutions that invested in rehabilitation early, to reduce the cost of continuing leaks, pipe repair, energy use and GHG emissions.

5.2 Case studies

The utility of the framework described in the previous sections will now be explored by two different case studies that have different water sources and many variables that need to be considered. These case studies are provided as an example of how the framework could be applied to optimize system operations. The first case study is a managed aquifer recharge (MAR) system that harvests stormwater from an urban creek for irrigation of reserves and sporting fields. This case study demonstrates the importance of analyzing the system by simulation prior to optimization in order to formulate the optimization problem. The second case study is a water supply system in a rural town that supplies potable water from multiple alternative water sources. This system is optimized for minimization of energy use of the many pumps used to transfer water from the various sources.

5.2.1 Ridge Park managed aquifer recharge – case study 1

Ridge Park is located in the Adelaide metropolitan area in South Australia, within the City of Unley local government area. The scheme supplies harvested stormwater to sports fields and recreational reserves in the local area for non-potable irrigation use. The scheme is designed to harvest up to 60 ML of stormwater per year, which occurs over the winter, while in summer the harvested water is used for irrigation. During winter, stormwater from Glen Osmond Creek, an urban waterway, is collected in the Harvest Pond created by a dam on the creek (Fig. 2). Water is then pumped to the Bioretention Basin which provides some treatment, and then pumped to a small treatment plant that includes UV and filtration. Once the water has been adequately treated, it is stored in an above ground tank next to the treatment plant and final pump station. From the Storage Tank, water is injected into an artesian, fractured rock aquifer for long term storage. In summer, when no water is being harvested, water is extracted from the Aquifer and to the Storage Tank, before being pumped or gravity-fed to irrigation points. The Ridge Park Reserve is irrigated by a pressurized irrigation line, as it is at higher elevation than the Storage Tank. Fraser Reserve is also connected to the pressurized system to ensure adequate pressures for irrigation. In total, the pressurized system supplies almost 15 ML of water per year. The remaining seven reserves are on a gravity-fed line which supplies a total demand of roughly 35 ML per year. The layout and details of the system are given in Fig. 3. For more detailed data on this case study, please see the Additional files 1, 2, 3 and 4.

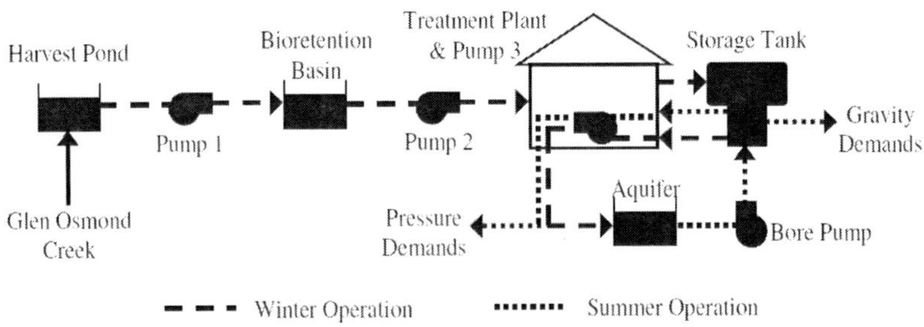

Figure. 2 Ridge Park managed aquifer recharge system process schematic

For existing systems, simulation analysis of the current operation is an important first step in formulating the optimization problem. Results of current operational simulations can highlight areas for improvement that can then be focused on in the optimization. The operation of the Ridge Park stormwater harvesting system was split between winter and summer operations and both were simulated in EPANET to determine current pump operational costs. Trigger levels (related to volumes in the three storages as shown in Table 2) control when the pumps in the Winter Harvesting and Injection system turn on and off (Table 2). The Bore Pump is also controlled by trigger levels in the Storage Tank. During summer, Pump 3 is controlled by the irrigation demands, which are on a schedule so that different reserves are irrigated on different nights (Table 3). Pump 3 is a VSP and is operated at 80% of full speed for injection (such that the flow is less than the 7 L/s maximum for injection) and 75% of full speed for irrigation (such that the target pressure downstream of the pump is achieved at the expected demand rates). Both systems were simulated for a period of 1 week in EPANET, with a 15 min hydraulic time step and 5 min reporting time step. Several week-long streamflow series for the available flow in Glen Osmond Creek at a daily resolution were used in the harvesting and injection model (Fig. 4). A peak/off-peak electricity price tariff applied to the entire system; a peak price of 25.53 c/kWh was applied from 7 am to 9 pm on weekdays, and an off-peak price of 15.26 c/kWh was applied over night and on weekends (tariff pattern and simulations started on a Sunday).

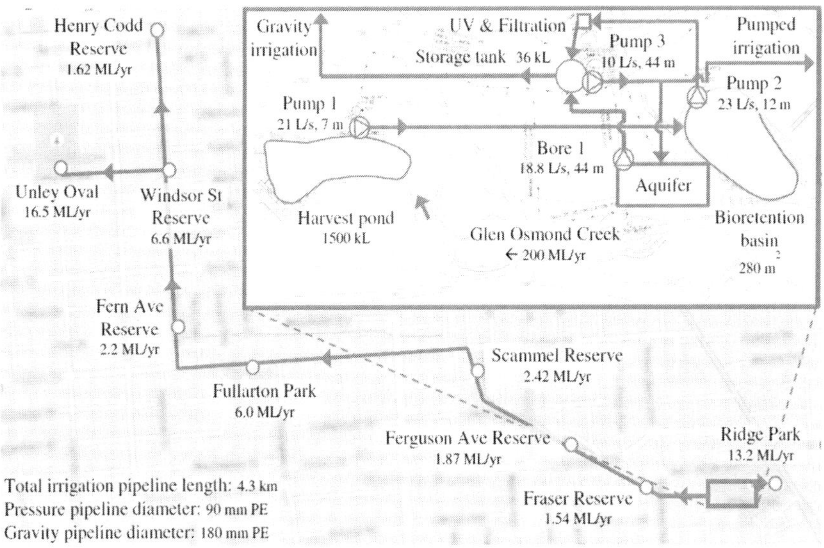

Figure. 3 Ridge Park managed aquifer recharge system layout and data

Table 2 Trigger levels for the Ridge Park system

Storage and trigger level type	Current setpoint		Start pump	Stop pump
	Volume (%)	Level (m)		
Harvest Pond High Level	80	1.6	1	-
Harvest Pond Low Level	50	1.0	-	1
Biofiltration Basin High	90	0.80	2	1

Level				
Biofiltration Basin Low Level	50	0.59	-	2
Storage Tank High Level	90	2.25	3	2, Bore
Storage Tank Low Level	70	1.75	Bore	3

Table 3 Irrigation demand schedule for the Ridge Park system

Reserve	Demand rate (L/s)	Duration/day (hr)	Start time	Irrigation days
Ridge Park 1	3.53	8.33	9:30 PM	Mon & Wed
Ridge Park 2	3.53	8.67	9:30 PM	Tues & Thurs
Fraser Reserve	1.41	5.83	9:30 PM	Mon & Wed
Ferguson Ave Reserve	2.00	5.00	9:30 PM	Tues & Thurs
Scammell Reserve	2.15	6.00	10:00 PM	Tues & Thurs
Fullarton Park 1	3.85	1.67	10:00 PM	Mon & Wed
Fullarton Park 2	3.85	6.67	10:00 PM	Tues & Thurs
Fern Ave Reserve	3.53	3.33	10:00 PM	Mon & Wed
Windsor St Reserve	2.20	8.00	8:30 PM	Tues & Thurs
Henry Codd Reserve	1.10	8.00	10:00 PM	Mon & Wed
Unley Oval	5.57	9.00	9:00 PM	Sun, Mon & Wed

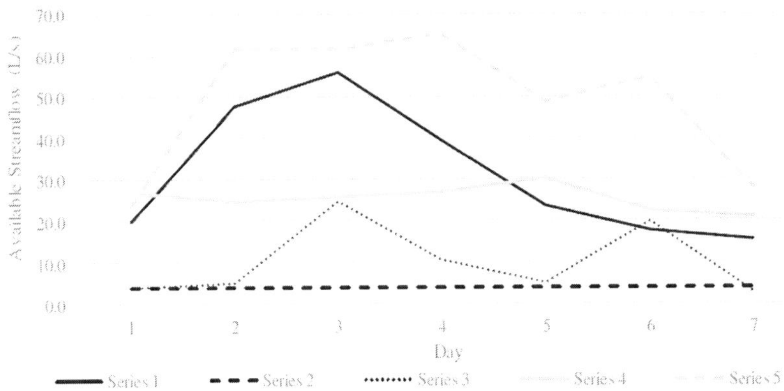

Figure. 4 Streamflow Series used for simulation of the winter harvesting and injection operation

5.3 Winter harvesting and injection system current pumping operation results

When there was adequate water available, such as in Streamflow Series 1, 4 and 5, the volume of water injected into the aquifer (by Pump 3) was a little over 3 ML per week (Table 4). This was significantly less than the volume available, which reflects the limited flow rate of Pump 3 (7 L/s for injection to the aquifer), as well as the water that would be lost to overflow when the inflow rate is greater than the flow rate of Pump 1 (approximately 22 L/s). The total pump energy cost estimate for the harvesting and injection system ranged from $163 to $267 per week, with an average of $235 per week. Pump 1 was the most cost-effective to run, while Pump 3 was the most expensive. Pumps 1 and 2 operated at similar times throughout the day, however, Pump 2 has much lower efficiencies, which increased its energy use. Pump 3 operated at a lower flow rate but much higher head than Pumps 1 and 2, and was more likely to be switched on for the entire day, which contributed to its higher cost of operation. Pumps 1 and 2 turned on and off very frequently, and operated at a much higher flow rate than Pump 3 (Fig. 5). The flow rate of Pump 3 in Fig. 5(c) reduced over the week as the headloss through the bore increased from assumed clogging of the bore. As the storages are relatively small, in particular the storage tank, it did not take long for them to be filled and emptied (Fig. 6), which contributed to the frequent pump switches. The current trigger levels in the Storage Tank are very close together (70 and 90% volume) as a result of possible pump priming issues that occurred during the commissioning of the system. These close together trigger levels also contributed to the short fill and empty times.

Table 4 Current operation results for the winter harvest and injection system

Streamflow Series	Available volume (ML/wk)	Cost (c/kL)			Volume injected (ML/wk)	Total cost ($/wk)
		Pump 1	Pump 2	Pump 3		
1	19.0	0.64	2.28	5.49	3.14	267
2	2.29	0.68	2.32	6.19	1.76	163
3	6.19	0.69	2.23	5.87	2.44	222
4	15.4	0.64	2.24	5.46	3.18	258
5	29.7	0.63	2.25	5.47	3.16	264
Average	14.5	0.66	2.26	5.70	2.74	235

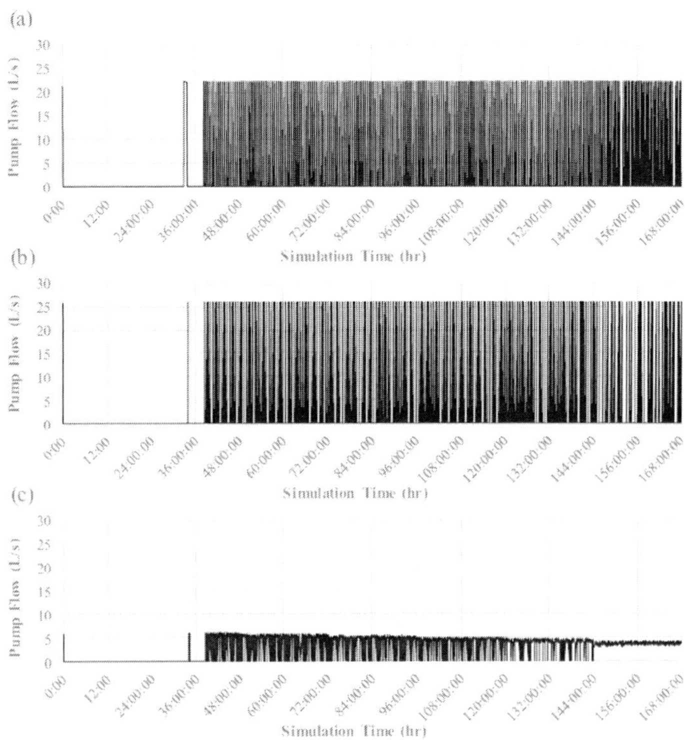

Figure. 5 Pump flows for Streamflow Series 2 for **A** Pump 1, **B** Pump 2 and **C** Pump 3 for a one week EPANET simulation

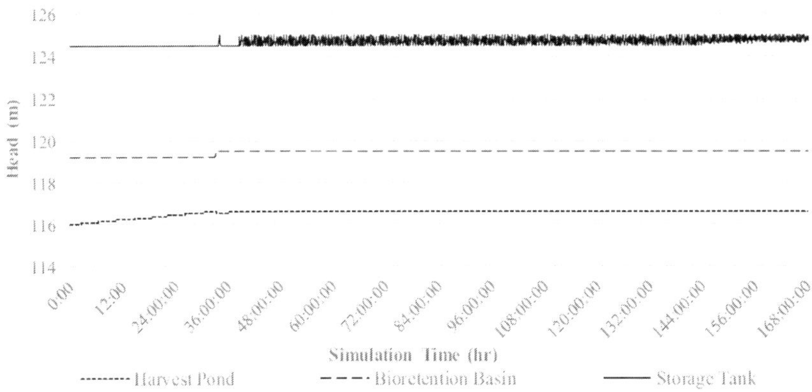

Figure. 6 Storage levels for Streamflow Series 2

5.4 Summer extraction and injection system current pumping operation results

Simulation of the irrigation system gave a total weekly pump energy cost of $90 (Table 5). The Bore Pump was more expensive overall, however, cost less per megaliter than Pump 3. This occurred because while the Bore Pump has a greater efficiency than Pump 3, it also has a higher flow and head, which increased the energy consumption. The higher volume pumped from the bore contributed to a lower cost rate than Pump 3. All of the pumping for this system occurred overnight (Fig. 7) when irrigation of all fields is allowed. The Bore Pump turned on and off very frequently when it was operating, again due to the small capacity of the Storage Tank which meant it did not take long for the pump to fill the operating volume (Fig. 8).

Table 5 Current operation results for the summer extraction and irrigation system

Pump	Volume (ML/wk)	Cost (c/kL)
Bore Pump	1.93	3.52
Pump 3	0.57	3.97
Total		$90.3/week

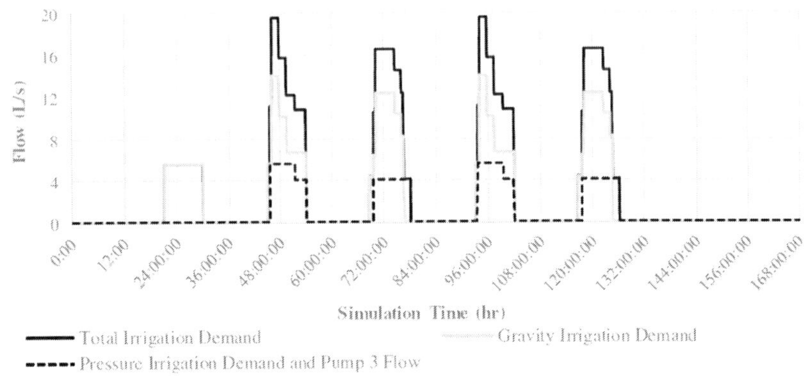

Figure. 7 Current demand rate and pump flows for the irrigation system

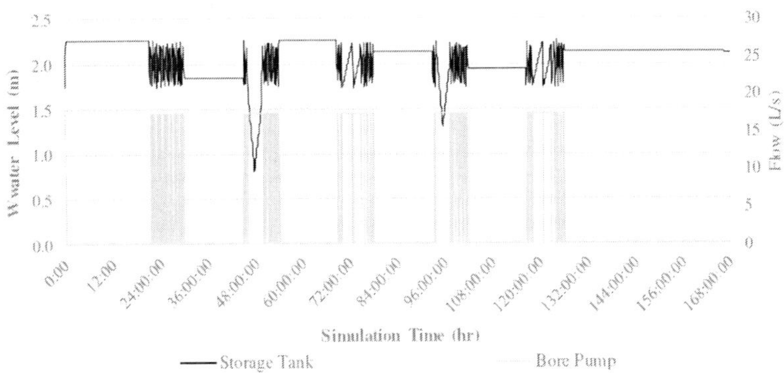

Figure. 8 Storage tank level and bore pump flow for the summer extraction and irrigation system

5.5 Optimization formulation

Initially, optimization of the Ridge Park system was considered to be an operational problem, however, results of the current operation simulation suggest that design decision variables need to be considered as well. Replacing Pumps 1 and 2 with models that would operate at much lower flow rates (to reduce the headlosses) and increasing the size of the Storage Tank will be considered along with operational decision variables (Table 6). These design decisions would aim to counter-act mismatched pump rates (Pumps 1 and 2 operating at a much higher rate than Pump 3) and small storage volumes that lead to frequent pump switches. Short-term operational decisions include trigger levels in the Harvest Pond,

Bioretention Basin and Storage Tank that will govern when pumps are turned on and off, a schedule for irrigation (that is, which reserves will be irrigated at which times), and VSP multipliers for Pump 3. In the current operation, VSP multipliers for Pump 3 were selected to ensure the required flow rate (for injection) and pressure (for irrigation) were achieved. With different levels in the Storage Tank considered, the VSP multipliers for Pump 3 can be altered, especially if efficiency is improved. If the pump priming issues discussed earlier were to be resolved, trigger levels that utilize the full height of the Storage Tank (rather than the 20% range in water elevation that is currently used) would be considered in the optimization. There are also long-term decision variables deciding when to switch between summer and winter operation and vice versa (Table 6). As the scheme injects to and extracts from the aquifer through the same bore, it is not possible to frequently switch between injecting and extracting water, therefore there will be only two switch times per year; one going into winter operation and one going into summer operation. The decision variables presented in Table 6 may all be considered together in an optimization problem, however, they could also be analyzed prior to optimization in a simulation sensitivity analysis. Simulating the system initially with the different pump models and storage tank sizes could help to decide if these actions are worthwhile considering in an optimization formulation. Engineering judgement may be sufficient to determine which pump model(s) would be best to replace Pumps 1 and 2, and therefore reduce the size of the optimization problem.

Table 6 Possible decision variables for the Ridge Park MAR scheme

Short-term winter harvesting and injection operation	
Pump 1 Off	Harvest Pond Level Low
	Bioretention Basin Level High
Pump 1 On	Harvest Pond Level High
Pump 2 Off	Bioretention Basin Level Low
	Storage Tank Level High
Pump 2 On	Bioretention Basin Level High
Pump 3 Off	Storage Tank Level Low
Pump 3 On	Storage Tank Level High
Pump 3 Speed	Storage Tank Level

Short-term summer extraction and irrigation operation	
Bore Pump Off	Storage Tank Level High
Bore Pump On	Storage Tank Level Low
Irrigation Schedule	Days of Irrigation at each Reserve
	Start Time of Irrigation at each Reserve
Pump 3 Speed	Required Demand Rate
Long-term operations	
Day to Switch Between	Summer to Winter
Seasonal Operational Regimes	Winter to Summer
System Design	
Storage Tank Size	Doubled, 5 times, 10 times current size
Pumps 1 and 2	Selection of pump curves with lower operating rates

Constraints on the system include an environmental flow for Glen Osmond Creek, an extraction limit from the Aquifer and meeting the weekly irrigation volumes for each reserve in the summer (Table 7). If there was not enough water harvested over the winter to supply the summer demands, a potable back-up supply is available at a cost. The main objective for this case study is to minimize the pump energy cost; there is also a secondary objective of minimizing the number of pump switches. To create an incentive for the optimization to find solutions that harvest more water, the cost objective includes the energy cost for the harvesting and distribution operation as well as the cost of purchasing potable water if the harvested volume is not enough to supply demand. The objective function is formulated as the cost per volume of water harvested as another means to ensure enough water is harvested from the system during winter to supply summer irrigation. During the conceptualization and design of this scheme, regulations from the South Australian Environmental Protection Authority (EPA), the Department for Environment, Water and Natural Resources (DEWNR) and the Department of Health (DoH) were considered. A license to recharge water into the aquifer was required from the EPA, while the DEWNR regulates how much water can be extracted from MAR schemes. DoH regulations informed the level of treatment implemented and the irrigation practices, which must limit the risk of public exposure.

Table 7 Possible constraints for the Ridge Park MAR scheme

Constraint	Value
Glen Osmond Creek Environmental Flow	>2 L/s
Aquifer Extraction in Summer	<80% of Injection Volume
Pressurized System Demands	>15 ML/year
Gravity System Demands	>37 ML/year

5.6 Orange integrated supply system – case study 2

Orange is a rural town roughly 250 km west of Sydney in the state of New South Wales, Australia. The water supply system serves a population of around 36,800 people with an average annual demand of approximately 5,400 ML. The majority of water supply is from the local surface water catchment, which culminates in the roughly 19,000 ML Suma Park reservoir (Fig. 9). Australia experienced severe drought between 2000 and 2010, and Orange was one of the hardest hit areas in New South Wales. Even with severe water restrictions almost halving the town's demand, Orange had less than two years of water supply heading into summer of 2009 and was relying only on surface water catchments (Montgomery Watson Harza, 2011). This prompted the Orange City Council to diversify their water supply, and they therefore developed two stormwater harvesting schemes and a long pipeline from an adjacent catchment, as well as re-opening several groundwater bores. Figure 2 shows a schematic process diagram of the system, which is described below, and Fig. 3 shows the layout (note that the 'Shearing Shed' Bore and 'Bore 5' in Fig. 2 are grouped as the 'Clifton Grove' Bores in Fig. 10). For more detailed data on this case study, please see the Additional files 1, 2, 3 and 4.

Water from the Ploughman's Creek Stormwater Scheme is treated through a series of wetlands, and then combined with water from the Blackman's Swamp Creek Stormwater Scheme. After treatment, this water can be used to top up Suma Park reservoir. Due to the severely low water supply levels during the drought, Emergency Authorization was initially given, and Council subsequently sought approval for use of the stormwater schemes on a permanent basis. Continuous water quality monitoring is undertaken to meet regulations of the Office for Water, the New South Wales Environmental Protection Authority and the Ministry of Health. To the authors' knowledge, this is the only system in Australia that has been approved to use harvested stormwater for potable supply. In order to use harvested stormwater for potable supply, the Council had to meet requirements of the Office for Water. The Macquarie pipeline transfers water from the adjacent Macquarie River catchment to Suma Park reservoir. It is 38 km long and requires more than 600 m of pumping head. Each of the three pumping stations has two pumps operating in parallel. Water from the groundwater bores is pumped first to balancing storages and then to Suma Park reservoir, with a combined licensed volume of 462 ML per year. Water from all of the sources is combined in Suma Park reservoir and treated at a water treatment plant before being delivered to consumers.

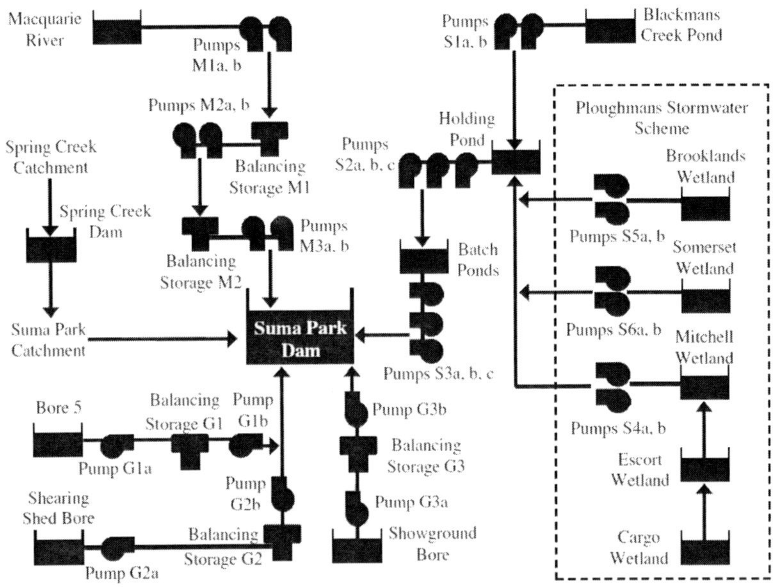

Figure. 9 Orange integrated supply system process schematic – inflow to Suma Park Dam

The Orange City Council is interested in optimizing the operation of this while delivering a secure yield from Suma Park Dam. In addition to the primary objective of minimizing energy cost, there are objectives of minimization of spill from Suma Park reservoir, minimization of (perceived) environmental impact, maximization of (perceived) water quality, and minimization of energy use. The Council has an explicit objective to minimize spill to ensure water and energy are not wasted by pumping from one of the three alternative sources to fill up Suma Park reservoir just before a rainfall event that would supply water from the natural catchment at no cost or energy output. As this system supplies potable demands, it is undesirable to apply water restrictions to consumers, thus minimizing time spent in restrictions is important. Objectives for the perceived environmental impact and water quality will be formulated as a preference ranking between the different sources based on community views of which sources are better for the environment and water quality. The constraints of the problem include environmental flows for the Macquarie River (downstream of the pumping station offtake point) and stormwater schemes, a water source license for the Macquarie River and extraction limits on the groundwater bores (Table 8).

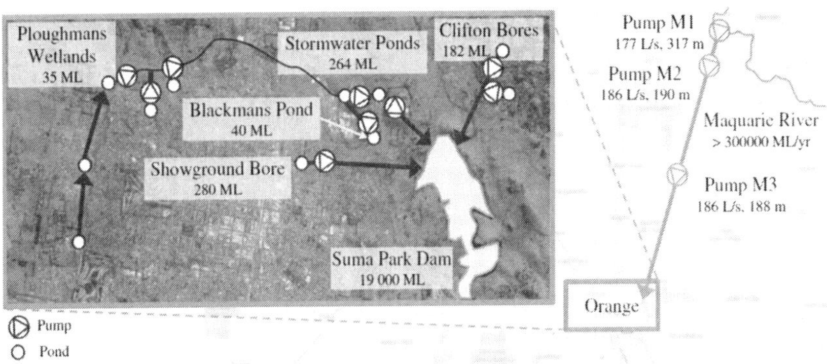

Figure. 10 Orange integrated supply system layout and data

Table 8 Constraints for the Orange integrated supply system

Constraint	Value
Macquarie River Environmental Flow	>108 ML/day
Blackmans' Creek Environmental Flow	>20 ML/day
Ploughmans' Creek Environmental Flow from Pump S4	>0.4 ML/day
Ploughmans' Creek Environmental Flow from Pump S5	>2 ML/day
Ploughmans' Creek Environmental Flow from Pump S6	>2 ML/day
Clifton Grove (Shearing Shed and Bore 5) Aquifer Extraction	<182 ML/year
Showground Aquifer Extraction	<280 ML/year
Macquarie River Extraction License	<12 ML/day

5.7 Energy optimization formulation

In this section, the developed framework is applied to the Orange case study to help set up the optimization procedure and identify the components and data to be modeled. Note that the model has been built taking into account all possible objectives of the system, however, the example of results presented here will focus on the minimization of energy consumption.

As all components of the system have already been constructed and considered sufficient for the operation of the system, there are no design decisions to consider, only operating decisions. For this case study, operating decisions consist of trigger levels in the various storages. These types of decision variables are chosen considering the control system available at each pump station (based on storage levels and not on time of the day) and the fact that the controls have to be defined for an operational horizon of 1 year or longer. As all of the pump stations have two or more pumps arranged in parallel, having different trigger level values may have a large impact on the operating point of the pumps and consequently their energy consumption. It is also likely that different trigger levels will be chosen for peak and off-peak electricity tariff periods when they are included in a cost optimization. For this system a peak/off-peak electricity tariff applies on weekdays, with weekends priced at the off-peak rate. A peak monthly electrical energy demand charge also applies to the Macquarie River pipeline pumping system. In order to assess the performance of different tank trigger levels, the infrastructure to be modeled includes the natural and urban catchments for the surface water and stormwater systems respectively, Suma Park reservoir, pipelines and pumps in the groundwater, Macquarie River and stormwater systems, and wetlands and storage ponds in the stormwater systems.

In general, the system could be modelled using hydrologic models, mass balance models, and/or hydraulic models. The choice of which model(s) will be used depends on the objectives and the processes to be modelled, on the available data and the computational times. In particular, hydrologic modeling is usually used to transform rainfall to runoff for the natural and urban catchments. For this case study, inflows inputs or approximate relationships between rain and flows were provided by previous studies by the Orange City Council. Hydraulic models are usually used for short term operations: pump energy costs can be computed accurately based on the hydraulic equations. Mass balance modeling is usually used for assessing the system in long term operations, as it can quickly compute the water available after evaporation and other losses in the system have occurred and after minimum environmental flows have been released. It cannot, however, take into account the non-linearity in the hydraulic equations and therefore assumptions need to be made in regard to the flow delivered by the pumps in the system. While hydraulic simulation would be most appropriate for the pumping stations in the system as they have multiple pumps and sometimes have connected pipelines, mass balance models would need to be used to compute the additional processes, such as evaporation and the release of minimum environmental flows that need to be taken into account given the long duration of the simulation. During an optimization process, simulating each potential solution using both a mass balance and a hydraulic model would increase considerably the computational time, particularly if data transfer between the two models was required. It is therefore suggested that the primary simulation tool should be a hydraulic solver. Rainfall-runoff modeling could be performed pre-optimization, and supplemental code added to a hydraulic model to account for functionality of a mass balance model. This would allow for consideration of the evaporation from and rainfall directly to reservoirs, changes to demands based on water restrictions and environmental flows that depend on the combined volume of two reservoirs (Spring Creek and Suma Park), infiltration losses when transferring water between reservoirs and peak power demand charges.

Another important issue to consider is what simulation time step should be used. Using a shorter time step will increase the accuracy of this hydraulic analysis and often results in feasible optimization times for storages that empty or fill in a day or two (as would likely be the case for the stormwater ponds and Macquarie pipeline balancing storages). Simulating the behavior of Suma Park dam is more challenging, however, as the variations in the water levels can have a period of several years. Thus, the computation times with a short time step become prohibitively long. A balance needs to be found between using a short enough time step for the detailed hydraulics and a long simulation time for the large storages without having a prohibitively large computational time. Given the data availability (there is 118 years of rainfall and inflow data available, with a daily time step) the time step chosen is 1 day.

Given that the time-step is automatically shortened by the hydraulic solver chosen (EPANET in this case), the model of the real system has been simplified in order to avoid excessive computational times. In particular, given that the levels in the balancing storages along the Macquarie pipeline vary rapidly, these storages were removed and the pipeline simulated with two parallel pumps, each representing the equivalent of the three stages of pumping (that is, the pump curves for Pumps M1a and b in Fig. 2 were adjusted such that they represented Pumps M2a, M3a and Pumps M2b, M3b as well). This simplification is considered acceptable as the pumps in series in the Macquarie pipeline will usually be operated at the same time, given that each pump will still be controlled also by the level of Suma Park Dam. Longer computational times were also caused by the small storages after the groundwater bores. The pumps used for extraction from the aquifers (Pumps G1a, G2a and G3a in Fig. 2) operate at relatively consistent rates, and as such they could be removed from the model and their energy use accounted for relative to the volume pumped from the second pump in each system (Pumps G1b, G2b and G3b respectively). To take into account the limited volume available from the groundwater bores, the storage tanks in the groundwater system each had a volume equivalent to a year's allocation for the respective bores. All of the stormwater pumps except for Pump S2c and Pump S3c, which are standby pumps and not in use, were included in the model. As well as the operating point of the pumps changing depending on the number of pumps used in parallel, there may be slight differences in efficiency and therefore energy use, and thus including all pumps here provided more accuracy.

All of the pumps included in the model were controlled using rule-based controls in EPANET, with conditions based on levels in one or more storages as well as time. Conditions based on downstream storages were considered as decision variables, while conditions based on upstream storages were fixed (Table 9). For the Macquarie pumps, there were also conditions based on the flow in the river to ensure that no water would be taken when there was not enough water available. There were four possible decision variables for each pump, a lower and upper trigger level in both the peak and off-peak time. For optimization of energy use, only two are required, as peak and off-peak tariffs are not considered. As the model was set up for other objectives including cost, which does use a peak and off-peak electricity tariff, the capability to choose different trigger levels in different periods was incorporated. A maximum of 15 pump switches per day per pump were allowed, and the end level of Suma Park Dam was constrained to 16 m (to be approximately the same as the start level). Based on license conditions, Macquarie River water can only be used when the Suma Park Dam level is below 90%, so choices for Pump M1a and M1b trigger levels in Suma Park Dam are more restricted than for other pumps.

Table 9 Decision variables and fixed controls for the Orange integrated supply system

Pump station action	Storage(s) controlling operation	Decision variable or fixed
Macquarie Pump M1a, M1b Off	Suma Park Dam Level High	Decision Variable
Macquarie Pump M1a, M1b On	Suma Park Dam Level Low	Decision Variable

Stormwater Pump S1a, S1b Off	Holding Pond Level High Blackmans Stormwater Pond Level Low	Decision Variable Fixed
Stormwater Pump S1a, S1b On	Holding Pond Level Low Blackmans Stormwater Pond Level High	Decision Variable Fixed
Stormwater Pump S2a, S2b Off	Batch Ponds Level High Holding Pond Level Low	Decision Variable Fixed
Stormwater Pump S2a, S2b On	Batch Ponds Level Low Holding Pond Level High	Decision Variable Fixed
Stormwater Pump S3a, S3b Off	Suma Park Dam Level High Batch Ponds Level Low	Decision Variable Fixed
Stormwater Pump S3a, S3b On	Suma Park Dam Level Low Batch Ponds Level High	Decision Variable Fixed
Stormwater Pump S4a, S4b Off	Holding Pond Level High Mitchell Wetland Level Low	Decision Variable Fixed
Stormwater Pump S4a, S4b On	Holding Pond Level Low Mitchell Wetland Level High	Decision Variable Fixed
Stormwater Pump S5a, S5b Off	Holding Pond Level High Brooklands Wetland Level Low	Decision Variable Fixed
Stormwater Pump S5a, S5b On	Holding Pond Level Low Brooklands Wetland Level High	Decision Variable Fixed
Stormwater Pump S6a, S6b Off	Holding Pond Level High Somerset Wetland Level Low	Decision Variable Fixed
Stormwater Pump S6a, S6b On	Holding Pond Level Low Somerset Wetland Level High	Decision Variable Fixed
Groundwater Pump	Suma Park Dam Level High	Decision Variable

G1 Off		
Groundwater Pump G1 On	Suma Park Dam Level Low	Decision Variable
Groundwater Pump G2 Off	Suma Park Dam Level High	Decision Variable
Groundwater Pump G2 On	Suma Park Dam Level Low	Decision Variable
Groundwater Pump G3 Off	Suma Park Dam Level High	Decision Variable
Groundwater Pump G3 On	Suma Park Dam Level Low	Decision Variable

5.8 Energy optimization results

Minimization of pump energy use over the longer term is presented here as an example of optimization of this system. Note that the system is simulated over 1 year, at a daily time step in EPANET. Additional computer code was added to the EPANET hydraulic simulation to take into account other process such as rainfall to and evaporation from storages. This code essentially adds a mass balance component to the hydraulic simulation. Historical rainfall for the catchments in the system was modelled in MUSIC hydrologic software to develop inflow series for the ponds and reservoirs. For this optimization the year with the closest to average rainfall was used, however, other years of rainfall were available and this optimization could be extended to consider other climate conditions.

NSGAII (Non-dominated Sorting Genetic Algorithm II) software was used for the optimization, with five random seeds, a population size of 50, 100 generations and probabilities of crossover and mutation of 0.8 and 0.02 respectively. In the best solution found, the system used a total of 793 MWh of energy over the entire year. Table 10 shows the volume of water pumped from each source to Suma Park Dam (and supplied from the local catchment) and the energy used by each of the pumps for the optimal solution. Pumping from the Macquarie is very energy intensive so this is only used at the very end of the simulation when the level in Suma Park Dam is very low, in order to achieve the end target level constraint (Figs. 11 and 12). Groundwater and stormwater sources are used initially to increase the level of Suma Park Dam to its maximum, and then not used again until around Day 160 when the level in the dam has dropped again. Only one of the Macquarie pumps is used, as, despite operating at a lower energy efficiency point, it uses less energy overall than operating two pumps in parallel. In dryer years, both pumps may need to be utilized in order to ensure supply to Suma Park Dam. Nearly all of the available groundwater license is used; G1 and G2 have a combined license of 180 ML/year, and G3 280 ML/year. Groundwater is more energy intensive than stormwater, however, it can be used at any time throughout the year, while stormwater is reliant of inflow. Most of the stormwater provided to Suma Park Dam came from the Blackman's Creek scheme (S1) rather than the Ploughman's Creek scheme (S4, S5 and S6). While the storage capacity of the Blackman's Creek scheme is much lower, the pump capacity and energy efficiency is much greater than in the Ploughman's Creek scheme, so it provides more water.

Table 10 Volume of water pumped/supplied and energy used in the optimal energy solution

Source	Pump	Volume (ML)	Energy (MWh)	Energy Rate (MWh/ML)
Macquarie River	M1a	0	0	0
	M1b	74	150	2.02
	Total	74	150	2.02
Groundwater[a]	G1	24	11	0.46
	G2	146	79	0.54
	G3	235	106	0.45
	Total	405	196	0.48
Stormwater	S1a	258	39	0.15
	S1b	479	71	0.15
	S2a	828	65	0.08
	S2b	237	21	0.09
	S3a	1022	170	0.17
	S3b	22	5.5	0.25
	S4a	178	41	0.23
	S4b	12	3.1	0.27
	S5a	24	4.8	0.20
	S5b	56	11	0.19

	S6a	60	11	0.18
	S6b	26	5.0	0.19
	Total[b]	1044	447	0.43
Spring Creek and Suma Park Catchment	-	3865[c]	-	-

[a]The energy consumption for the groundwater pumps includes both the transfer and bore pumps, i.e. the energy for Pump G1 includes G1a (not modelling in EPANET, energy use estimated from volume) and G1b (modelled in EPANET).

[b]The total volume supplied by the stormwater schemes is measured as the combined volume supplied by Pumps S3a and S3b (which discharge to Suma Park Dam), while the total energy is the total of all pumps.

[c]This is the volume supplied by the natural catchment for the town's consumption, the total inflow from the catchment is greater than this however some is used to provide environmental flows and some spills.

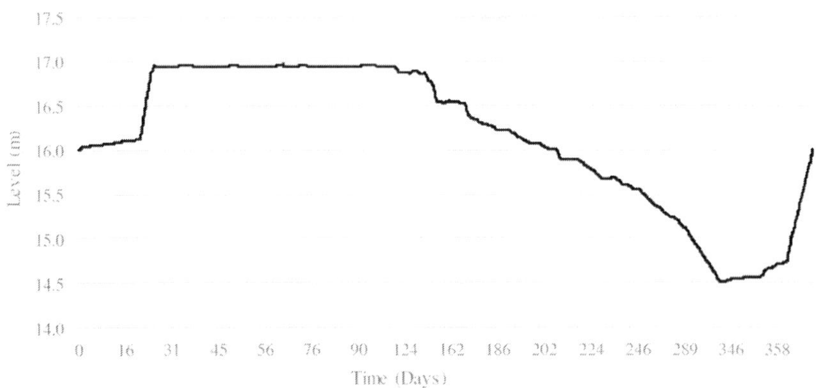

Figure. 11 Variation in Suma Park Dam level for the energy optimal solution

6. CONCLUSIONS

A generalized framework for the optimization of the design and operation of water supply and distribution systems has been developed and two case study systems have been used as examples of how to apply it. The framework is comprised of several components; the options component describes the design and operational decision variables for the optimization, the infrastructure component covers the infrastructure aspects of the system that need to be modeled and their data requirements, the analysis component includes the simulation of the system and evaluation against the objectives and constraints, and finally the government policy component describes the regulations that may affect other aspects of the framework. These components fit within an optimization algorithm structure, which firstly generates potential solutions using the decision variables in the options component, models the system according to the infrastructure component and evaluates potential solutions using the analysis component. The evaluation of potential solutions then feeds into the solution space which informs how the decision variables are changed in the next set of potential solutions. Sensitivity analysis of parameters will significant uncertainty should be undertaken to ensure robust solutions. The framework also applies to simulation of systems prior to or without optimization.

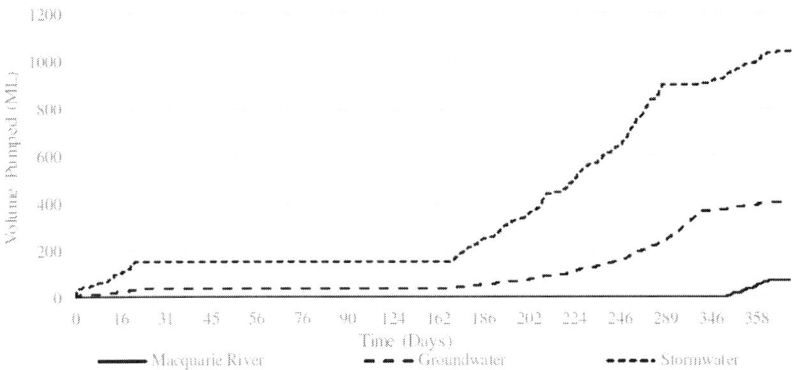

Figure. 12 Volume pumped from each source to Suma Park Dam for the energy optimal solution

The Ridge Park MAR Scheme Case Study harvests stormwater from an urban creek and stores it in an aquifer, to be extracted at a later time and used as non-potable supply for irrigation of sporting fields and reserves. For this case study, and similar ones, the simulation of the system may be simplified by splitting the system into two parts, one for the components of the system used in winter operation (harvesting and injection) and one for the components used in summer operation (extraction and irrigation). This system highlighted the importance of simulation and analysis prior to optimization, in order to focus the formulation of the optimization problem.

The Orange Integrated Supply System Case Study uses multiple water sources; natural catchment water, harvested stormwater, imported water and groundwater to supply potable demands. For this case study, finding an appropriate combination of simulation models and time step and simulation duration is important in order to provide accuracy in representing both long- and short-term operations without excessive computational times. Optimization of pump energy use for this system indicated that the groundwater and stormwater supplies are more desirable to supplement natural inflows than the imported water from the Macquarie River, which required a lot of energy to transfer water over a long distance and against a high elevation head.

The framework is generalized, and so could be applied to other water supply and distribution systems, particularly those using non-traditional water sources, to optimize their design and operation. While the framework attempts to cover all aspects of water supply in a generalized manner, it does have some limitations. Along with the supply of water, there will always be a need to manage wastewater. Apart from considering recycled wastewater as a source, this framework does not cover wastewater systems in terms of collection, transport, treatment and potential discharge of wastewater into the environment. Treatment of raw water supplies is included in the framework, however, the details of such treatment and measurement of water quality throughout a water distribution system are not focused on as much as the design and operation of the systems. A difficulty of applying this framework will be the definition of the boundary of a system and which aspects should be analyzed. Currently, there does not exist commercial software that has all of the capabilities considered in the framework (i.e. both hydrologic and hydraulic simulation). This means that specialist simulation models may need to be developed for particular systems (as was done for the second case study). Future developments in simulation software may reduce the difficultly of combining hydrologic, mass balance and hydraulic considerations and remove the need for specialist tools built for individual systems. In the future, the framework should be tested with other case study systems to fully investigate its benefits.

ACKNOWLEDGEMENTS

The support of the Commonwealth of Australia through the Cooperative Research Centre program is acknowledged. This research was part of the CRC for Water Sensitive Cities Project C5.1 (Intelligent Urban Water Networks) and was supported by funding for post-doctoral research and a PhD top-up scholarship. The authors would like to thank Martin Haege

(Geolyse), Chris Devitt (Orange City Council) and Matt Sanderson (The City of Unley) for providing information on the case study systems presented in this paper.

REFERENCES

1. Ashbolt S, Maheepala S, Perera BJC (2014) A framework for short-termoperational planning for water grids. Water Resour Manag 28:2367–2380.doi:10.1007/s11269-014-0620-4
2. Basupi I, Kapelan Z (2015) Flexible water distribution system design under futuredemand uncertainty. J Water Resour Plan Manag 141(4). doi:10.1061/(ASCE)WR.1943-5452.0000416
3. Becker N, Lavee D, Katz D (2010) Desalination and alternative water-shortagemitigation options in Israel: a comparative cost analysis. J Water Resour Prot2:1042–1056. doi:10.4236/jwarp.2012.212124
4. Beh EHY, Dandy GC, Maier HR, Paton FL (2014) Optimal sequencing of watersupply options at the regional scale incorporating alternative water supply sources and multiple objectives. Environ Model Softw 53:137–153.doi:10.1016/j.envsoft.2013.11.004
5. Beh EHY, Maier H, Dandy GC (2015) Adaptive, multiobjective optimal sequencing approach for urban water supply augmentation under deep uncertainty.Water Resour Res 51. doi:10.1002/2014WR016254
6. Blinco LJ, Simpson AR, Lambert MF, Auricht CA, Hurr NE, Tiggemann SM, Marchi AGenetic algorithm optimization of operational costs and greenhouse gas emissions for water distribution systems. In: 16th Conference on WaterDistribution System Analysis, Bari, Italy, 14–17 July 2014. Procedia Engineering, pp 509–516. doi:10.1016/j. proeng. 2014.11.246
7. Bogardi JJ, Dudgeon D, Lawford R, Flinkerbusch E, Meyn A, Pahl-Wostl C,Vielhauer K, Vörösmarty C (2012) Water security for a planet under pressure:
8. Curr Opin Environ Sustain 4:35–43. doi:10.1016/j.cosust.2011.12.002Cai X, Zeng R, Kang WH, Song J, Valocchi AJ (2015) Strategic planning for drought mitigation under climate change. J Water Resour Plan Mana141(9). doi:10.1061/(ASCE)WR.1943-5452.0000510
9. Campisano A, Modica C (2012) Optimal sizing of storage tanks for domesticrainwater harvesting in Sicily. Resour Conserv Recycl 63:9–12. doi:10.1016/j.resconnrec. 2012. 03.007
10. Chung G, Lansey K, Blowers P, Brooks P, Ela W, Stewart S, Wilson P (2008) Ageneral water supply planning model: evaluation of decentralized treatment.
11. Environ Model Softw 23:893–905. doi:10.1016/j.envsoft.2007.10.002Cook C, Bakker K (2012) Water security: debating an emerging paradigm.Glob Environ Chang 22:94–102. doi:10.1061/j.gloenvcha.2011.10.011
12. Dandy GC, Simpson AR, Murphy LJ (1996) An improved genetic algorithm for pipenetwork optimization. J Water Resour Res 32(2):449–458. doi:10.1029/95WR02917
13. Dawadi S, Ahmad S (2013) Evaluating the impact of demand-side management onwater resources under changing climate conditions and increasing population.J Environ Manag 114:261–275. doi:10.1016/j.jenvman.2012.10.015
14. Deb K, Pratap A, Agarwal S, Meyarivan T (2002) A fast and elitist multiobjectivegenetic algorithm: NSGA-II. Trans Evol Comput 6(2):182–197
15. Dobbie MF, Brown RR (2012) Risk perceptions and receptivity of Australiaurban water practitioners to stormwater harvesting and treatment systems.Water Sci Technol Water Supply 12(6):888–894. doi:10.2166/ws.2012.069
16. Fearnside PM (2002) Time preference in global warming calculations: a proposalfor a unified index. Ecol Econ 41(1):21–31. doi:10.1016/S0921-8009(02)00004-6

17. Fielding KS, Gardner J, Leviston Z, Price J (2015) Comparing public perceptions of alternative water sources for potable use: the case of rainwater, stormwater,desalinated water, and recycled water. Water Resour Manag 29:4501–4518.doi:10.1007/s11269-015-1072-1
18. Friedman K, Heaney JP, Morales M, Palenchar M (2014) Analytical optimizationof demand management strategies across all urban water use sectors.Water Resour Res 50:5475–5491. doi:10.1002/2013WR014261
19. Gollier C, Weitzman ML (2010) How should the distant future be discountedwhen discount rates are uncertain? Econ Lett 107:350–353. doi:10.1016/j.econlet.2010.03.001
20. Hwang AH-S, Valeo C, Draper D (2006) Public perceptions and attitudes towardstormwater recycling for irrigation. Can Water Resour J 31(3):185–196.doi: 10.4296/cwrj3103185
21. IPCC (2000) In: Nakicenovic N, Swart R (eds) Special Report on EmissionsScenarios. Cambridge
22. IPCC (2007) Synthesis Report: Contribution of Working Groups I, II and III to the Fourth Assessment Report of the Intergovernmental Panel on Climatechange Climate Change 2007. IPCC, Geneva
23. Londoño Cadavid C, Ando AW (2013) Valuing preferences over stormwatermanagement outcomes including improved hydrologic function. Water Resour Res 49:4114–4125. doi:10.1002/wrcr.20317
24. Makropoulos CK, Natsis K, Liu S, Mittas K, Butler D (2008) Decision support for sustainable option selection in integrated urban water management. Environ Model Softw 23:1148–1460. doi:10.1016/j.envsoft.2008.04.010
25. Marchi A, Dandy GC, Maier HR (2016) Integrated approach for optimizing thedesign of aquifer storage and recovery stormwater harvesting schemes accounting for externalities and climate change. J Water Resour Plan Manag 142(4). doi:10.1061/(ASCE)WR.1943-5452.0000628
26. McArdle P, Gleeson J, Hammond T, Heslop E, Holden R, Kuczera G (2011)Centralised urban stormwater harvesting for potable reuse. Water Sci Technol 63(1):16–24. doi:10.2166/wst/2011.003
27. Miller S, Shermer H, Semiat R (2015) Energy and environmental issues in desalination. Desalination 366:2–8. doi:10.1016/j.desal.2014.11.034
28. Milly PCD, Betancourt J, Falkenmark M, Hirsch RM, Kundzewicz ZW, Lettenmaier DP,Stouffer RJ (2008) Stationary is dead: whither water management? Science319:573–574. doi:10.1126/science.1151915
29. Montgomery Watson Harza (2011). Orange drought relief connection conceptinvestigation report, report prepared for the Orange City Council. http://watersecurity.orange.nsw.gov.au/client_images/1046362.pdf
30. Mpelasoka FS, Chiew FHS (2009) Influence of rainfall scenario constructionmethods on runoff projections. J Hydrometeorol 10(5):1168–1183.doi:10.1175/2009jhm1045.1
31. Muga HE, Mihelcic JR (2008) Sustainability of wastewater treatment technologies.J Environ Manag 99:437–447. doi:10.1016/j.jenvman.2007.03.008
32. Nagal R (2015) Making direct potable reuse a reality. J Am Water Works Assoc 107(7): 76–82. doi:10.5942/jawwa.2015.107.0103
33. Naylor T, Moglia M, Grant AL, Sharma AK (2012) Self-reported judgements ofmanagement and governance issues in stormwater and greywater systems.J Clean Prod 29–30:144–150. doi:10.1016/j.jclepro.2012.02.003
34. Nicklow J, Reed P, Savic D, Dessalegne T, Harrell L, Chan-Hilton A, Karamouz M,Minsker B, Ostfeld A, Singh A, Zechman E (2010) State of the art for geneticalgorithms and beyond in water resources planning and management. J WaterResour Plan Manag 136(4):412–432. doi:10.1061/(ASCE)WR.1943-5452.0000053
35. Oron G, Adel M, Agmon V, Friedler E, Halperin R, Lesham E, Weinberg D (2014)Greywater use in Israel and worldwide: standards and prospects. Water Res58:92–101. doi:10.1016/j.watres.2014.03.032

36. Paton FL, Maier HR, Dandy GC (2013) Relative magnitudes of sources ofuncertainty in assessing climate change impacts on water supply security forthe southern Adelaide water supply system. Water Resour Res 49:1643–1667.doi:10.1002/wrcr.20153
37. Paton FL, Dandy GC, Maier HR (2014) Integrated framework for assessing urbanwater supply security of systems with non-traditional sources under climatechange. Environ Model Softw 60:302–319. doi:10.1016/j.envsoft.2014.06.018
38. Rambaud SC, Torrecillas MJM (2005) Some considerations on the social discount rate.Environ Sci Policy 8:343–355. doi:10.1016/j.envsci.2005.04.003
39. Reis LFR, Walters GA, Savic D, Chaudhry FH (2005) Multi-reservoir operationplanning using hybrid genetic algorithm and linear programming (GA-LP:an alternative stochastic approach. Water Resour Manag 19:831–848.doi:10.1007/s11269-005-6813-0
40. Rodriguez C, van Buynder P, Lugg R, Blair P, Devine B, Cook A, Weinstein P (2009)Indirect potable reuse: a sustainable water supply alternative. Int J EnvironRes Public Health 6(3):1174–1209. doi:10.3390/ijerph6031174
41. Roshani E, Filion YR (2014) Water distribution system rehabilitation under climatechange mitigation scenarios in Canada. J Water Resour Plan Manag 141(4).doi:10.1016/(ASCE).WR.1943-5452.0000450
42. Rozos E, Makropoulos CK (2013) Source to tap urban water cycle modelling.Environ Model Softw 41:139–150. doi:10.1016/j.envsoft.2012.11.015
43. Seifi A, Hipel K (2001) Interior-point method for reservoir operationwith stochastic inflows. J Water Resour Plan Manag 127(1):48–57.doi:10.1061/(ASCE)0733-9496(2001)127:1(48)
44. Stern NH (2006) Stern review: the economics of climate change. CambridgeUniversity Press, Cambridge
45. Stokes CS, Simpson AR, Maier HR (2014) The cost-greenhouse gas emissionnexus for water distribution systems including the consideration of energygenerating infrastructure: an integrated conceptual optimization frameworkand review of literature. Earth Perspectives 1(9). doi:10.1186/2194-6434-1-9
46. Umapathi S, Chong MN, Sharma AK (2013) Evaluation of plumbed rainwater tanks in households for sustainable water resources management: a real-timemonitoring study. J Clean Prod 42:204–214. doi:10.1016/j.jclepro.2012.11.006
47. Vrugt JA, Robinson BA (2007) Improved evolutionary optimization from geneticallyadaptive multimethod search. Proc Natl Acad Sci U S A 104(3):708–711.doi:10.1073/pnas.0610471104
48. Weitzman ML (2007) A review of the Stern review on the economics ofclimate change. J Econ Lit 45:703–724. doi:10.1257/jel.45.3.703
49. West C, Kenway S, Hassall M, Yuan Z (2016) Why do residential recycled waterschemes fail? A comprehensive review of risk factors and impact onobjectives. Water Res 102:271–281. doi:10.1016/j.watres.2016.06.044
50. Wu W, Maier HR, Simpson AR (2010a) Single-objective versus multiobjectiveoptimization of water distribution systems accounting for greenhouse gasemissions by carbon pricing. J Water Resour Plan Manag 136(5):555–565.doi:10.1061/(ASCE)WR.1943.5452.0000072
51. Wu W, Simpson AR, Maier HR (2010b) Accounting for greenhouse gasemissions in multiobjective genetic algorithm optimization of waterdistribution systems. J Water Resour Plan Manag 136(2):146–155.doi:10.1061/(ASCE)WR.1943.5452.0000020
52. Wu W, Simpson AR, Maier HR (2012a) Sensitivity of optimal tradeoffs betweencost and greenhouse gas emissions for water distribution systems toelectricity tariff and generation. J Water Resour Plan Manag 138:182–186.doi:10.1061/(ASCE)WR.1943. 5452.0000169
53. Wu W, Simpson AR, Maier HR, Marchi A (2012b) Incorporating variable-speedpumping in multi-objective genetic algorithm optimization of waterdistribution systems to reduce greenhouse gas emissions. J Water ResourPlan Manag 138(5):543–552. doi:10. 1061/(ASCE)WR.1943.5452.0000195
54. Wu W, Maier HR, Simpson AR (2013) Multiobjective optimization of waterdistribution systemsaccounting for economic cost, hydraulic reliability,and greenhouse gas emissions. Water Resour Res 49(3):1211–1225.doi:10.1002/wrcr.20120

CHAPTER 3

Stakeholder perceptions of water systems and hydro-climate information in Guanacaste, Costa Rica

Matthew Babcock Gabrielle Wong-Parodi, Mitchell J. Small and Iris Grossmann

AUTHORS' CONTRIBUTIONS

MB conceived and designed the data collection, performed the data collection, analyzed the data, and wrote the paper. GWP, MS, and IG conceived and designed the data collection and wrote the paper. All authors read and approved the final manuscript.

ABSTRACT

In the face of changing environmental and socio-economic drivers, access to, understanding of, and the use of probabilistic climate forecasts and other sources of scientific hydro-climate information are important for informed decision making in the water sector. This paper characterizes and compares local perceptions of the water system and hydro-climate information in the seasonally dry province of Guanacaste, Costa Rica. Semi-structured interviews were conducted with a total of 40 participants from 7 water-related groups. Interview results were used to compare mental models of the drivers of water systems and water scarcity mitigation/adaptation options, and relate them to stakeholder information needs, accuracy ratings, and use. Our results suggest that: 1) while there appear to be similar perceptions of the drivers of rainfall and groundwater, there is a gap between groups in the use of forecasts, the awareness of management options, and the level of detailed understanding of how the water system works; 2) there are potential mismatches between the information presented in rainfall forecasts and the stated and/or salient information needs of some stakeholders, specifically in the case of groundwater resources; 3) there appear to be different perceptions of forecasts even among groups that rate the accuracy of such forecasts the same; and 4) there appears to be a relationship between the use of forecasts and certain types of management actions such as long-term planning. Our findings warrant further investigation and confirmation and may contribute to the development of communications that help stakeholders make informed decisions about freshwater management in semi-arid regions.

Keywords: Stakeholder, perceptions, Climate change, Water management, Hydro-climate information

1. INTRODUCTION

1.1 Background

Successful freshwater resources management is a complex task that involves diverse stakeholder groups and decision making at multiple spatial and temporal levels. Increasingly, changes in climatic patterns, socio-economic pressures, and other large-scale drivers of water systems have created (and are expected to continue to create) contexts in which older water management practices and decision-making strategies are not adequate to meet the demand for freshwater in multiple sectors. Experts, national governments, and international agencies have developed and promoted scientific hydrologic modeling and forecasting tools for use in integrated water resource management, and as part of adaptation plans for water systems under changing climate conditions (Stern and Easterling 1999). Scientific hydro-climate information refers to information that is derived from statistical and modeling analyses that incorporate uncertainty, unlike earlier or traditional approaches. However, it has been shown that many water managers from different sectors and geographic areas do not incorporate this information into their decision-making but rather continue to rely on traditional methods that were developed during (or which assume) a more stationary environment (Rayner et al. 2005;

Kirchhoff 2012; Orlove et al. 2004). Without using such information, stakeholders' may not be making *informed* decisions – where the costs, benefits and uncertainties of choices are understood well enough to enable decisions to be made in accordance with values and preferences. This lack of informed decision making may contribute to maladaptive decisions for managing socio-ecological systems (SES), especially in arid and semi-arid areas of the developing world where water stress is already elevated and where institutional and physical resource constraints leave such systems more vulnerable to rapid social and ecological change.

A number of specific factors influence the use of scientific climate information in both wealthier and developing nations, including but not limited to: whether the information relates to local conditions and the scale at which the user is operating, the perceived reliability of forecasts and trust in the forecasting agency, the timing of the forecasts, the socio-economic status of the potential user, the specific sector of work the user was involved in, political differences between providers and users, forecast content and the form of communication, institutional standard operating procedures, and the perceived risk from current and future threats to water resources (O'Conner et al. 2005; Rayner et al. 2005; Orlove et al. 2004; Moser and Luer 2008; Letson et al. 2001; Hansen et al. 2004; Lemos et al. 2002). Key factors in the developing nation context include whether forecasts are perceived as appropriate and trustworthy, whether there is access to such forecasts, and whether there are resources to act on that information (Letson et al. 2001; Orlove et al. 2004; Lemos et al. 2002). Furthermore, each of these factors may influence decision making differentially depending on the decision contexts faced by stakeholders acting in different water use sectors (e.g., agriculture, domestic use, energy production, recreational, environmental protection). While it is clear that a number of factors influence the use of climate information, recent research suggests that not all factors are equally influential across all stakeholders. Indeed, a number of additional factors inherent to the decision maker (e.g., who they are; level of expertise/education) and external factors such as context (e.g., water sector decision) and goal (e.g., agricultural production, energy production, recreation, environmental protection) can influence whether a decision maker has access to and can or will make use of such information. Recognizing this, recent work calls for more work on identifying the decision making processes of stakeholder groups with respect to questions of climate adaptation (Kirchhoff et al. 2013; Jain et al. 2015). Therefore, in this paper we ask: What do different groups of water end users think of scientific hydro-climate information and how, if at all, they use this type of information in their decision making.

1.2 Research questions and framework for comparing stakeholder perceptions

Here we explore the perceptions of multiple stakeholder groups in Guanacaste, Costa Rica, that are facing increasing water stress in the face of a changing climate (Kuzdas 2012; van Eeghan 2011). To assess stakeholder perceptions in a systematic fashion, we developed a simplified framework (Fig. 1) that incorporates concepts from Theory of Planned Behavior (Ajzen 1991), the Planned Risk Information Seeking Model (Kahlor 2010), and Protective Motivation Theory (Rogers et al. 1983).[1] This framework takes as its end points the actual implementation of water management actions and/or seeking out and using hydro-climate information. In this framework, a stakeholder's decision to implement management actions is affected by the perception of a threat or opportunity that requires the action, the perception that they have the ability to implement the action, and/or the perception that there is a social pressure to perform that action. Similarly, whether they seek new information and/or use such information is affected by their perceived need for that information, ability to use it, and any subjective norms related to the use of that information. The use of new information in turn can affect the implementation of the management action. Underlying the stakeholder's perceptions of threats/opportunities, information needs, ability, and social norms are their perceptions of the water resources related social-ecological system and of the different aspects of the hydro-climate information. Our focus therefore is on the following main questions:

1) How do various stakeholder groups perceive the freshwater system in their local and regional areas? Specifically how do they perceive the system drivers, states, and responses within the system? What are the similarities and differences between stakeholder groups?

2) How do various stakeholder groups perceive and use hydro-climate knowledge and specific sources of information? Specifically, how do they perceive the information sources; the attributes of the information, including trustworthiness and accuracy; and whether and how the information is used for informing decisions? What are the similarities and differences between stakeholder groups?

The next section describes the study area and the development and implementation of the interview protocol. Section 3 describes the qualitative and quantitative results of the interviews, and Section 4 discusses the suggested implications of these results for water management policy in the region and how they may inform future work.

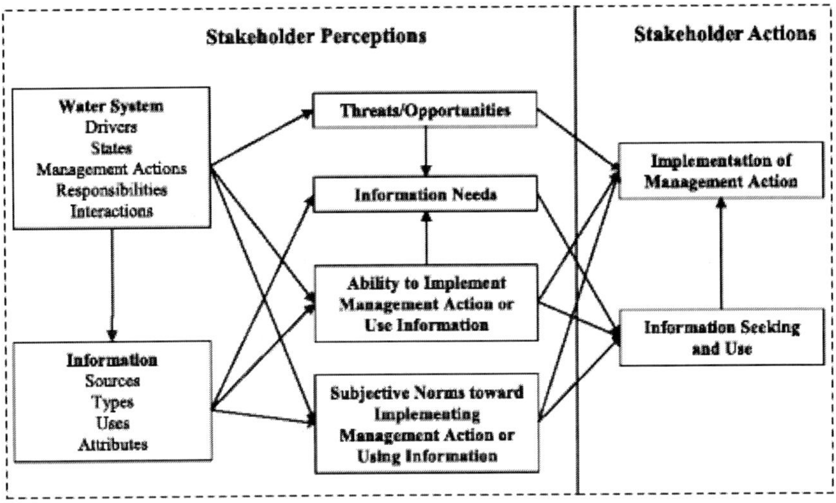

Figure. 1 Simplified stakeholder perceptions and decision making framework

2. METHODS

2.1 Study area and stakeholders

2.1.1 Freshwater resources in Guanacaste, Costa Rica

This study is part of the FuturAgua Project in Guanacaste, Costa Rica, a multidisciplinary, multinational research effort supported by the G8 Belmont Forum to study climate change and freshwater security in developing nations (FuturAgua 2015). Guanacaste is a seasonally dry tropical province, with a yearly rainfall pattern that is typically comprised of a 6-month dry season from late November until May, a smaller rainy season from May to July, a mid-summer drought in July/August, and the main rainy season from August to November. This pattern is significantly affected by the status of the El Niño Southern Oscillation and the North Atlantic Oscillation climate systems. Climate change forecasts and models, such as those included in the Fourth and Fifth Assessment Report of the IPCC, predict changes to the annual cycle of precipitation and increased temperatures, both of which may additionally stress water supplies in the region (Rauscher et al. 2008, 2011; Karmalkar et al. 2011; Ryu and Hayhoe 2014; Neelin et al. 2006; Steinhoff et al. 2014).

Costa Rica guarantees a healthy environment to its citizens in the national constitution and has passed water related laws that establish that freshwater resources cannot be privately owned. The Water Directorate of MINAE, the Costa Rican Ministry of the Environment and Energy, manages concessions of groundwater and river water for use by municipalities, hydroelectric power facilities, and private entities such as farms, off-grid households, and resorts. Within the last 30 years, municipal population growth, changing agricultural activity, increased hydro-electric power production, increased tourism developments, and continuing environmental protection interests all have placed increasing demands on freshwater resources, and there has been a recent history of inter-stakeholder group conflict over water issues (Ramírez-Cover 2008). These conflicts have been shown to occur in part due to underrepresentation of local stakeholders in decision making. In addition, there is a lack of credible or available scientific measures of water quality and quantity, and without these measures the ability to distinguish between the physical lack of water resources and misallocation of such resources has proven difficult (Kuzdas 2012 and van Eeghan 2011).

2.1.2 Stakeholder groups

For this study we separate stakeholders into the following groups: government agencies, large farmers, small farmers, hydroelectric system managers, tourism businesses, village water committees called ASADAs, and the public. The government agencies that make decisions at a local and regional level about water resources or are impacted by such decisions include MINAE as mentioned above, the Ministry of Aqueducts (AyA), the Ministry of Agriculture and Livestock (MAG), and the Ministry of Subterranean Water and Irrigation (SENARA). AyA is mandated to provide potable water to all citizens in the country for domestic use. In the larger towns and cities, AyA manages the water systems, whereas in the smaller less-connected towns the water systems are managed by local water committees or ASADAs. The executive council of each ASADA are volunteers that are voted in by the users every 2–3 years (some ASADAs also pay the administrators and technicians that work on the systems). The volunteer councils are legally responsible for maintaining the water systems and have the authority to collected water use fees, but typically have less technical expertise then the central AyA offices. Almost all municipal water systems in Guanacaste source their water from groundwater or rivers.

The Ministry of Agriculture and Livestock (MAG) is mandated to provide technical assistance to Small Farmers and this outreach includes assistance with irrigation and climate adaptation efforts. Small farmers are either tenants of Large Farms or family enterprises who either raise cattle or grow a variety of crops for local and sometimes export markets (rice and sugar, but also peppers, coffee and vegetables). Large Farms are large estates used either for cattle grazing or for the growing of cash crops such as rice, sugar, and/or melons and employ agronomic engineers as well as large numbers of laborers/tenant farmers. Large and Small farms typically use a mix of direct rainfall, groundwater, and river water depending upon their location and crop. Depending on the farm type they have a mix of irrigation methods installed on their property (this is more widespread on Large Farms, but Small Farms may also have their own systems). In special irrigation districts, SENARA is responsible for providing irrigation water to Small and Large farmers.

Hydroelectric power generation in Guanacaste comes from the ArCoSa system (3 plants in series for a total capacity of 360 MW) operated by the Costa Rican Electricity Institute or ICE, and a two plant system run by the rural electrification cooperative, COOPGUANACASTE. These systems are located in the mountainous region along the eastern border of the Province which receives a larger amount of yearly rainfall and use a mix of reservoir and river water.

2.1.3 Available hydro-climate information

The main source of climate forecast information in the region is the Costa Rican National Meteorological Institute (Spanish acronym, IMN). The IMN provides daily and weekly weather forecasts through its website (IMN 2016). The IMN also provides for free on this website monthly climate reports that review the past months precipitation and temperature data and project future precipitation by region typically up to three months ahead (IMN also less frequently releases predictions for the next 6–12 months). Internet coverage in Guanacaste is relatively good and many access the internet through mobile devices (this is more true of younger generations). Additionally all of this information is also transmitted through local and national public media (TV, radio, and newspapers). The IMN also provides more detailed historical data to the public and other government agencies for a fee. The other government agency that has direct access to climate measurements is ICE, though typically ICE does not share this information. ICE also has information regarding reservoir levels that is used in the management of hydroelectric power stations. Many Large Farmers have their own meteorological equipment and have access to NOAA forecast information. Streamflow and groundwater data are more difficult to come by and this lack of information about how much water exists in certain aquifers has been identified as a factor in local water conflicts (Kuzdas 2012 and van Eeghan 2011).

2.2 Interview protocol

In order to elicit stakeholder perceptions, a variation on the mental models approach (Morgan et al. 2002) was employed. This approach includes the use of a formative semi-structured interview that aims to more broadly and openly elicit perceptions from participants. The results of this interview are then used to inform the development of surveys to confirm the prevalence of interview results and test hypotheses generated from the original interview (Klima et al. 2012). Typically this approach has been used to compare risk perceptions and facilitate risk communication between experts and laypeople (Morgan et al. 2002; Hansen et al. 2004). It has also been used to compare climate and adaptation perceptions across experts (Otto-Banaszak et al. 2011). In this study, the approach is used to help compare perceptions of water systems and climate information across multiple stakeholder groups.

Drawing from previous literature and input from other FuturAgua researchers during the winter and spring of 2014, the English language interview protocol was developed. It was then translated into Spanish by the lead author and edited for language by two native Spanish speakers (a coordinator from the FuturAgua project and a member of the local advisory group located in Nicoya, one of the main towns in Guanacaste). In May 2014, the protocol was pilot tested for understanding with two different members of the Nicoya advisory group (an environmental ministry employee and a university professor) prior to the start of the field interviews. The protocol was structured into three main sections: 1) open ended questions about stakeholder perceptions of the social-ecological system (SES); 2) open ended questions about perceptions of water system information and sources and closed questions rating the accuracy of mentioned sources; 3) specific questions about forecasts and climate change (the full protocol can be found in Additional file 1).

2.3 Interview participants and process

A total of 40 participants were interviewed from 7 different stakeholder groups: Agencies ($n = 10$, including government employees of AyA, MAG, MINAE, and SENARA), ASADAs ($n = 7$), Small ($n = 6$) and Large ($n = 4$) Farmers, Hydroelectric power managers ($n = 3$), Tourism-centered businesses ($n = 4$), and members of the Public ($n = 6$). Participants were recruited through a variety of strategies. Members of the Nicoya advisory group, other FuturAgua researchers, or government agency contacts suggested most of the participants and named them as either knowledgeable or interested individuals. Some ASADA members were contacted based on a list of contact information provided by the Aqueduct ministry (AyA) in Nicoya. Other ASADA members were recruited using snowball sampling [34], in which ASADA group interview participants were asked to name other ASADA members to be contacted. All participants from the Tourist and Public stakeholder groups were directly recruited as a convenience sample (Berg 2001) by the lead author in the street, shops, restaurants, businesses or hotels. All potential participants contacted were interviewed with the exception of two (1 hydroelectric project manager and 1 small farmer - both due to scheduling issues). The mix of convenience and snowball sampling in these types of studies is standard practice in the field (see Kirchhoff 2012 and Orlove et al. 2004 as examples), however one possible issue with proceeding in this manner is that the results may not include the views of individuals who live farther away from others or those who have less societal connections.

Interviews were conducted in Guanacaste during June and July 2014[2]. The interviews were recorded and conducted in Spanish except for one (an English-speaking hotel owner who was from the United States and did not want to be recorded). The interviews were performed one-on-one, though occasionally in some interviews there were interruptions and additional comments made by others (family members, neighbors, and in some cases one of FuturAgua's local advisers). Interviews lasted between 25 and 90 min, with most being approximately 45 min long. Participants were not monetarily compensated. After the interview, each participant was given a FuturAgua mug as a thank you gift (participants were not informed of the gift in advance of the interview).

The median age of interview participants was 54.5 years. Overall, 55 % of the participants had at least a college degree. The percentage of stakeholders that had such a degree of education within the Large Farmer, Agency, and Hydroelectric groups was 90 % or above, whereas the percentage in the other groups was 50 % or below. Only 17 % of the participants were female, which, while very low in terms of the general public and elected government positions, is closer to the percentage that are in ASADAs (20 % based on AyA records) or that work for MINAE (25 %) in Guanacaste. Recognizing that the intent of studies employing the mental models approach with in-depth interviews is to discover concepts and suggest hypothesis (not to test them), a sample that includes participants from the targeted groups was sought, but it did not need to be representative.

2.4 Coding and analysis

All interviews were transcribed (Spanish to Spanish) either directly by the lead author or by transcribers recruited through Amazon Mechanical Turk, an Internet platform for crowdsourcing short "human intelligence" tasks (Amazon Mechanical Turk 2015). All Mechanical Turk transcriptions were checked for errors and corrected by the lead author. Interview transcripts were translated into English by the lead author and then, as a quality check, several interview transcripts were back translated from English to Spanish by native Spanish speakers from the broader FuturAgua research team.

The lead author used multiple iterations of an open-coding procedure (Strauss 1987) to inductively find common and interesting themes from the interview transcripts for further analysis. For the one interview that was not recorded, the lead author's notes from the interview were used as the transcript for coding purposes. The codes were separated into groups concerning drivers, states, and uses of the water system, actions taken to mitigate or adapt to water scarcity, and information sources and attributes (a full list of the sub-codes under these categories can be found in Additional file 2).

QDA Miner Lite software (QDA Miner Lite 2015) was used to "tag" excerpts with one or more codes, allowing the grouping of similar quotes and descriptive comparisons of pairs of codes. A second rater coded a subset of 11 of the transcripts, and there was 69 % agreement between the two raters as to whether a specific code was mentioned in a specific transcript. Literature on inter-rater reliability suggests that a percent agreement of 69 % indicates "substantial agreement" (Landis and Koch 1977). The first coder conducted the interviews and thus likely had a more nuanced perspective of the transcripts and allocated more codes than the second coder. These statements have been added to the manuscript.

Binary frequencies of mention (mentioned in transcript versus not mentioned) were determined for each participant for each single sub-code (e.g., DRIVER-ELNINO) and for select pairs of sub-codes (e.g., did the transcript mention both INFOSOURCE-FORECAST and INFOATTRIBUTE-USED?). Pairing sub-codes allowed frequency counts of interactions such as Driver/State pairings. The percentage of participants within each stakeholder group that mentioned a certain sub-code or pair of sub-codes at least once was then calculated and used for comparing across stakeholder groups.

Transcript excerpts that mentioned the numerical rating of information sources were collected and analyzed with simple descriptive statistics. No inferential statistics were performed as the participants were not randomly selected and there were only a small number of participants in each of the different stakeholder groups.

3. RESULTS

3.1 How various stakeholder groups perceive the freshwater system in their local and regional areas

3.1.1 Views on drivers of rainfall amount and duration and groundwater levels

The main sources of water for stakeholder use mentioned by interview participants from all groups were direct rainfall and/or groundwater (accessed through wells and/or aqueducts). Table 1 shows the perceived drivers of the amount and distribution of rainfall mentioned by interview participants. Global warming-related climate change was the rainfall driver identified by the highest percentage of members within the Agency, ASADA, and Tourism groups. Climate change and the effects of the El Niño Southern Oscillation (ENSO) were both mentioned by the highest percentage of members within the Small Farmer, Large Farmer, and Hydroelectric groups. Climate change was not mentioned at all by the Public as a rainfall driver, but ENSO, Deforestation, and God were tied for most mentions within this group. Climate change and ENSO were the most mentioned drivers across groups, followed by. Deforestation, Geological drivers, and unspecified Natural cycles.

Table 1 Drivers of rainfall and groundwater levels mentioned by members of different stakeholder groups

	% of stakeholder group members mentioning specific driver						
	Agency (10)	**ASADA (7)**	**Sm. Farmers (6)**	**Lg. Farmers (4)**	**Tourism (4)**	**Hydroelectric (3)**	**Public (6)**
Rainfall drivers							
Climate Change	**40 %**	**43 %**	**50 %**	**50 %**	**50 %**	**67 %**	0 %
ENSO	20 %	29 %	**50 %**	**50 %**	0 %	**67 %**	**17 %**
Nature	10 %	14 %	0 %	25 %	25 %	0 %	0 %

Deforestation	0 %	0 %	33 %	25 %	25 %	33 %	**17 %**
Geological/Geographical	20 %	0 %	33 %	**50 %**	25 %	33 %	0 %
Ozone Destruction	0 %	0 %	0 %	0 %	25 %	0 %	0 %
God	0 %	0 %	0 %	0 %	25 %	0 %	**17 %**
Groundwater drivers							
Rainfall	20 %	**29 %**	**50 %**	**50 %**	0 %	0 %	0 %
Climate Change	20 %	**29 %**	0 %	0 %	0 %	0 %	0 %
ENSO	20 %	14 %	0 %	25 %	0 %	0 %	0 %
Geological/Geographical	10 %	**29 %**	33 %	25 %	0 %	0 %	**33 %**
Deforestation	**30 %**	14 %	0 %	0 %	0 %	0 %	0 %
Population Growth	20 %	0 %	17 %	0 %	0 %	0 %	0 %
Tourism Use	20 %	14 %	17 %	25 %	0 %	0 %	0 %
Large Farm Use	**30 %**	14 %	0 %	0 %	0 %	0 %	0 %
Other Population	0 %	14 %	17 %	0 %	0 %	0 %	0 %
Misuse	20 %	0 %	0 %	0 %	0 %	0 %	0 %
Damage	10 %	0 %	0 %	0 %	0 %	0 %	0 %
God	0 %	14 %	0 %	0 %	0 %	0 %	0 %

Bolded values represent the drivers mentioned by the highest percentage of members within the group

() indicate the number of participants in each group

Specific driver definitions can be found in Additional file 2

Table 1 also shows the percentage of each stakeholder group that mentioned different drivers of groundwater levels. Hydroelectric managers did not mention (and were not asked about) groundwater drivers during the interviews, most likely because the source of water for their facilities comes from rivers and reservoirs. Tourism group members also did not mention groundwater drivers, which is of potential interest as the source of water for some of these members is groundwater and some of the political conflict in the region has been over tourist use of aquifers (Kuzdas 2012 and van Eeghan 2011). The most mentioned driver of groundwater resources by several groups was the amount and timing of rainfall. This association clearly has a strong physical basis, especially for shallow (surface) aquifers and wells. The broadest range of drivers were mentioned by the Agency and ASADA group members, which may be a function of the fact that these two groups mentioned groundwater the most out of any other group in general.

Participants mentioned both perceived direct and indirect drivers of rainfall and groundwater levels. A partial explanation for these results comes from the fact that different participants expressed different levels of specificity and sophistication when discussing these issues. For example, the two quotes below show different levels of detail and knowledge about the role El Niño plays:

"Depends if it is El Niño or La Niña, and it is a little complex and everything but the El Niño is because of the heating of the Equatorial Pacific Ocean... the climate variability that is a normal process. El Niño has always existed, La Niña also...It's that the storms on the Pacific side, when it rains a lot 5 days, 7 days, 15 days (is when) the Pacific wind tries to go toward the mountain system and it rains a lot. But this year less will be seen, it is expected that 45 % less storms in the Pacific and the wind is hitting here makes it that the humid breeze from the Pacific backs off and it doesn't rain." (Agency 10)

"Yes. Well they say (there is drought) because of the El Niño phenomena. I don't know when it will be. I don't know." (Public 5)

In general, the participants from the Large Farmer, Hydroelectric, and Government Agency groups were more specific and more confident in describing the relationship between global climate change, the El Niño system, and rainfall than were participants from the other groups.

3.1.2 Views on the state of water resources and appropriate management responses to lack of water resources

Most participants mentioned that they had enough water at the time of the interview to meet their needs. But, many participants also mentioned that they knew of other areas that did not, or expressed concern and/or uncertainty over whether they would have enough in the future. Table 2 shows the management options mentioned by each stakeholder group for addressing current or future water shortages (additional information about each option can be found in Additional file 2). At least one member of each group mentioned changing the mentality of users and society to be more environmentally friendly, reforestation actions, and finding new water supplies as responses to not having enough water. Table 2 also shows that the more frequently mentioned (and therefore assumed to be most practiced) management options varied by stakeholder group. For example, for Agency, Large Farmer, and Hydroelectric group members, the most mentioned options were grouped under the code "Modify planning" which encompasses changing longer term operations such as the design of future municipal projects or hydropower facilities, phasing out crop types, or expanding agricultural enterprises

Table 2 Management actions for water shortages mentioned by members of different stakeholder groups

Management actions	**% of stakeholder group members who mentioned specific management action**						
	Agency (10)	**ASADA (7)**	**Sm. Farmer (6)**	**Lg. Farmer (4)**	**Tourism (4)**	**Hydroelectric (3)**	**Public (6)**
Nothing can be done	10 %	29 %	0 %	25 %	25 %	33 %	**50 %**
Store water	10 %	0 %	0 %	0 %	**50 %**	0 %	33 %
Increase efficiency	50 %	29 %	17 %	75 %	25 %	0 %	33 %
Standard operations	30 %	**57 %**	0 %	25 %	**50 %**	0 %	33 %
Use improved tech.	30 %	14 %	33 %	**100 %**	0 %	0 %	0 %
Buy crop insurance	0 %	0 %	17 %	25 %	0 %	0 %	0 %
Modify planning	**90 %**	0 %	17 %	**75 %**	25 %	**100 %**	17 %

Change mentality	30 %	**57 %**	33 %	50 %	25 %	33 %	17 %
Make new law/rule	20 %	14 %	0 %	0 %	0 %	0 %	17 %
Protect watershed	40 %	**57 %**	33 %	25 %	25 %	67 %	0 %
Find new supply	70 %	43 %	**50 %**	50 %	25 %	33 %	33 %
Reforest	40 %	71 %	33 %	25 %	25 %	67 %	17 %
Megaprojects	40 %	0 %	0 %	0 %	0 %	0 %	0 %

Bolded values represent the drivers mentioned by the highest percentage of members within the group

() indicate the number of participants in each group

Specific action definitions can be found in Additional file 2

Buying insurance as a response to a lack of water resources was only mentioned by farmers, who have experience with the various institutional crop insurance products offered by the government and through cooperatives. One Large farmer (Lg Farmer 1) mentioned an interesting complicating policy factor related to the timing of the rainy season: in order to get crop insurance, farmers have to plant before a certain date. If this date is set too early in a year with a late rainy season then the insurance system may incentivize farmer decisions that result in additional crop loss.

Modification of planning activities (encompassing activities such as changing planting times and power generation schedules) was mentioned by a majority of members of government agencies, hydroelectric managers and large farmers and a minority of other groups.

3.2 Stakeholder perceptions and use of hydro-climate information

3.2.1 Perceived information needs

A total of 26 participants directly answered the question, "What would you like to know about the state of water resources that you don't already know?" The most common answer (9 participants) was related to how much water was in the aquifer of interest to that participant (including how much was in the aquifer now, how the groundwater flowed in or out, and how much would there be in the future). The main reason given for wanting this information was so that they would know if there was going to be enough water for their own use and the use by others. Other direct answers included, in order of frequency, information about groundwater quality (what was the quality and how to treat the water - 5 participants), information about climate change (forecasts and sources – 4 participants), what the rain forecast was going to be (3), information about surface water quality (3), and what the impact of climate would be on plants (2).

3.2.2 Awareness and use of different types of forecasts

During the interviews rainfall forecasts of four different types were mentioned: basic daily/weekly radio/TV/internet forecasts, seasonal 3-month/6-month forecasts created by the IMN or by ICE, year-long forecasts, and multi-year forecasts. Figure 2 shows that only a few participants mentioned multi-year forecasts, members of the Public group only mentioned daily/weekly forecasts, and the highest percentages of use of forecasts were for seasonal forecasts.

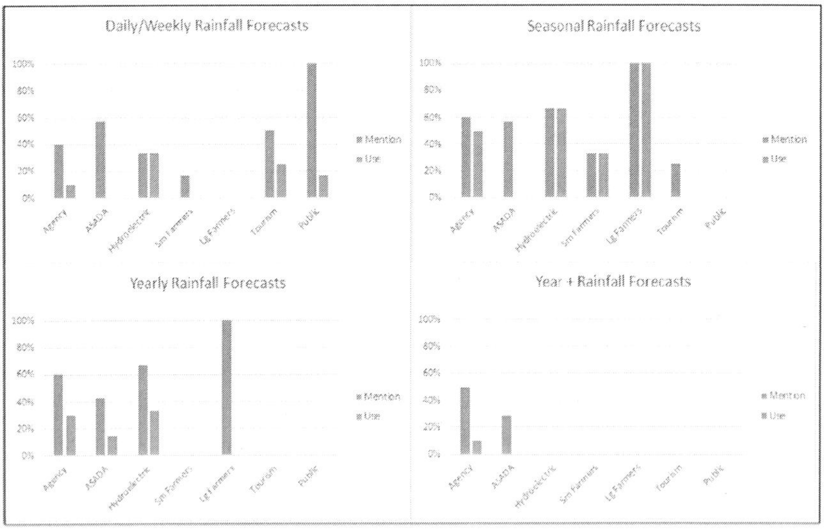

Figure. 2 Percentage of each stakeholder group that mentioned forecasts of different types

3.2.3 Perceptions of rainfall forecast accuracy

The most consistently mentioned information source related to water resources in Guanacaste was the short-term (weeks) and long-term (months/years) forecasts of rainfall provided by the IMN. The forecast accuracy was judged on a scale from 0 (completely inaccurate) to 10 (fully accurate). Based on participants' comments, the main reasons for high ratings were that the participant thought that the forecast matched what the participant actually experienced and that the people in charge of making the forecasts were skilled professionals/experts. As some participants put it:

"Like I said it is a trusted source, they take a lot into account. Maybe we don't make decisions only based on what they say but yes it forms an important part of the decision we make on our farm." (Lg Farmer 1)

"Okay based on the last 3 years' experience I would say very good precision for example from 1 to 10 I would put a 9. At least in that they have said, "Here comes a dry period" and it is certain that there have been dry years. For example, this year...they predicted that May, June, and July were going to be dry but the percentage in June was drier than they thought it was going to be. Nevertheless yes they were right that there were dry months. I really... I have trust in the IMN." (Agency 10)

Even participants who had a positive opinion on the accuracy of the forecasts usually also discussed some limitation to them:

"I would give between a 7 and 8. For example it fails. So they say in these 3 months it is going to rain 100 mm, 25-50-25, but maybe it rains 100 but 5-70-25, you understand that, or the total rain is more or less certain. I believe it is impossible that they do it so... In this they are not certain but it works very well." (Lg Farmer 1)

"No, I give them a 7 about there, yes, yes, yes. But for the 3 month forecasts. They that make these (forecasts) take data from the different meteorological stations. So, to 3 months it is close enough with the projection, but equally they also are very responsible to say that for example the projections for 1 year are not as confident because in reality they are variable or so uncertain that it is a projection nothing more." (Hydroelectric 1)

Participants often mentioned a mismatch between the spatial and temporal scales of the forecasts and the information they need to assess their water management decisions, and many participants mentioned that the farther into the future the forecast goes, the less accurate it will be. The reason for the spatial discrepancy that was most mentioned was that forecasts are given on the scale of the entire region, not on the scale of the sub-region, town, or farm level. Some participants from the Large Farmer and Agency groups stated that while 3-month-ahead average rainfall forecasts were useful, having a forecast that could describe the distribution of rainfall over those 3 months would be even more helpful.

For those assigning low accuracy ratings, the main reasons given included that the participants felt the forecast did not match what they actually experienced or more generally that reality was too variable for the forecasts to be accurate. As one participant said:

"They are not trustworthy. Sometimes they say it is going to rain in the afternoon and it is dry, sometimes they hit the target and sometimes they don't." (ASADA 7, rated accuracy of 5)

This last quote and the quote from Agency 10 are representative of several that basically say, "Sometimes what the forecast predicts occurs, and sometimes it doesn't." The results show that whether this statement is used to justify a high rating or a low rating varies across stakeholder groups – it seems that large farmers, government agencies, and hydroelectric managers give high ratings and other groups give low ratings (see Fig. 3).

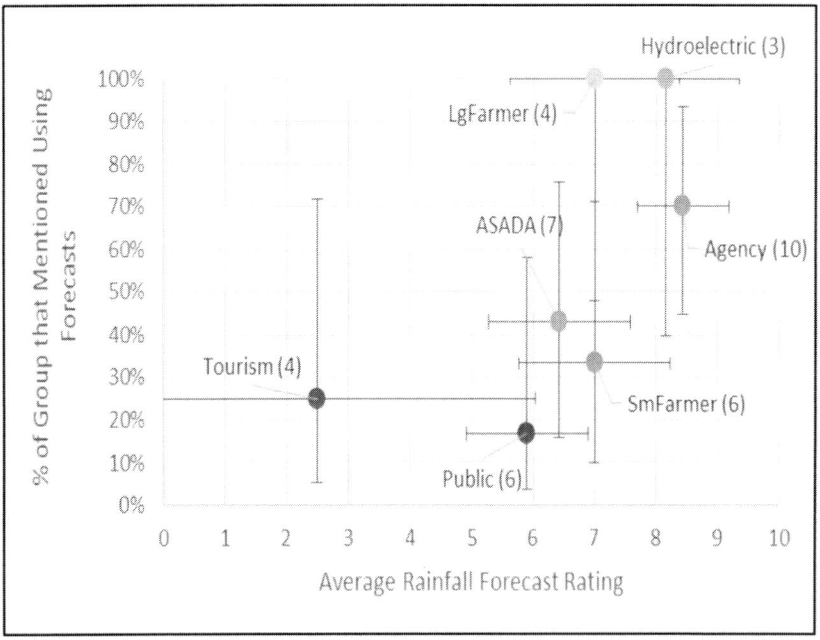

Figure. 3 Percentage of group that mentioned using forecasts vs. Average Rainfall Forecast Accuracy Rating. *Horizontal error bars* are 2 SD. Vertical error bars are 95 % CI on binomial p distribution. The (#) is the total sample size in the group used to calculate the % of group that mentioned using forecasts. The average rainfall accuracy is averaged only over those participants in each group that gave a numerical rating (and therefore can be less than (#))

3.3 Use of rainfall forecasts

Figure 3 also shows the average rainfall forecast accuracy rating calculated for each stakeholder group compared against the percentage of group members that mentioned using the forecasts. The results suggest that higher average accuracy ratings for rainfall forecasts are associated with a greater stated use of such forecasts, though significant variation is present within groups. This tendency would be in agreement with previous results: Orlove et al. (2004), in their study of the fishing sector in Peru, similarly found that their results, "supports the long-established claim that perceived accuracy influences forecast use" and that, "subpopulations differ significantly in their use of this information."

The results presented in Fig. 3 also may suggest that Large Farmers, Hydroelectric Managers, and Government Agency group members use rainfall forecasts more than ASADAs, Small Farmers, and members of the Public and Tourism industry. It also tentatively suggests that while Small Farmers and Large Farmers provided similar accuracy ratings for the rainfall forecasts, the percentage of group members mentioning using forecasts may be much higher among Large Farmers than among Small Farmers. This may be due to the Large Farmers potentially having more access or awareness of different types of rainfall forecasts.

More widespread use of forecasts within a group may be related to whether those forecasts are used to inform the modification of planning activities (changes in planting or equipment schedules, changes in energy production, etc.). Figure 4 shows that again there appears to be a distinction between Large Farmers/Hydroelectric Managers/Government Agencies and the other stakeholders, with the groups reporting the use of forecasts also reporting planning modification options for their water systems. These results could suggest either or both directions of influence: that these groups are responding with modifying planning activities because they use forecasts, or that they seek out and use forecasts because they are already modifying longer-term activities.

4. DISCUSSION AND CONCLUSIONS

4.1 Main research findings and study limitations

Our results suggest that most stakeholder groups are similar in their identification of climate change and ENSO as the major drivers of rainfall and groundwater resources, though they differ in their ability to explain the interactions between drivers and water resource states. Stakeholder groups also differ in the types of management responses they are most aware of, with some groups (Large Farmers, Hydroelectric, and Agency members) emphasizing longer-term planning. The timing of the rainy season and the amount of rainfall during the rainy season were identified as important factors in the water system, though individual stakeholders differed to a large extent in their awareness of different rainfall forecasts, their rating of forecast accuracy, and their mentioned use of forecasts. Participants from the Large Farmer, Hydroelectricity, and Government Agency groups rated forecast accuracy as higher and mentioned using forecasts to a larger extent than other groups.

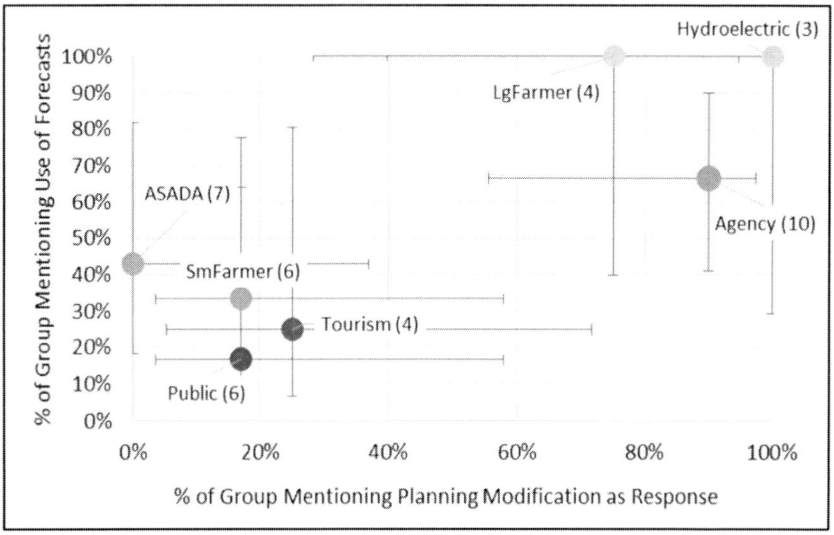

Figure. 4 Percentage of Group Mentioning Planning Modification Type Responses vs. Average Rainfall Forecast Accuracy Rating. *Horizontal* and *vertical error bars* are 95 % CI on binomial p distribution

As this study is based on a limited number of interviews, the results presented above should be considered tentative and the actual prevalence of these beliefs and actions requires further testing with larger populations. Also, the results are all self-reported and therefore in some cases may not be accurate. In addition to working with larger populations, the use of other corroborating evidence (census and ministry information and observations, network analysis etc.) will help increase the confidence in these results and their usefulness.

4.2 Implications for policy and future research

Our results tentatively suggest a split between two groups of stakeholders in the sophistication and specificity of descriptions of the interactions between water resource drivers, states, and management actions, in the perception and use of rainfall forecasts, and in the variation of opinions expressed on these issues within each group. This is important because as shown in the framework in Fig. 1, these perceptions impact whether a specific threat is perceived and whether the stakeholder perceives they have the information and the ability to act in the face of this threat. Stakeholder groups we call the Large type (the Agency, Large Farmer, and Hydroelectric stakeholder groups) seem to express a clearer and more descriptive understanding of the physical drivers of rainfall and groundwater, use rainfall forecasts to a larger extent, and have less variation in their opinions than groups from the Small type (Public, Tourist, Small Farmer, and ASADA)[3]. In addition to differences in education levels, one general distinction between Large and Small groups that may factor into these results is that the Large groups most likely have more direct institutional connections and longer term relationships with the National Meteorological Institute. For example, while the IMN releases its 3-month and El Niño forecasts publically on their website, and such forecasts are mentioned in the public media, only members of the Large Farmer, Hydroelectric, and Agency groups mentioned having direct meetings and specialized presentations with the IMN. These types of connections have been identified in previous literature as beneficial to the use of forecasts, specifically the "Interaction" part of Lemos et al.'s "Fit, Interplay, and Interaction" model (Lemos et al. 2012). "Interaction" opportunities that improve the use of forecasts such as co-production, long-term relationships, and two-way communication, appear to be more prevalent between Large Farmers, Agency, Hydroelectric groups and the IMN. The split in perceptions of the water system between Large and Small stakeholders may become problematic in cases where Large and Small groups interact, as in the case where the Ministry of Agriculture (MAG) is mandated to provide education and guidance on climate change and water management to small farmers. A similar relationship exists between the ASADAs and the Ministry of Aqueducts (AyA). An important implication for policy is that the educational material and communication practices of Large groups, should be developed and tested with Small groups to ensure understanding and usability of information to enhance the management of freshwater resources (Wong-Parodi et al. 2014).

Previous literature has emphasized the importance of providing information that helps answer the questions stakeholders actually have to deal with/are aware of (Moser and Luer 2008; Lemos et al. 2012; Fischhoff 2013). For example, in Moser and Luer's study of coastal managers (2008), they argue that, "the overarching message emerging from the information needs identified by coastal managers is that climate change science still needs to be translated into types of information that are salient to the manager". Our results suggest that rainfall and climate forecasts are not being translated into information about the more salient water resource concern, which in this case may be groundwater (based on the interest expressed in obtaining additional information about it as mentioned in the results section). In our results, while rainfall was perceived as a major driver of groundwater, information on groundwater levels and flows was directly identified as a need more often than climate and rainfall forecasts. If groundwater levels are information that people want and need in the study area, then perhaps communicators of climate information need to be able to do a better job of translating climate forecasts into a type of groundwater forecast (the feasibility of which depends greatly on how well the groundwater, surface water, and precipitation interactions are monitored and understood). If modeling of the groundwater levels or storage is too difficult to achieve this translation, then an alternative would be to try to understand better and communicate to stakeholders how increased groundwater withdrawals during dry periods can exacerbate the effects of rainfall deficits on groundwater (a total of 7 interview participants from the Agency, ASADA, Small Farmer, and Large Farmer groups mentioned the relationship between the lack of rain and increased withdrawals as causes for concern). Future exploration and testing of stakeholder's understanding of the relationship between rainfall and groundwater levels may also assist in designing more salient and effective communication of climate forecast information.

An unexpected finding from the interviews that suggests an opportunity for further policy analysis and improvement is that crop insurance mandates regarding when planting must start may incentivize farmers not to adjust their planting schedules to changing rainfall patterns, resulting in crop loss. This finding suggests future work should include attempts to determine how important insurance is to farmers in the region (e.g., through follow-up surveys). If found to be important, then the process by which the providers of crop insurance determine their cutoff dates should be reviewed and possibly updated.

Looking across Fig. 1 at how perceptions of information may influence the ability to use such information (and the subsequent use of that information), the differences between groups in the rating of forecast accuracy, whether forecasts are used, and how they are used suggest some additional questions about stakeholder decision making that could be tested in the Guanacaste and other contexts as part of future work. While it could be that the distinction between Large and Small type group stems solely from the inability of smaller groups to act due to the lack of opportunity or resources ("nothing can be done" was the most mentioned response for the Public), the fact that members of the ASADA, Small Farmer, and Tourism groups did mention actions that they take which could potentially be enhanced by forecast information (see Table 2) may indicate that other aspects are important. For example, signal detection theory, combined with an understanding of how stakeholders differ in their prior beliefs and motivational biases, could be used to better explain the finding that groups which basically agree on the fact that forecasts are not entirely certain rate the forecast accuracy differently (Green and Swets 1966; Small et al. 2014). It could be that the two groups see the distinction between how well the forecast matches reality the same way (same sensitivity), but one group's decision point for calling something "accurate" (different biases) and having a subsequent positive feeling about the forecast is higher or lower than the other's decision point. Another explanation for the discrepancy could be that some groups simply understand the forecast information or the underlying ideas of uncertainty less than others. Future studies that directly test such understanding may help determine if this is the more important determinate of accuracy perception. A different avenue to pursue based on our results involves exploring why the 3-month/6-month rainfall forecasts were the most well known and most used of the various types of forecasts used. For example, it could be that the perceived ability to use these types of forecasts is high because these forecasts fit the already established management schedules of different stakeholder groups (planting seasons, energy production projections, etc.).

CONCLUSIONS AND NEXT STEPS

Our findings support existing literature that those who have more resources (e.g., economic resources, organizational connections, etc.) are also those who tend to use forecasts (Lemos et al. 2002; Kirchhoff et al. 2013). Similarly, those who rate forecasts as more accurate also somewhat tend to use forecasts more as mentioned above (Orlove et al. 2004). We also find that other factors may be important such as recognition of groundwater levels in people's understanding of water availability, the potential for crop insurance to provide perverse incentives, and the differences in perceptions of forecast accuracy between different stakeholder groups. Indeed, these findings warrant further investigation and confirmation and may contribute to the development of communications that help stakeholders make informed decisions about freshwater management in Guanacaste and other semi-arid regions.

FOOTNOTES

[1] There have been attempts to integrate these various factors into summary frameworks such as Lemos et al.'s model of "fit, interplay, and interaction" which concentrates on the interaction between climate information providers and end-users (Lemos et al. 2012). Others have investigated these factors (and factors related to adaptation decision making and activity in general) using modified versions of established decision making, information seeking, and behavior frameworks such as the Theory of Planned Behavior (TPB) (Ajzen 1991), the Risk Information Seeking and Processing (RISP) model (Griffin et al. 1999), the Planned Risk Information Seeking Model (PRISM) (Kahlor 2010), and Protective Motivation Theory (PMT) (Rogers et al. 1983). For example, Truelove and colleagues use a modified version of the PMT they named the Risk, Coping, and Societal Appraisal (RCS) framework to study adaptation to climate change in the farming sector (Truelove et al. 2015) and Yang and colleagues have used RISP and PRISM to investigate and compare the seeking out of climate information between different groups (Yang et al. 2014a, b). These frameworks all tie into the idea that in order to improve informed decision making, producers of scientific information must identify the specific decisions, perceptions, and decision environments faced by those stakeholders (Fischhoff 2013).

[2] It should be noted that while most often this time of the year would have been the beginning of the wet season in Guanacaste, in 2014 this period was very dry and the government was in the process of issuing El Niño alerts (which may have primed some participants to bring up El Niño during the interview).

[3] It should be noted that 2 of the 3 Hydroelectric group participants also work at government agencies.

ACKNOWLEDGEMENTS

The work presented in this article was funded through FuturAgua, a multinational, multidisciplinary Belmont Forum-G8 Initiative Collaborative Research project. Support for Carnegie Mellon's participation in FuturAgua was provided by the US National Science Foundation (ICER1342941-NSF). We would like to thank the other members of FuturAgua for their research and administrative support. We also especially would like to thank the study participants from Guanacaste for their participation.

REFERENCES

1. Ajzen I (1991) The theory of planned behavior. Organ Behav Hum Dev 50:179–211 Amazon Mechanical Turk. http://www.mturk.com Accessed 30 Mar 2015.
2. Berg BL (2001) Qualitative research methods for the social sciences. Allyn and Bacon, Boston
3. Fischhoff B (2013) The sciences of science communication. Proc Natl Acad Sci U S A 10(3):14033–14039
4. FuturAgua Project Website. http://www.futuragua.ca/ubc.home/ Accessed 30 Mar 2015.
5. Green DM, Swets JA (1966) Signal detection theory and psychophysics. Wiley, New York
6. Griffin RJ, Dunwoody S, Neuwirth K (1999) Proposed model of the relationship of risk information seeking and processing to the development of preventive behaviors. Environ Res 80:S230–S245
7. Hansen JW, Marx SM, Weber EU (2004) The role of climate perceptions, expectations, and forecasts in farmer decision making: the Argentine Pampas and South Florida. International Research Institute for Climate Prediction Technical report 04-01. The Earth Institute at Columbia University.
8. Instituto Meteorológico Nacional de Costa Rica website. https://www.imn.ac.cr/web/imn/inicio Accessed 20 Apr 2016.
9. Jain M, Naeem S, Orlove B, Modi V, DeFries R (2015) Understanding the causes and consequences of differential decision-making in adaptation research: adapting to a delayed monsoon onset in Gujarat, India. Global Environ Chang 31:98–109
10. Kahlor L (2010) PRISM: a planned risk information seeking model. Health Commun 25(4):345–356
11. Karmalkar AV, Bradley RS, Diaz HF (2011) Climate change in Central America and Mexico: regional climate model validation and climate change projections. Clim Dyn 37:605–629
12. Kirchhoff CJ (2012) Understanding and enhancing climate information use in water management. Clim Change 119:495–509
13. Kirchhoff CJ, Lemos MC, Dessai S (2013) Actionable knowledge for environmental decision making: broadening the usability of climate science. Annu Reb Env Resour 38:3.1–3.22
14. Klima K, Bruine de Bruin W, Morgan MG, Grossman I (2012) Public perceptions of hurricane modification. Risk Anal 32(7):1194–1206
15. Kuzdas C (2012) Unpacking Water Conflict in Guanacaste, Costa Rica. GWF Discussion Paper 1242, Global Water Forum, Canberra, Australia.
16. Landis JR, Koch GG (1977) The measurement of observer agreement for categorical data. Biometrics 33:159–174
17. Lemos MC, Finan TJ, Fox RW, Nelson DR, Tucker J (2002) The use of seasonal climate forecasting in policymaking: lessons from Northeast Brazil. Clim Change 55:479–507
18. Lemos MC, Kirchhoff CJ, Ramprasad V (2012) Narrowing the climate information usability gap. Nat Clim Chang 2:789–794
19. Letson D, Llovet I, Podesta G, Royce F, Brescia V, Lema D, Parellada G (2001) User perspectives of climate forecasts: crop producers in Pergamino Argentina. Clim Res 19:57–67
20. Morgan MG, Fischhoff B, Bostrom A, Atman CJ (2002) Risk communication: a mental models approach. Cambridge University Press, New York
21. Moser SC, Luer AL (2008) Managing climate risks in California: the need to engage resource managers for successful adaptation to change. Clim Change 87:S309–S322
22. Neelin JD, Munnich M, Su H, Meyerson JE, Holloway CE (2006) Tropical drying trends in global warming models and observations. Proc Natl Acad Sci U S A 13(16):6110–6115
23. O'Conner R, Yarnal B, Dow K, Jocoy CL, Carbone GJ (2005) Feeling at risk matters: water managers and the decision to use forecasts. Risk Anal 25(5):1265–1275

24. Orlove B, Broad KF, Petty AM (2004) Factors that influence the use of climate change forecasts: evidence from the 1997/98 El Nino event in Peru. Bull Am Meteorol Soc 84(11):1735–1743
25. Otto-Banaszak I, Matczak P, Wesseler J, Wechsung F (2011) Different perceptions of adaptation to climate change: a mental model approach applied to the evidence from expert interviews. Reg Environ Chang 11:217–228
26. QDA Miner Lite (Version 1.4) http://provalisresearch.com/products/qualitative-data-analysis-software/freeware/ Accessed 15 Jan 2015.
27. Ramírez-Cover A (2008) Conflictos socioambientales y recursos hídricos en Guanacaste: Una descripción desde el cambio en el estilo de desarrollo (1997-2006). Anuario de Estudios Centroamericanos 33–34:359–385
28. Rauscher SA, Giorgi F, Diffenbaugh NS, Seth A (2008) Extension and intensification of the Meso-American mid-summer drought in the twenty-first century. Clim Dyn 31:551–571
29. Rauscher SA, Kucharski F, Enfield DB (2011) The role of regional SST warming variations in the drying of Meso-America in future climate projections. Am Meteorol Soc 24:2003–2016
30. Rayner S, Lach D, Ingram H (2005) Weather forecasts are for wimps: Why water resource managers do not use climate forecasts. Clim Change 69:197–227
31. Rogers RW, Cacioppo JT, Petty R (1983) Cognitive and physiological processes in fear appeals and attitude change: A revised theory of protection motivation. In: Social psychophysiology: A sourcebook. p 153–177.
32. Ryu JH, Hayhoe K (2014) Understanding the sources of Caribbean precipitation biases in CMIP3 and CMIP5 simulations. Clim Dyn 42:3233–3252
33. Small MJ, Güvenç Ü, DeKay ML (2014) When can scientific studies promote consensus among conflicting stakeholders? Risk Anal. doi:10.1111/risa.1223
34. Steinhoff DF, Monaghan AJ, Clark MP (2014) Projected impact of twenty-first century ENSO changes on rainfall over Central America and northwest South America from CMIP5 AOGCMs. Clim Dyn. doi:10.1007/s00382-014-2196-3
35. Stern P, Easterling W (eds) (1999) Making climate forecasts matter. National Academy Press, Washington DC
36. Strauss AL (1987) Qualitative analysis for social scientists. Cambridge University Press, New York
37. Truelove HB, Carrico AR, Thabrew L (2015) A socio-psychological model for analyzing climate change adaptation: a case study of Sri Lankan paddy farmers. Global Environ Chang 31:85–97
38. van Eeghan SE (2011) Water conflicts in Costa Rica? Sardinal: a case study on the emergence of a water conflict in the context of high speed growth in (residential) tourism. MS Thesis. Utrecht University, Department of Geosciences.
39. Wong-Parodi G, Fischhoff B, Strauss B (2014) A method to evaluate the usability of interactive climate change impact decision aids. Clim Change 126(3-4):485–493
40. Yang ZJ, Kahlor LA, Li H (2014a) A United States-China comparison of risk information-seeking intentions. Commun Res 41:935–960
41. Yang ZJ, Rickard LN, Harrison TM, Seo M (2014b) Applying the risk information seeking and processing model to examine support of climate change mitigation policy. Sci Commun 36:296–324

CHAPTER 4

Characterising the present-day stress regime of the Georgina Basin

Bailey, E. Tenthorey & B. Ayling

Geoscience Australia, GPO Box 378, Canberra, ACT 2601, Australia

ABSTRACT

The onshore Georgina Basin in northern Australia is prospective for unconventional hydrocarbons; however, like many frontier basins, it is underexplored. A well-connected hydraulic fracture network has been shown to be essential for the extraction of resources from the tight reservoirs that categorise unconventional plays, as they allow for economic flows of fluid from the reservoir to the well. One of the fundamental scientific questions regarding hydraulic stimulation within the sub-surface of sedimentary basins is the degree to which local and regional tectonic stresses act as a primary control on fracture propagation. As such, an understanding of present-day stresses has become increasingly important to modern petroleum exploration and production, particularly when considering unconventional hydrocarbon reservoirs. This study characterises the regional stress regime in the Georgina Basin using existing well data. Wellbore geophysical logs, including electrical resistivity image logs, and well tests from 31 petroleum and stratigraphic wells have been used to derive stress magnitudes and constrain horizontal stress orientations. Borehole failure features interpreted from wellbore image and caliper logs yield a maximum horizontal stress orientation of 044°N. Integration of density log data results in a vertical stress gradient of 24.6 MPa km^{-1}. Leak-off and mini-fracture tests suggest that this is the minimum principal stress, as leak-off values are generally shown to be at or above the magnitude of vertical stress. The maximum horizontal stress gradient is calculated to be in the range of 31.3–53.9 MPa km^{-1}. As such, a compressional stress regime favouring reverse/reverse–strike-slip faulting is interpreted for the Georgina Basin.

Keywords: Georgina Basin, *in situ* stress, present-day stress, stress magnitudes, hydraulic stimulation, central Australia, Arthur Creek Formation, image logs, leak-off tests, borehole failure

1. INTRODUCTION

The Australian continent hosts a highly variable stress field (Hillis & Reynolds, 2003; Rajabi, Tingay, King, & Heidbach, 2016a; Rajabi, Tingay, & Heidbach, 2016b) where stress patterns are not oriented sub-parallel to absolute plate motion as they are in many other continents (Rajabi et al., 2016b). Knowledge of contemporary stresses can provide important information for understanding large-scale processes, such as intraplate deformation, geodynamic and neotectonic processes, through to localised processes, with implications for the mining, petroleum, and geothermal industries (Bell, 1996; Hillis & Reynolds, 2003; Rajabi et al., 2016b). Patterns of stress orientations in the Australian continent have been analysed in the Australian Stress Map project (Hillis & Reynolds, 2000, 2003; Reynolds, Coblentz, & Hillis, 2002) and in subsequent studies (e.g. Bailey, King, Holford, & Hand, 2016; King, Hillis, & Reynolds, 2008; Rajabi et al., 2016a) (Figure 1). However, data from some locations, including the Georgina Basin in northern Australia, have not yet been analysed and/or incorporated into the most recent version of the Australian Stress Map (Hillis & Reynolds, 2003; Rajabi et al., 2016a) (Figure 1).

The Georgina Basin is an onshore Australian basin considered to be prospective for conventional and unconventional oil and gas (Ambrose, Kruse, & Putnam, 2001; Munson, 2014). The Georgina Basin is a frontier basin, and with limited seismic data coverage and only »60 wells targeting prospective formations drilled to date, it is underexplored (Bennett, Philipchuk, & Freeman, 2010; Munson, 2014; PetroFrontier, 2012, 2014; Smith et al., 2013). Conventional hydrocarbon exploration and stratigraphic drilling in the basin commenced in the 1960s; however, commercial flows of oil and gas are yet to be achieved. Recent exploration has primarily targeted source rocks in the Arthur Creek Formation and Thorntonia Limestone in the southern Georgina Basin for unconventional oil and gas.

The extraction of oil and gas from tight reservoirs such as shale requires hydraulic stimulation to create fracture pathways to enable adequate flows of fluid from the reservoir to the well (Bell, 1990; Bell & Babcock, 1986). Local and regional tectonic stresses act as primary controls over the generation and propagation of hydraulic fractures; thus, an understanding of present-day stresses has become increasingly important to modern petroleum exploration and production (Bell, 1996; Hillis & Reynolds, 2003; Zoback, 2007) (Figure 1). While critical for predicting hydraulic fracture patterns associated with reservoir stimulation, the orientations and magnitudes of present-day stresses also have implications for borehole stability, water-flood design, and fault reactivation (Addis, Last, Boulter, Roca-Ramisa, & Plumb, 1993; Caillet, 1993; Hillis & Reynolds, 2003; Hillis & Williams, 1993a, 1993b; Horn, 1991; Mildren, Hillis, Fett, & Robinson, 1994).

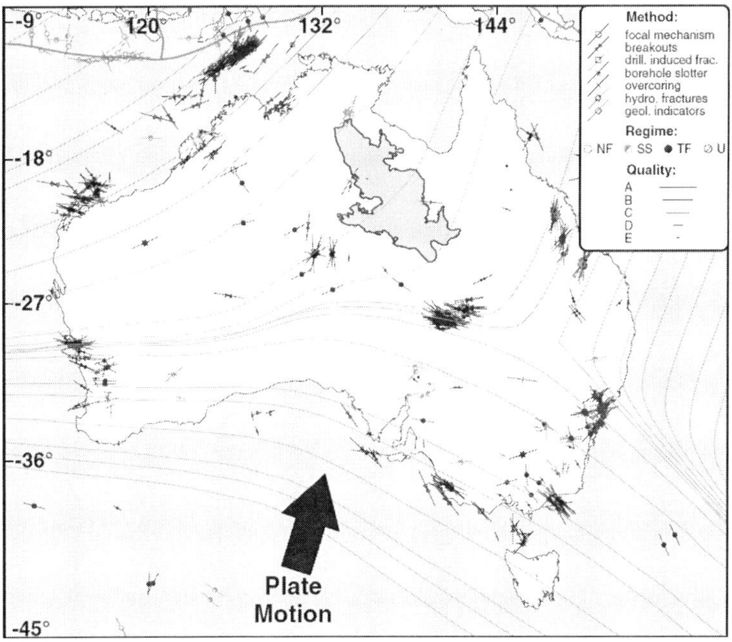

Figure 1. Present-day stress pattern of Australia (Heidbach et al., 2010; Hillis & Reynolds, 2003; Rajabi et al., 2016a). Different symbols show the orientation of maximum horizontal stress, and different colours show different types of stress indicators and the stress regimes (TF, thrust faulting; SS, strike–slip faulting; NF, normal faulting; black, unknown). The length of the lines indicates the quality of data. Solid grey lines show stress trajectories of Hillis and Reynolds (2003). Note the lack of the present-day stress indicators in the Georgina Basin (modified from Rajabi et al., 2016a).

Lithospheric stresses can be defined in terms of the orientation and magnitude of three orthogonal principal stresses, namely a maximum ($\sigma 1$), minimum ($\sigma 3$), and intermediate ($\sigma 2$). One stress is generally vertical owing to the mass of overburden and is referred to as sV (Anderson, 1951; Sibson, 1974). The two remaining principal stresses are generally confined to the horizontal plane and are referred to as the maximum (σH) and minimum (σh) horizontal stresses. The relative magnitudes of these three principal stresses define a tectonic regime in which a type of faulting will dominate (Zoback et al., 1989), although preexisting structures can exhibit hybrid failure modes under any given regime (Anderson, 1951; Heidbach & Hohne, 2008; Sibson, 1977). With one stress assumed to be vertical, knowledge of one of the horizontal stress orientations (σH or σh) enables the orientations of all three stresses to be determined (Bell, 1996; Tingay, Hillis, Morley, Swarbrick, & Okpere, 2003).

The aim of this study was to characterise the regional stress regime within the Georgina Basin and the variation in stress orientations within the region (Figures 1, 2). Owing to limited public availability of appropriate, detailed geophysical log data, lithological controls on stress regime within the basin are not evaluated in this study, and we focus on the regional stress regime. Data from 31 petroleum and stratigraphic wells were used to derive stress parameters from which a whole-of-basin representation of the present-day stress regime was constructed. Stress magnitudes were constrained using wellbore geophysical logs and tests. Vertical stress (σv) magnitudes were derived from checkshot-calibrated density logs, minimum horizontal stress (sh) magnitudes were constrained using data from leak-off tests (LOTs) and formation integrity tests (FITs). Maximum horizontal stress (σH) magnitudes were constrained using frictional limits calculations (Jaeger & Cook, 1979) and estimates assuming shear failure of pre-existing fractures during LOTs (Couzens-Schulz & Chan, 2010). Electrical resistivity-based image logs from six wells were used in conjunction with four-arm caliper (4AC) logs to assess σH orientations within the Georgina Basin. Common wellbore-failure features, such as borehole breakouts (BOs) and drilling-induced tensile fractures (DITFs), have been shown to be reliable indicators of stress orientations within sedimentary basins and can be identified on both electrical resistivity-based image and caliper logs (Brudy & Zoback, 1999; Dart & Zoback, 1989; Hillis & Reynolds, 2000).

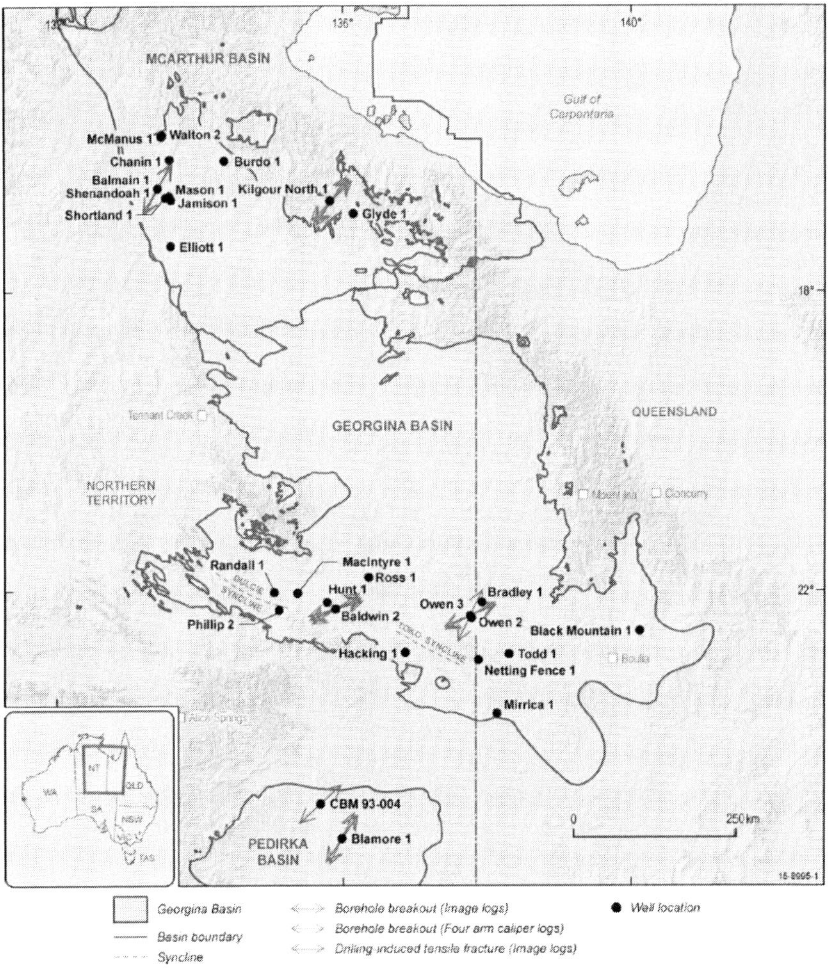

Figure 2. Regional map of the Georgina Basin. Interpreted maximum horizontal stress orientations from wells featuring four-arm caliper or electrical resistivity image logs are displayed as scaled arrows based on the world stress map quality ranking (Heidbach et al., 2010) assigned to that data point (Table 1). Two distinct domains of the basin are generally recognised: a southern depocentre south of latitude 21_S and a central-northern quiescent platform (Munson, 2014).

2. GEOLOGICAL SETTING

The Georgina Basin is a polyphase Neoproterozoic to Paleozoic intracratonic basin and is the largest remnant of the Neoproterozoic to early Paleozoic Centralian Superbasin (Ambrose et al., 2001; Greene, 2010; Munson, 2014; Walter, Veevers, Calver, & Grey, 1995). The basin covers an area of approximately 333 000 km2 throughout the central-eastern Northern Territory and western Queensland (Figure 2). The basin is bounded by the Tennant Region to the west, the McArthur Basin, South Nicholson Basin and Lawn Hill Platform to the northeast, and the Mount Isa Province to the east. The basin was once contiguous with the Amadeus Basin to the south; however, uplift of the Arunta Block led to their separation (Ambrose et al., 2001; Munson, 2014). Unconformably overlying the Proterozoic crystalline basement, the Georgina Basin hosts up to 5 km of Cryogenianto Devonian-aged sedimentary rocks that have experienced several major deformation events, most notably the Paleozoic Alice Springs Orogeny and the Neoproterozoic Petermann Orogeny (Ambrose et al., 2001; Greene, 2010; Munson, 2014). The basin is commonly separated into two distinct domains:

1) a southern depocentre south of latitude 21_S that incorporates the primary sedimentary succession of Cryogenian to Devonian rocks; and

2) a central-northern quiescent platform that includes some Ediacaran sediments and a thin middle Cambrian platform succession (Munson, 2014). The dominant petroleum prospectivity is associated with the southern Georgina Basin, and in particular in the Dulcie and Toko synclines (Figure 2).

The main interval of interest for petroleum exploration in the Georgina Basin has historically been the middle Cambrian succession, which contains high-quality marine source rocks (Ambrose et al., 2001, Boreham & Ambrose, 2005, 2007; Dunster, Kruse, Duffett, & Ambrose, 2007; Munson, 2014; Tiem, Horsfield, & di Primo, 2011). Of these, within the Arthur Creek Formation (Figure 3) oil shows increase in abundance with target (Ambrose & Putnam, 2007; Boreham & Ambrose, 2005; Munson, 2014). The Thorntonia Limestone, a transgressive dolostone that underlies the Arthur Creek Formation and features interbeds of carbonaceous mudstone, is also considered to host high-quality source rocks with unconventional prospectivity (Ambrose et al., 2001; Boreham & Ambrose, 2005; Dunster et al., 2007; Munson, 2014).

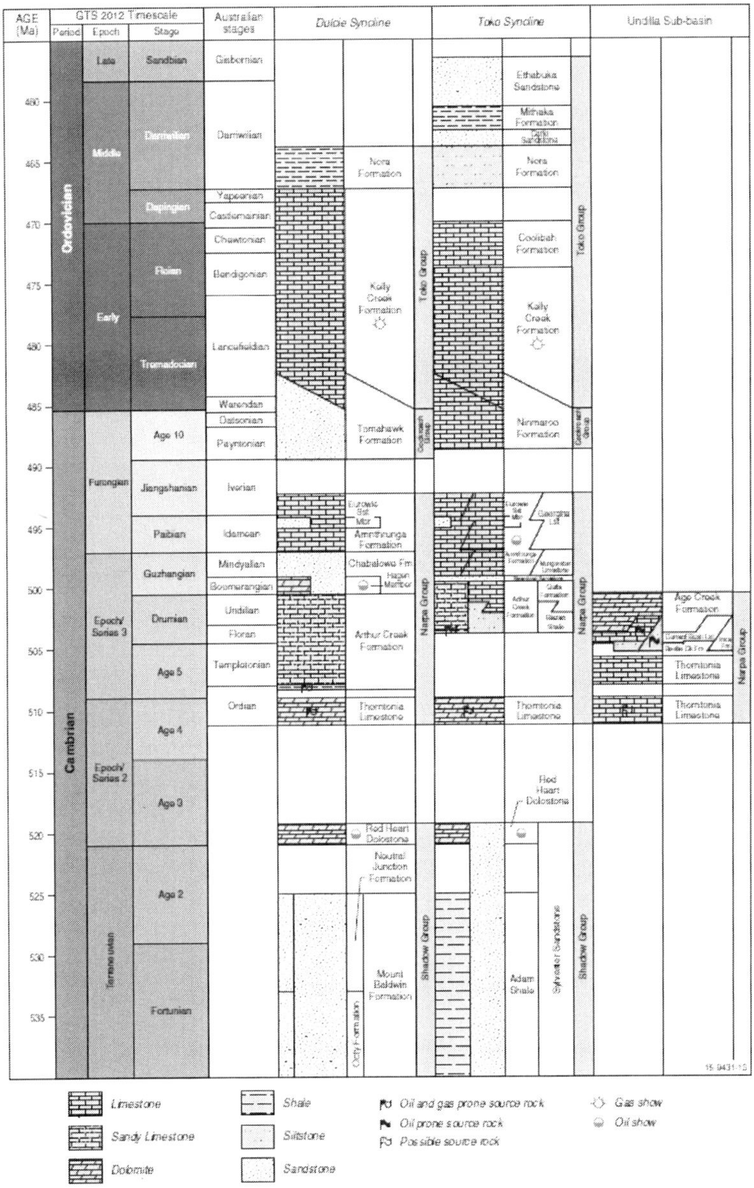

Figure 3. Stratigraphy of the Georgina Basin. Recent exploration has focused on the potential unconventional resources of the middle Cambrian Arthur Creek Formation and Thorntonia Limestone.depth (Ambrose et al., 2001). The basal hot shale facies of the lower Arthur Creek Formation is thought to have the greatest potential as both a traditional source rock and an unconventional

3. METHOD AND RESULTS

3.1 Data sources and data evaluation

Data were sourced from open-file well-completion reports [FITs, LOTs, drill stem tests (DSTs), mini-fracture tests] and associated well-log data (density, velocity and resistivity image logs). In addition, image-log and other digital data were obtained for two recently drilled wells in the basin (Owen 3 and Baldwin 2).

Data were evaluated for quality and reliability, processed to remove erroneous values and compiled. Of the 60 wells identified within the study area, 31 wells were found to contain relevant parameters for the analysis and met the criteria for reliability. These wells are detailed in the Supplementary Papers (Table A1). Several wells from the McArthur Basin, which underlies the Georgina Basin in the north, and from the northern Pedirka Basin, which lies to the south of the Georgina Basin, were used to validate and extend the interpretation of regional sH azimuths (Figure 2).

3.2 Vertical stress magnitude

The magnitude of vertical stress (σv) is defined as the stress applied by the mass of overburden above a specific depth and is calculated through an integration of measured formation bulk density to the depth of interest (Bell, 1996; Brown & Hoek, 1978; Engelder, 1993; Tingay et al., 2003). Where no overlying water column is present, the following equation applies:

$$\sigma_V = \int_0^Z \rho(z) g \, dz \tag{1}$$

where $p(z)$ is overburden density at depth z, and g is gravitational acceleration (Bell, 1996; Engelder, 1993; King, Neubauer, Hillis, & Reynolds, 2010; Tingay et al., 2003). Density logs were used to calculate σV (Bell, 1996; King et al., 2010; Tingay et al., 2003). In wells that are not run to surface, a top-of-log stress value can be calculated from well velocity surveys using the Gardner, Gardner, and Gregory (1974) empirically derived velocity–density transform (Bell, 1996; Tingay et al., 2003), allowing a vertical stress profile to be constructed (King et al., 2010; Zoback, 2007).

Density logs from 23 wells within the Georgina Basin produce a mean σv gradient of 24.6 MPa km^{-1} (SD D 0.84 MPa km^{-1}). Calculated magnitudes for σv at 1 km depth range from 22.8 to 26.1 MPa km^{-1} (Figures 4, 5; Table 2).

3.3 Minimum horizontal stress magnitude

Hydraulic fracture tests, such as mini-fracture tests and LOTs, allow for direct determination of σ_H magnitude (Bell, 1990; Breckels & van Eckelen, 1982; Dickey, 1986; Nelson, Hillis, & Mildren, 2006; White, Traugott, & Swarbrick, 2002).

In the Georgina Basin, two different sets of LOT results were observed: those where leak-off occurs below the calculated value of σ_v, and those where leak-off occurs at or above the calculated value of σ_v (Figure 4). Nine LOTs and two minifracture tests from nine wells were analysed in this study (Table 3). An additional 24 FITs from 22 wells were used to provide estimates of lower bounds to σ_3 (Figure 4) (Supplementary Papers, Tables A1, A3). Calculated values of σ_3 from LOTs range from 2.0 MPa at 0.09 km to 41.4 MPa at 1.27 km; calculated values of σ_3 from the two mini-fracture tests range from 27.6 MPa at 0.90 km to 70.7 MPa at 2.54 km (Table 3).

Five LOTs in four wells had leak-off pressures that were below the calculated value of σ_v. One of these wells, Bradley 1, was discounted owing to reported poor LOT conditions.

3.4 Maximum horizontal stress magnitude

Several methods are used to estimate σ_H magnitudes from petroleum wellbore data. The most commonly used is the frictional limits method (Jaeger & Cook, 1979). A maximum limit to horizontal stress magnitude can also be calculated when assuming shear failure of fractures during LOTs (Couzens- Schultz & Chan, 2010). The magnitude of σ_H can also be constrained by observed breakout width (Barton, Zoback, & Burns, 1988; Vernik & Zoback, 1992) and through the circumferential stress relationships around a vertical well (Bell, 1996; Hubbert & Willis, 1957); however, these approaches require higher-quality data than were available for this study. Detailed one-dimensional mechanical earth models can be constructed using poroelastic stress equations that take tectonic strain into consideration (Brooke_Barnett et al., 2015; Mildren, Clarke, Holford, & Titus, 2013; Thiercelin & Plumb, 1994). However, these rely on the availability of extensive wireline log suites and laboratory rock property measurements for calibration (Brooke _Barnett et al., 2015) that are not presently publicly available in the Georgina Basin. Thus, we applied the first two approaches to estimate σ_H within the Georgina Basin.

(1) Frictional limits theory (Jaeger & Cook, 1979) assumes that the maximum stress is limited by the rock's frictional properties, and that this stress is horizontal. These assumptions point to a maximum value for σ_H magnitude (Sibson, 1974; Zoback & Healy, 1984), as represented by the following equations:

$$\sigma_H = f(\mu) \times (\sigma_3 - P_P) + P_P \tag{2}$$

$$f(\mu) = \left[(1+\mu^2)^{\frac{1}{2}} + \mu\right]^2 \tag{3}$$

and where m is the coefficient of internal friction, assumed to be 0.6 (Byerlee, 1978; Zoback & Healy, 1984.

Figure 4. Plot of depth vs stress (to a cutoff of 2 km depth) illustrating the principal stress magnitudes interpreted from data for the 26 Georgina Basin wells used in this study. Vertical stress profiles are plotted for each well with the range of values shown. Horizontal stresses are presented as a range of possible values limited by their frictional limits, with the estimated range for maximum horizontal stress highlighted in green. These values are calculated using Equations 2 and 4, respectively. Minimum horizontal stress is interpreted to be closer to vertical stress magnitude than maximum horizontal stress magnitude, as significant anisotropy must exist between the minimum and maximum horizontal stress in order to form borehole breakouts. Therefore, a reverse-faulting (or transitional strike-slip to reversefaulting) stress regime is interpreted for the Georgina Basin.

Assuming a hydrostatic pore pressure, and that $\sigma_V = \sigma 3$, an average upper frictional limit boundary of 53.9 MPa km^{-1} (1 SD D 2.6 MPa km^{-1}) was calculated for the Georgina Basin (Table 2; Figure 4).

(2) Couzens-Schultz and Chan (2010) provide a method for estimating the maximum upper limits of σ_H when shear failure is suspected in LOTs (see Discussion for treatment of suspect LOT data). Estimation of σ_H can be achieved using the following equation:

$$\sigma_{H,lim} = \sigma_V + \frac{2\left(\sigma_V - \left(LOP - \frac{C_0}{\mu}\right)\right)\sin^{\circ}\emptyset}{(1 - \sin^{\circ}\emptyset)} \tag{4}$$

Where $\sigma_{H,lim}$ is the upper limit of σ_H, C_0 is cohesion (assumed to be negligible), and μ is the coefficient of friction (assumed to be 0.6) (Couzens-Schulz & Chan, 2010). Using LOTs with suspected shear failure (Equation 4), σ_H magnitudes ranging from 2.8 MPa at 0.09 km depth to 4.7 MPa at 0.13

km depth are predicted (Table 4). This gives an extrapolated σ_H gradient of 34.7 MPa km−1 (Figure 4). The two methods applied to estimate σ_H magnitude yield estimates that range from 34.7 MPa km^{-1} to 53.9 MPa km^{-1}. It should be noted that the maximum value derived from frictional limits theory is likely to overestimate σ_H.

3.5 Maximum horizontal stress orientation

Two main wellbore-failure features are commonly used as stress indicators: borehole breakouts (BO) and DITF (Figure 6). As rock is removed during drilling, stress is concentrated around the wellbore: circumferential stress is maximised at the minimum horizontal stress azimuth and minimised at the maximum horizontal stress azimuth (Gough & Bell, 1981; Haimson & Herrick, 1986; Maloney & Kaiser, 1989; Zheng, Kemeny, & Cook, 1989; Zoback, Moos, Mastin, & Anderson, 1985) (Figure 6a). Borehole failure through the creation of BOs and DITFs is a result of stress perturbations that exist under such conditions (Kirsch, 1898; Plumb & Hickman, 1985; Zheng et al., 1989; Zoback et al., 1985). BOs form in the direction of σ_H, and DITFs occur in the direction of σ_H (Figure 6a). These features and their orientations can be identified using Figure 5. Plot of depth vs stress illustrating the range of stress gradients calculated for wells with rock density data. either four-arm caliper data (Figure 6b) or electrical resistivity image logs (Figure 6c, d).

Seven wells were identified within the study area (Figure 2) that had resistivity image logs, 4AC logs or both (Figure 7) (Supplementary Papers, Table A1). As a consequence of the low data density within the Georgina Basin, wells in the nearby Pedirka Basin and underlying McArthur Basin were included in the interpretation to establish the regional consistency of σ_H azimuths (Figure 2). Several types of electrical resistivity image logs were used: Formation Micro-Imager (FMI), Formation Micro-Scanner, Compact Micro-Imager (CMI) and Simultaneous Acoustic and Resistivity Imager (STAR) (Supplementary Papers, Table A2). Glyde 1 hosted no identifiable stress indicators. However, the remaining six interpreted wells hosted an appropriate population of stress indicators (Figure 7) to qualify as D quality or higher according to the World Stress Map guidelines (Table 1). The only exception to this was the BOs interpreted from the Baldwin 2 image log. Using the valid stress orientations from borehole failure data (Figure 2), a mean σH orientation of 044_N was interpreted for the study area (Figure 7). Application of the Rayleigh test (Mardia, 1975) showed that on a regional scale these data represent a non-random distribution and so are likely to be representative of the regional σH orientation. However, the orientation of σH varies subtly between locations, and this is likely controlled by local factors such as stress perturbations around existing structures (e.g. Barton & Zoback, 1994). In wells that deviate from vertical,

BO and DITF azimuths may be perturbed from the true orientation of σ_H and σ_H, respectively (Peska & Zoback, 1995). This is a result of the borehole axis not being aligned parallel to one of the principal stresses in highly deviated wells (Mastin, 1988; Peska & Zoback, 1995). Well deviation ranges from 1.5_ to 18.5_ in the studied wells (Table 1), with the majority of wells being deviated <10_. Additionally, the two wells

Table 1. Interpreted maximum horizontal stress orientations from four-arm caliper logs and resistivity image logs (using both borehole breakouts and drillinginduced tensile fractures).

Well	Log	Indicator	Quality	Orientation
Baldwin 2	FMI	BO (4AC)	C	58.0
		BO (Image)	E	74.8
		DITF	D	66.9
Blamore 1	STAR	BO (4AC)	–	–
		BO (Image)	D	26.2
		DITF	C	34.7
CBM 93 004	CMI	BO (4AC)	C	48.3
		BO (Image)	–	–
		DITF	–	–
Kilgour North 1	FMI	BO (4AC)	B	46.8
		BO (Image)	C	45.8
		DITF	D	33.1
Owen 3	CMI	BO (4AC)	B	26.3
		BO (Image)	D	62.6
		DITF	D	65.4
Shenandoah 1	FMI	BO (4AC)	D	30.9
		BO (Image)	–	–
		DITF	D	30.7

Quality rankings are based on the World Stress Map Project criteria (Heidbach et al., 2010). Further details on the factors contributing to a given quality ranking can be found in Supplementary Papers, Table A3. Tools: FMI, Formation Micro-Imager; CMI, Compact Micro-Imager; STAR, Simultaneous Acoustic and Resistivity Imager. Indicators: BO, Borehole Breakouts; DITF, Drilling-Induced Tensile Fractures; 4AC, Four-Arm Caliper.

Table 2. Calculated stress magnitudes for the wells featuring density logs in the Georgina Basin.

Well	S_V gradient (MPa km^{-1})	Frictional limits (MPa km^{-1})
Northern Georgina Basin		
Altree 1	24.7	54.3
Balmain 1	23.5	50.3
Burdo 1	24.1	52.4
Chanin 1	24.4	53.3
Elliot 1	24.0	52.1
Glyde 1	24.7	54.0
Mason 1	22.8	48.4
Shenandoah 1	23.5	50.3
Shortland 1	23.4	50.1
Walton 2	23.7	50.9
Southern Georgina Basin		
Baldwin 1	25.4	56.3
Baldwin 2	25.7	57.2
Bradley 1	24.3	52.9
Hacking 1	24.4	53.1
Hunt 1	25.1	55.5
MacIntyre 1	26.1	58.4
Mirrica 1	25.2	55.6
Owen 2	25.1	55.5
Owen 3	25.1	55.5
Phillip 2	25.5	56.8
Randall 1	24.5	53.5
Ross 1	25.0	55.0
Todd 1	25.8	57.6

Results are separated by well location within the basin. Vertical stress gradient (S_v) as calculated using Equation 1 and frictional limits to maximum horizontal stress as calculated using Equation 2 are presented.

3.6 Pore pressure

Pore pressure Formation pressure can be extrapolated from production tests such as the DST and through wireline tools such as the Modular Formation Dynamics Test (MDT) (Bradley, 1975; Horner, 1951; Kuchuk, 1998). Formation test data were obtained for 17 wells in the Georgina Basin, providing 39 estimates of formation pressure (Figure 8). Pore pressures were observed to be at or near the hydrostatic value for saline groundwater, and this was also indicated by the mud weights applied when drilling in the Georgina Basin, which closely reflect the hydrostatic gradient (Figure 8). While a hydrostatic gradient is assumed for the remainder of this study based on these results, overpressures may exist within the basin.

Table 3. Leak-off tests and resultant estimate of least principal stress magnitude for each well in which leak-off was achieved in the Georgina Basin. Quality rankings are as per King et al. (2008). Depth is given as true vertical depth as measured from the Kelly bushing (m TVDkb). LOT, leak-off test; Minifrac, Mini-Fracture Test; FCP, fracture closure pressure.

Well	Test	Quality	Depth (m TVDkb)	Leak-off pressure (MPa)	Leak-off gradient (MPa km^{-1})
Bradley 1	LOT	D	820.40	10.2	12.5
Bradley 1	LOT	D	117.80	1.71	14.5
MacIntyre 1	LOT	D	125.30	2.6	20.8
Randall 1	LOT	D	128.00	2.8	21.9
Phillip 2	LOT	D	88.10	2.0	22.8
Shenandoah 1	Minifrac	Reported FCP gradient	2539.80	70.5	27.8
Shenandoah 1	Minifrac	Reported FCP gradient	1640.00	45.5	27.8
Shenandoah 1	Minifrac	Reported FCP gradient	2537.00	70.7	27.9
Shenandoah 1	Minifrac	Reported FCP gradient	1754.00	49.2	28.1
Shenandoah 1	Minifrac	Reported FCP gradient	1873.50	53.5	28.6
Baldwin 2	MiniFrac	Reported closure Pressure	895.00	27.6	30.9
Baldwin 2	LOT	C	423.00	13.2	31.1
Owen 3H	LOT	D	1273.00	41.4	32.5
Ross 1	LOT	D	101.50	3.3	32.6
Baldwin 1	LOT	D	102.50	5.1	49.8

Quality rankings are as per King et al. (2008). Depth is given as true vertical depth as measured from the Kelly bushing (m TVDkb). LOT, leak-off test; Minifrac, Mini-Fracture Test; FCP, fracture closure pressure.

Table 4. Leak-off tests (LOT) with particularly low leak-off values recalculated using the Couzens-Schultz and Chan (2010) relationships (Equations 4 and 5), assuming shear reactivation rather than the creation of a new tensile fracture in estimating a minimum and maximum limit to the horizontal stress magnitudes.

Well	Test	Quality	Depth (m TVDkb)	$S_{h,lim}$ MPa	$S_{h,lim}$ gradient (MPa km^{-1})	$S_{H,lim}$ MPa	$S_{H,lim}$ gradient (MPa km^{-1})
MacIntyre 1	LOT	D	125.25	2.8	23.5	4.7	38.7
Randall 1	LOT	D	128.00	2.9	23.6	3.8	31.3
Phillip 2	LOT	D	88.14	2.1	25.7	2.8	34.1

Depth is given as true vertical depth as measured from the kelly bushing (m TVDkb).

studied *i*th deviations of >10_, Owen 3 and Kilgour North 1, are deviated <10_ for the majority of their length. As all of the wells studied are approximately vertical (well deviations <20_ from vertical), we consider that our interpretations of BO and DITF provide reliable indications of σH and σH, respectively (Mastin, 1988).

4. DISCUSSION

4.1 Diagnosis of stress regime

In determining the present-day stress regime, it is essential to determine which principal stress is the minimum. The LOT data show two types of results: values that are at or above the value of σ_V, and values that are significantly lower than the value of σ_V (Figure 4). Values at or above σ_V suggest that σ_V D σ_3, and that horizontal tensile fractures are being generated (Zoback, 2007). This would place the Georgina Basin in a dominantly reverse-faulting stress regime (Anderson, 1951). However, some data are below the vertical stress gradient, which suggests two possible scenarios: (1) fractures optimally oriented for shear failure were reactivated, or (2) vertical tensile fractures were generated (e.g. under a much reduced σh).

Such dichotomous LOT results have been observed in active thrust belts; in some wells leak-off pressures are close to the vertical stress, and in others much less than vertical stress (e.g. Couzens-Schultz & Chan, 2010). It has been suggested that pre-existing fractures intersecting the borehole can affect LOTs. Where cohesion is low, fractures may experience shear slip allowing for leak-off, giving abnormally low values of sH (Bailey et al., 2016;

Chan, Hauser, Couzens- Schultz, & Gray, 2014; Couzens-Schultz & Chan, 2010; Gjønnes, Cruz, Horsrud, & Holt, 1998; King et al., 2012). We consider the first scenario to be more likely, i.e. data that are below σ_V are reflecting shear reactivation of optimally oriented fractures. This is supported by FIT data that demonstrate values obtained without reaching leak-off are close to σ_V (Figure 4). The magnitude of σ_H is, therefore, likely to be equal to or higher than the magnitude of σ_V.

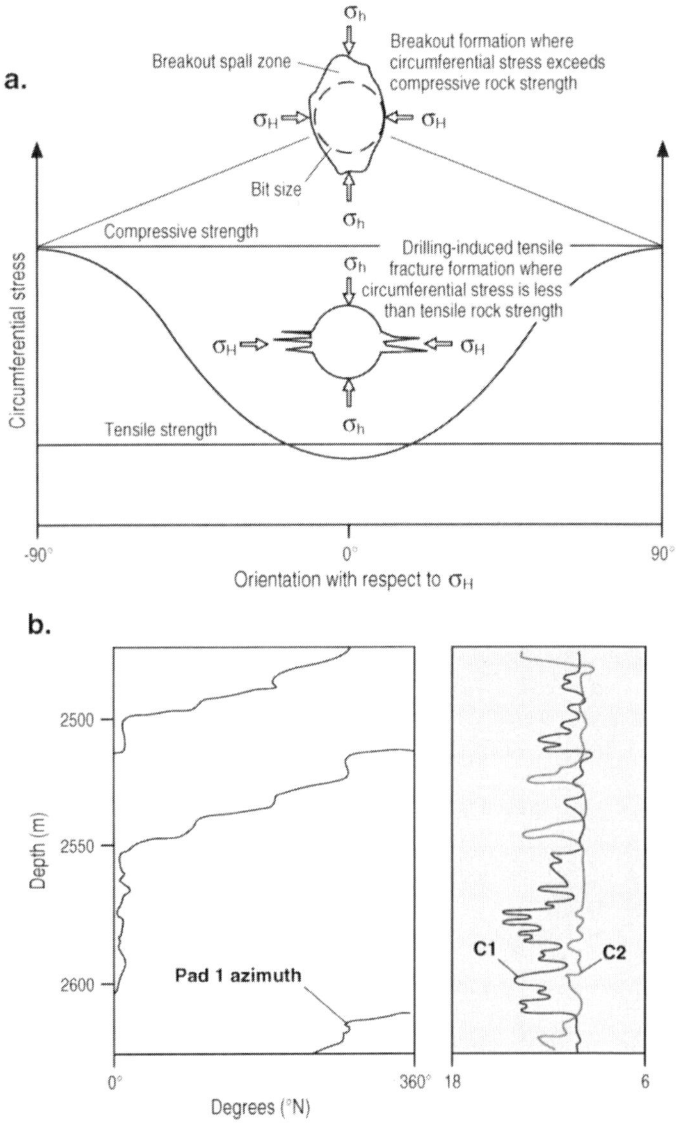

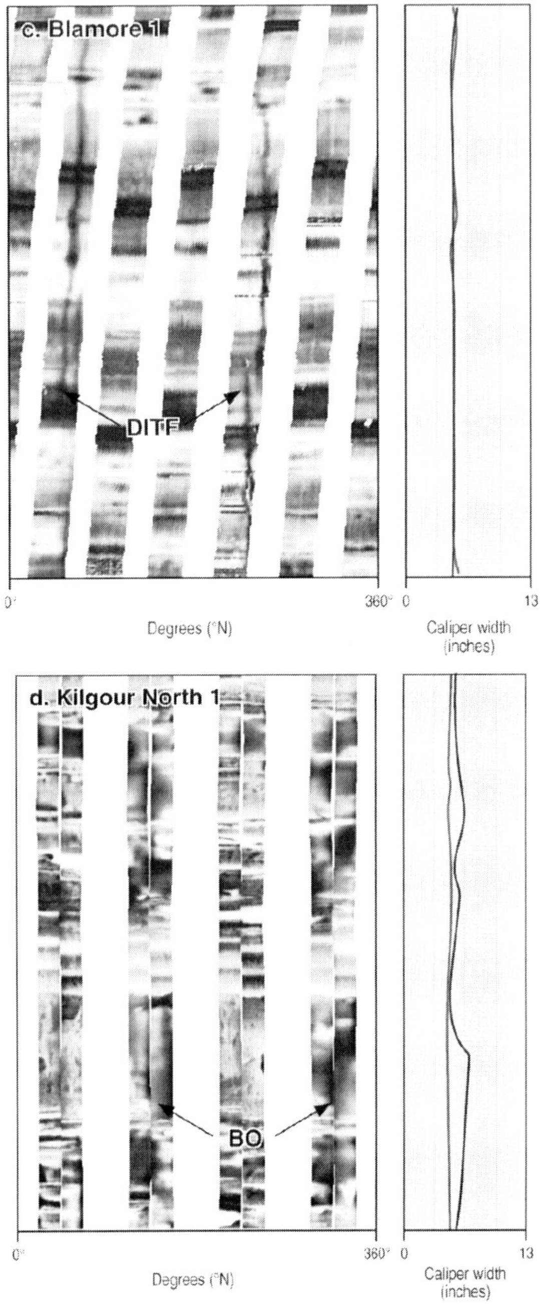

Figure 6. (a) Azimuth of borehole breakouts and drilling-induced tensile fractures with respect to the circumferential stress around the wellbore

(Hillis & Reynolds 2003). (b) Four-arm caliper logs showing borehole breakouts, with associated pauses in pad rotation (Reynolds & Hillis, 2000). (c) Section of STAR image from Blamore 1 showing drilling-induced tensile fractures. (d) Section of FMI image from Kilgour North 1 showing borehole breakouts.

Couzens-Schultz and Chan (2010) propose the following equation to calculate a lower-bound estimate for the two horizontal stress magnitudes based on shear reactivation of existing fractures:

$$\sigma_{H.lim} = \sigma_V - \frac{2\left(\sigma_V - \left(\text{LOP} - \frac{C_0}{\mu}\right)\right)\sin\theta}{(1+\sin\theta)} \tag{5}$$

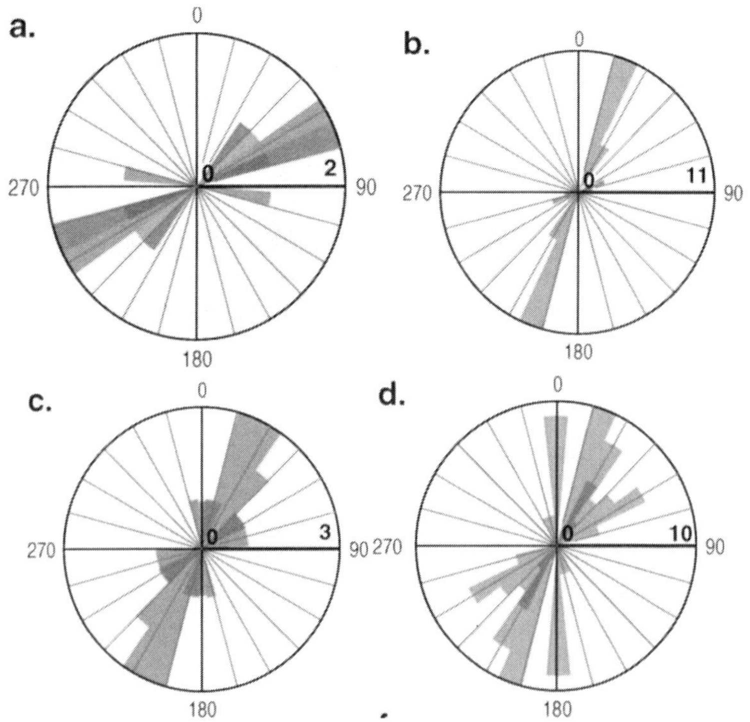

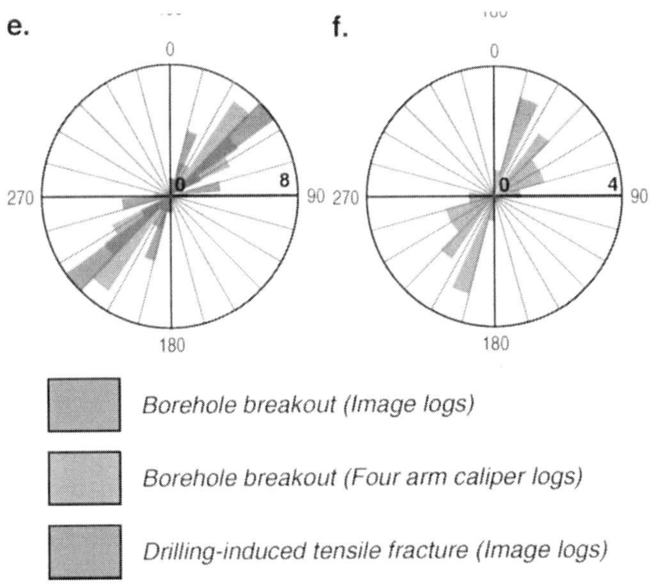

Figure 7. Rose diagrams of maximum horizontal stress derived from each well. Green roses represent 4AC interpreted breakout azimuths, red roses represent image log interpreted breakout zimuths, and blue rose outlines represent image log interpreted DITF azimuths. Further statistical data can be found in the Supplementary Papers. Each diagram features an individual scale.

where σ_H,lim is the lower limit of σH (Couzens-Schulz & Chan, 2010). The five data points with anomalously low LOT values (Table 4) were recalculated using Equation 5. Values of $\sigma_{h,lim}$ from this method range from 2.1 MPa at »0.09 km to 2.9 MPa at 0.13 km, giving gradients that are approximately equal to the calculated gradient of σ_v (Tables 2, 4), and in line with estimates from the other LOTs.

Calculated stress magnitudes demonstrate that the Georgina Basin is subject to a dominantly reverse-faulting stress regime (Table 5). It is possible that $\sigma_3 = \sigma_2$ ($\sigma_V = \sigma_h$), meaning that the Georgina Basin may host a transitional strike-slip to reverse-faulting stress regime. Given that enough stress anisotropy exists between the horizontal stresses for the formation of BOs and DITFs, it is unlikely that $\sigma_H = \sigma_h$ (Table 5).

The orientations of the principal stresses appear to be consistent throughout the basin, with a non-random mean σ_H orientation of 044_N. This orientation is obtained from widespread data; however, it agrees with previously modeled σH orientations (Dyksterhuis, M€uller, & Albert, 2005; Muller & Dyksterhuis, 2005; Reynolds et al., 2002) as well as the predicted orientations of the Australian Stress Map (Hillis & Reynolds, 2000, 2003) (Figures 1, 2). We propose that σ_H is greater or equal to σ_v thereby placing the Georgina Basin in a reverse/reverse–strike-slip stress regime (Zoback et al., 1989) (Table 5).

4.2 Lithological controls over stress

At local scales, it is well demonstrated that lithological transitions can affect in situ stress magnitudes, primarily owing to variations in rock properties. Stress magnitudes in units with low Poisson ratio are generally greater than those with high Poisson ratios (Thiercelin & Plumb, 1994). Variation of stress magnitudes by lithology can be beneficial for hydraulic stimulation, primarily as a means by which fracture growth can be controlled: where a 'low-stress' unit is located between two 'high stress' units, fractures formed in the centre unit can be easily compartmentalised by limiting injection pressures (e.g. Zoback, 2007). Brooke-Barnett et al. (2015) demonstrate an example in Queensland's Surat basin where mechanical stratigraphy exerts a control over tectonic stresses, with lowerrigidity rocks supporting lower tectonic stresses than highrigidity rocks.

Analysis of stress magnitudes derived from LOTs presented in this study reveals no distinct correlation between lithology and either high or low values for leak-off, so we suggest that the anomalously low LOT values observed are due to LOTs having been performed in the presence of pre-existing fractures as described above (Couzens-Schulz & Chan, 2010). The formation of DITFs is generally observed in fine-grained (siltstone and shale) intervals and can be a result of a strike-slip stress regime (Zoback, 2007). Hence, as DITFs in this study similarly appear to be restricted to fine-grained intervals in the Georgina Basin data, this may be an example of mechanical stratigraphy and may indicate intervals of strike-slip faulting within the basin. However, while some lithological control may be implied in our data, this study has primarily focused on the regional and basin-scale trends in stress regime of the Georgina Basin. For field-scale applications, these possible lithological variations in stress regime should be further investigated.

4.3 Distribution of stresses

Although data coverage is sparse, the Georgina Basin does exhibit some trends in the distribution of stress. Contour maps of the calculated values for the principal stresses were created to highlight general trends (Figure 9), although it should be recognised that the paucity of data in the northern part of the basin means that basin-wide trends must be interpreted with caution.

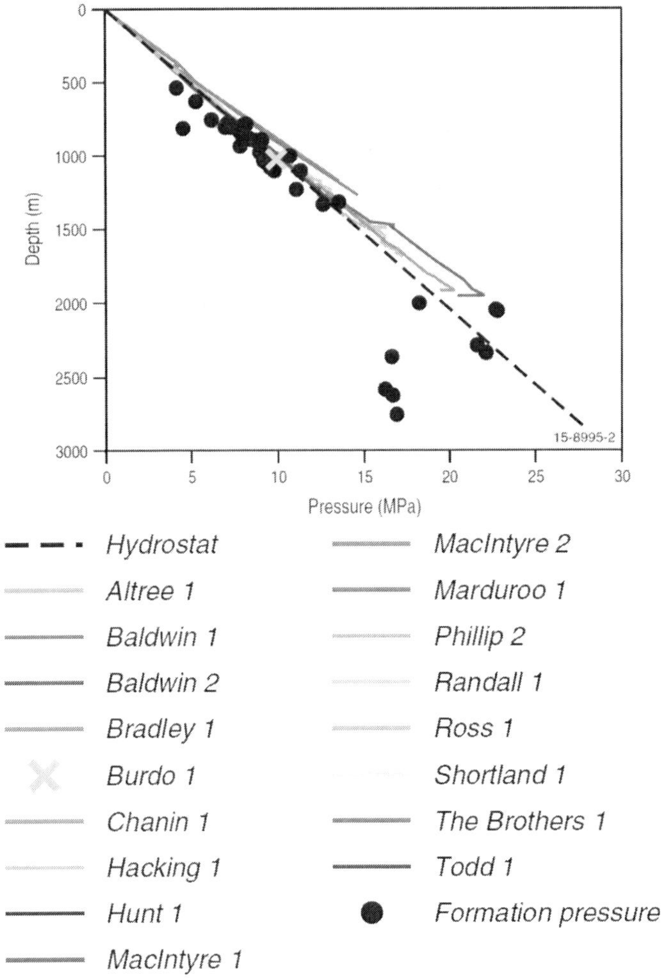

Figure 8. Pore pressure profile for Georgina Basin. Point data sourced from DSTs and modular formation dynamics tests (MDT) undertaken in wells within the basin and plotted over the recorded mud weight profiles for various wells. Hydrostatic pressure is represented as a dashed line.

Table 5. Summary table of all interpreted stress data demonstrating that the Georgina Basin likely hosts a reverse-faulting stress regime.

Stress orientation	Vertical stress (σ_v) magnitude (MPa km^{-1})	Minimum horizontal (σ_h) stress magnitude (MPa km^{-1})	Maximum horizontal (σ_H) stress magnitude (MPa km^{-1})	Interpreted stress regime
044°	22.8–26.1	Between σ_v and σ_H	53.86	Reverse

Vertical stress magnitudes are given for 1 km depth.

High vertical stresses are common, most notably in the southern Georgina Basin in the Toko and Dulcie synclines (Figure 9a). This trend may be explained by extensive uplift, which is thought to have occurred along the basin's southern margin during the Paleozoic Alice Springs Orogeny and the Neoproterozoic Petermann Orogeny (Ambrose et al., 2001; Hand & Sandiford, 1999; Munson, 2014). Uplifted sedimentary sections are over-compacted relative to current burial depths, leading to higher rock densities (King et al., 2010; Tingay et al., 2003). Conversely, in the northern part of the basin (including the Beetaloo Sub-basin and the underlying McArthur Basin), σ_V magnitudes are lower, implying less uplift-related over-compaction of the sedimentary section at current burial depths (Munson, 014). As σ_H magnitudes calculated through frictional limits are based on the values of σ_V, they exhibit similar trends, with higher σ_H magnitudes observed in the southern Georgina Basin (Figure 9b).

Analysis of LOT results highlights few distinct spatial trends, primarily owing to the low data density (Figure 2). However, lower pressures appear to be necessary to achieve leak-off in the eastern part of the Dulcie Syncline (Figures 2, 9c). As proposed above, this may be due to the presence of pre-existing flaws in the tested rock (i.e. shear reactivation of existing fractures), rather than a reduction in σ_3 magnitudes. As such, it could indicate higher natural fracture densities in the eastern Dulcie Syncline compared with the rest of the basin. Given the low data density and the presence of low LOT results in other parts of the basin, it is possible that the clustering of low LOT values is coincidental, and further data are required to identify any reliable trends in fracture density distribution.

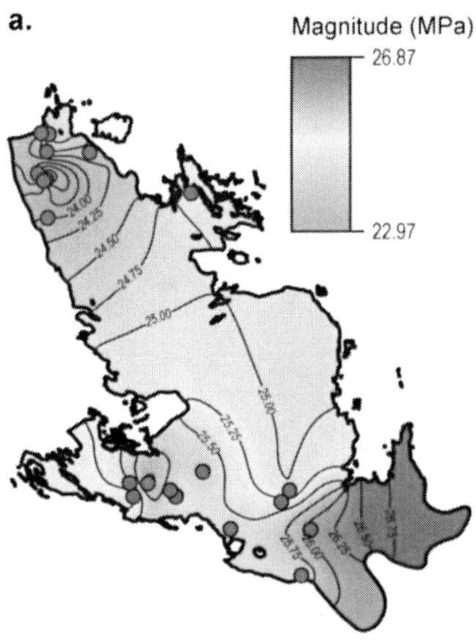

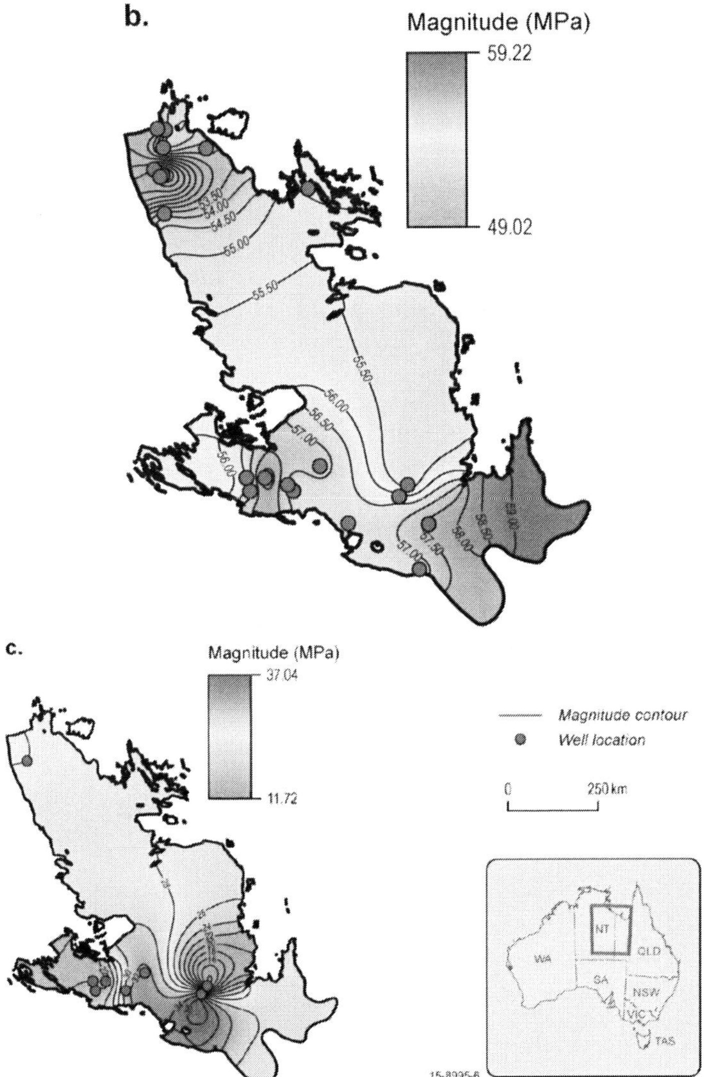

Figure 9. Contour maps showing the distribution of: (a) vertical stress gradients within the Georgina Basin; (b) interpreted frictional limits to maximum horizontal stress within the Georgina Basin; and (c) leak-off test results within the Georgina Basin. Moderately high values in the eastern portion of the basin are an artefact of low data density and software extrapolation.

4.4 Implications for hydraulic stimulation

Hydraulic stimulation is commonly undertaken to improve fluid recovery, particularly in tight reservoirs in shale gas, shale oil, tight gas and coal seam gas plays (Fjaer, Holt, Horsrud, Raaen, & Risnes, 2008; Yew & Weng, 2014). Created through injection of high-pressure fluids into a target formation, hydraulic fracture formation and propagation are primarily controlled by the present-day stress regime, rock properties, rock heterogeneity and pore pressures (Pe_ska & Zoback, 1995). The orientation and magnitudes of the three principal stresses commonly influence the size and orientation of fractures formed through hydraulic stimulation (Anderson, 1951; Hubbert & Willis, 1957). We have interpreted the Georgina Basin to be in a reverse/ reverse-strike-slip regime. In normal and strike-slip faulting stress regimes, tensile fractures (such as those induced through hydraulic fracturing) are observed to form vertically, as $\sigma_3 = \sigma_h$ (Anderson, 1951; Hossain, Rahman, & Rahman, 2000; Sibson, 1990; Zoback, 2007). Studies have found that in a reverse-faulting stress regime, and at distance from the well bore, tensile fractures will form horizontally as $\sigma_3 = \sigma_V$ (Hossain et al., 2000; Hubbert & Willis, 1957; Sibson, 1990; Zoback, 2007). In reverse-faulting stress regimes, induced tensile fractures are observed to form vertically at the wellbore wall owing to hoop stress effects (Zoback, 2007), before rotating to horizontal as they propagate out of the perturbed stress field surrounding the wellbore (Baumg€artner & Zoback, 1989; Hossain et al., 2000; Zoback, 2007).

4.5 Implications of stress field for well stability

Stress field anisotropy can cause wells to become unstable and form BOs and DITFs (King et al., 2012; Zoback, 2007), resulting in unacceptable reservoir and wellbore damage (Zoback, 2007). Boreholes are therefore drilled along a trajectory that minimises stress anisotropy (Zhou, Hillis, & Sandiford, 1996). In a reverse-faulting stress regime, the greatest anisotropy has been demonstrated to occur between σ_V and σ_H thus, horizontal wells drilled towards σ_h are proposed to be the least stable, as they are subject to the greatest stress anisotropy (King et al., 2012). In the Georgina Basin, σ_V and σ_H are interpreted to be of similar magnitudes (»25–26 MPa km−1). In similar settings, studies have shown that horizontal to sub-horizontal wells oriented in the direction of σ_H will be exposed to the least stress anisotropy and, accordingly, will be the most stable (Zhou et al., 1996; Zoback, 2007). However, such wells would have minimal hoop stress differential and may therefore require larger pressures to initiate tensile hydraulic fracturing (Zoback, 2007). This is because having some hoop stress at the wellbore effectively reduces the magnitude of the smaller stress, thereby promoting tensile fracturing at lower hydraulic pressures (Zoback, 2007).

4.6 Georgina Basin stress regime in context

Our data are consistent with other studies of onshore Australian basins, which suggest reverse to strike-slip faulting stress regimes (e.g. the Cooper-Eromanga, McArthur and Bowen basins; Hillis, Enever, & Reynolds, 1999; Mildren et al., 2013; Reynolds et al., 2002; Reynolds, Mildren, Hillis, & Meyer, 2006), largely as a result of the high compressive tectonic stresses in the Australian continent (Cloetingh & Wortel, 1986; Heidbach et al., 2010; Reynolds et al., 2002). In comparison, the major producing shales in the USA, such as the Barnett, Bakken, Woodford and Marcellus, are mostly Devonian to Carboniferous (Abousleiman et al., 2007; Arthur, Bohm, & Layne, 2009; Fisher et al., 2002; Smith & Bustin, 2000) and are found in basins characterised by extensional stress regimes (Heidbach et al., 2010; Zoback & Zoback, 1980). The Neoproterozoic to Devonian sediments of the Georgina Basin have had a complex evolution, having been buried and exhumed during several major tectonic events (Ambrose et al., 2001; Munson, 2014; Summons, Powell, & Boreham, 1988). As a result, their physical properties are likely to differ from the American equivalents.

5. CONCLUSIONS

Geomechanical datasets from the Georgina Basin indicate a dominantly reverse-faulting stress regime with a regionally consistent maximum horizontal stress orientation of approximately 044_N. Vertical stress is interpreted to be the least principal stress, as the majority of leak-off test results report values at, or above, the vertical stress owing to overburden. The integration of calibrated density logs yields an estimated vertical stress gradient of 24.6 MPa km^{-1}. It is, therefore, likely that vertical stress is being evaluated rather than minimum horizontal stress, as is generally assumed. Assessment of LOTs that appear to have failed through shear reactivation rather than tensile initiation provides stress estimates that further support this interpretation. Frictional limits calculations provide an upper bound for the maximum horizontal stress gradientof approximately 53.9 MPa km^{-1}, while estimates of maximum horizontal stress magnitude from the assumption of shear failure range between approximately 31.0 MPa km^{-1} and 39.0 MPa km^{-1}.

The minimum horizontal stress likely falls between the magnitude of vertical and maximum horizontal stress; thus, a reverse/reverse–strike-slip faulting stress regime is interpreted for the Georgina Basin.

ACKNOWLEDGEMENTS

The authors would like to acknowledge the NTGS, PetroFrontier and Statoil for their assistance in locating and providing well data. Particular thanks are reserved for Steve Abbott and Andrew McPherson at Geoscience Australia, for their invaluable input in improving this paper, and Silvio Mezzomo and David Arnold are thanked for their assistance in preparing the figures for publication. Thanks to Mojtaba Rajabi, for providing current stress map data. We would also like to thank the reviewers, Scott Mildren and Emma Tavener, for the time and effort spent improving this paper. This paper is published with the permission of the CEO, Geoscience Australia.

REFERENCES

1. Abousleiman, Y. N., Tran, M. H., Hoang, S., Bobko, C. P., Ortega, A., & Ulm, F. J.(2007). Geomechanics field and laboratory characterization of the Woodford Shale: The next gas play. SPE Annual Technical Conference and Exhibition. Richardson TX: Society of Petroleum Engineers.
2. Addis, T., Last, N., Boulter, D., Roca-Ramisa, L., & Plumb, D. (1993). The quest for borehole stability in the Cusiana field, Colombia. *Oilfield Review, 5*, 33–43
3. Ambrose, G. J., Kruse, P. D., & Putnam, P. E. (2001). Geology and hydrocarbon potential of the southern Georgina Basin, Australia. *APPEA Journal, 41*(1), 139–163.
4. Ambrose, G. J., & Putnam, P. E. (2007). Carbonate ramp facies and oil plays in the Middle/Late Cambrian, southern Georgina Basin, Australia. In T. J. Munson & G. J. Ambrose (Eds.), Proceedings of the Central Australian Basins Symposium (CABS), Alice Springs, Northern Territory, 16–18 August, 2005. Darwin NT:Northern Territory Geological Survey, Special Publication 2, 236–253.
5. Anderson, E. M. (1951). *The dynamics of faulting*. Edinburgh, UK: Oliver & Boyd.
6. Arthur, J. D., Bohm, B., & Layne, M. (2009). Hydraulic fracturing considerations for natural gas wells of the Marcellus Shale. *Gulf Coast Association of Geological Societies Transactions, 59*, 49–59.
7. Bailey, A. H., King, R. C., Holford, S. P., & Hand, M. (2016). Incompatible stress regimes from geological and geomechanical datasets: Can they be reconciled? An example from the Carnarvon Basin, Western Australia. *Tectonophysics, 683*,405–416.
8. Barton, C. A., & Zoback, M. D. (1994). Stress perturbations associated with active faults penetrated by boreholes: Possible evidence for near-complete stress drop and a new technique for stress magnitude measurement. *Journal of Geophysical Research: Solid Earth, 99*, 9373–9390.
9. Barton, C. A., Zoback, M. D., & Burns, K. L. (1988). In situ stress orientation and magnitude at the Fenton Hill geothermal site, New Mexico, determined from wellbore breakouts. *Geophysical Research Letters, 15*, 467–470.
10. Baumgärtner, J., & Zoback, M. D. (1989). Interpretation of hydraulic fracturing pressure – time records using interactive analysis methods. *International Journal of Rock Mechanics and Mining Sciences & Geomechanical Abstracts, 26*, 461–469.
11. Bell, J. S. (1990). Investigating stress regimes in sedimentary basins using information from oil industry wireline logs and drilling records. In A. Hurst & M. A. Lovell (Eds.), *Geological applications of wireline logs* (Vol. 48, pp. 305–325).London, UK: Geological Society of London, Special Publication.

12. Bell, J. S. (1996). Petro Geoscience. 1. In situ stresses in sedimentary rocks. 1. Measurement techniques. *Geoscience Canada, 23*(2), 85–100.
13. Bell, J. S., & Babcock, E. A. (1986). The stress regime in the western Canadian sedimentary basin and implications for hydrocarbon production. *Bulletin of Canadian Petroleum Geology, 34*, 364–378.
14. Bennett, P. J., Philipchuk, P., & Freeman, A. (2010). Arthur Creek "hot shale": A Bakken unconventional oil analogy in the Georgina Basin of Northern Territory, Australia. AAPG International Conference and Exhibition, Calgary, Canada.
15. Boreham, C. J., & Ambrose, G. J. (2005). Cambrian petroleum systems in the southern Georgina basin, Northern Territory, Australia. Central Australian Basins Symposium: Petroleum and minerals potential (pp. 16–18). Alice Springs, NT.
16. Boreham, C. J., & Ambrose, G. J. (2007). Cambrian petroleum systems in the southern Georgina Basin, Northern Territory, Australia. In T. J. Munson & G. J.Ambrose (Eds.), Proceedings of the Central Australian Basins Symposium (CABS), Alice Springs, Northern Territory,16–18 August, 2005 (pp. 254–281). Darwin, NT:Northern Territory Geological Survey, Special Publication 2.
17. Bradley, J. S. (1975). Abnormal formation pressure. *AAPG Bulletin, 59*(6), 957–973.
18. Breckels, I. M., & Van Eekelen, H. A. M. (1982). Relationship between horizontal stress and depth in sedimentary basins. Journal of Petroleum Technology,*34*(09), 2191–2199.
19. Brooke-Barnett, S., Flottmann, T., Paul, P. K., Busetti, S., Hennings, P., Reid, R., &Rosenbaum, G. (2015). Influence of basement structures on in situ stresses over the Surat Basin, southeast Queensland. *Journal of Geophysical Research: Solid Earth, 120*(7), 4946–4965.
20. Brown, E. T., & Hoek, E. (1978). Trends in relationships between measured in-situ stresses and depth. *International Journal of Rock Mechanics and Mining Sciences & Geomechanics Abstracts, 15*(4), 211–215.
21. Brudy, M., & Zoback, M. D. (1999). Drilling-induced tensile wall-fractures: Implications for determination of in-situ stress orientation and magnitude.*International Journal of Rock Mechanics and Mining Sciences, 36*, 191–215. Retrieved from http://dx.doi.org/10.1016/S0148-9062(98)00182-X
22. Byerlee, J. (1978). Friction of rocks. *Pure and Applied Geophysics, 116*, 616–626.
23. Caillet, G. (1993). The caprock of the Snorre Field, Norway: A possible leakage by hydraulic fracturing. *Marine and Petroleum Geology, 10*, 42–50
24. Chan, A. W., Hauser, M., Couzens-Schultz, B. A., & Gray, G. (2014). The role of shear failure on stress characterization. *Rock Mechanics and Rock Engineering, 47*,1641–1646.
25. Cloetingh, S., & Wortel, R. (1986). Stress in the Indo-Australian plate.*Tectonophysics, 132*, 49–67.
26. Couzens-Schultz, B. A., & Chan, A. W. (2010). Stress determination in active thrust belts: An alternative leak-off pressure interpretation. *Journal of Structural Geology, 32*, 1061–1069. Retrieved fromhttp://dx.doi.org/10.1016/j.jsg.2010.06.013
27. Dart, R. L., & Zoback, M. L. (1989). Wellbore breakout stress analysis within the central and eastern continental United States. *The Log Analyst, 30*(1), 12–25.
28. Dickey, P. A. (1986). *Petroleum development geology* (3rd ed.). Tulsa, Ok: Pennwell Books.
29. Dunster, J. N., Kruse, P. D., Duffett, M. L., & Ambrose, G. J. (2007). Geology and resource potential of the southern Georgina Basin. Darwin NT: Northern Territory Geological Survey, Digital Information Package DIP007.
30. Dyksterhuis, S., Müller, R. D., & Albert, R. A. (2005). Paleostress field evolution of the Australian continent since the Eocene. *Journal of Geophysical Research: Solid Earth, 110*, B05102, doi:10.1029/2003JB002728.
31. Engelder, T. (1993). *Stress Regimes in the Lithosphere*. Princeton, New Jersey: Princeton University Press.
32. Fisher, M. K., Wright, C. A., Davidson, B. M., Goodwin, A. K., Fielder, E. O., Buckler,W. S., & Steinsberger, N. P. (2002). Integrating fracture mapping technologies to optimize stimulations in the Barnett Shale. SPE Annual Technical Conference and Exhibition Society of Petroleum Engineers, January 2002.

33. Fjaer, E., Holt, R. M., Horsrud, P., Raaen, A. M., & Risnes, R. (2008). Mechanics of hydraulic fracturing. *Developments in Petroleum Science, 53*, 369–390.
34. Gardner, G. H. F., Gardner, L. W., & Gregory, A. R. (1974). Formation velocity and density–the diagnostic basics for stratigraphic traps. *Geophysics, 39*, 770–780.
35. Gjønnes, M., Cruz, A. M., Horsrud, P., & Holt, R. M. (1998). Leak-off tests for horizontal stress determination? *Journal of Petroleum Science and Engineering, 20*,63–71.
36. Gough, D. I., & Bell, J. S. (1981). Stress orientations from oil-well fractures in Alberta and Texas. *Canadian Journal of Earth Sciences, 18*(3), 638–645.
37. Greene, D. C. (2010). Neoproterozoic rifting in the southern Georgina Basin, central Australia: Implications for reconstructing Australia in Rodinia. *Tectonics,29*, TC5010, doi:10.1029/2009TC002543.
38. Haimson, H. C., & Herrick, C. G. (1986). Borehole breakouts–a new tool for estimating in situ stress? ISRM International Symposium International Society for Rock Mechanics.
39. Hand, M., & Sandiford, M. (1999). Intraplate deformation in central Australia, the link between subsidence and fault reactivation. *Tectonophysics, 305*, 121–140.
40. Heidbach, O., & Höhne, J. (2008). CASMI—a visualization tool for the World Stress Map database. *Computers & Geosciences, 34*, 783–791.
41. Heidbach, O., Tingay, M., Barth, A., Reinecker, J., Kurfess, D., & Mueller, B. (2010).Global crustal stress pattern based on the World Stress Map database release 2008. *Tectonophysics, 482*(1–4), 3–15. doi:10.1016/j.tecto.2009.07.023
42. Hillis, R. R., Enever, J. R., & Reynolds, S. D. (1999). In situ stress field of eastern Australia. *Australian Journal of Earth Sciences, 46*, 813–825.
43. Hillis, R. R., & Reynolds, S. D. (2000). The Australian stress map. *Journal of the Geological Society, 157*, 915–921.
44. Hillis, R. R., & Reynolds, S. D. (2003). In situ stress field of Australia. *Geological Society of America Special Papers, 372*, 49–58.
45. Hillis, R. R., & Williams, A. F. (1993a). The stress field of the North West Shelf and wellbore stability. *APEA Journal, 33*, 373–385.
46. Hillis, R. R., & Williams, A. F. (1993b). The contemporary stress field of the Barrow-Dampier Sub-Basin and its implications for horizontal drilling.*Exploration Geophysics, 24*, 567–576.
47. Horn, M. K. (1991). Play concepts for horizontal drilling. Geological aspects of horizontal drilling. American Association of Petroleum Geologists, Continuing Education Course Note Series, *33*, 189–323.
48. Horner, D. R. (1951). Pressure build-up in wells: Proceedings of the Third World Petroleum Congress. The Hague, Section II, 503–521
49. Hossain, M. M., Rahman, M. K., & Rahman, S. S. (2000). Hydraulic fracture initiation and propagation: Roles of wellbore trajectory, perforation and stress regimes. *Journal of Petroleum Science and Engineering, 27*, 129–149.
50. Hubbert, M. K., & Willis, D. G. (1957). Mechanics of hydraulic fracturing.*Petroleum Transactions AIME, 210*, 153–163.
51. Jaeger, J. C., & Cook, N. G. W. (1979). *Fundamentals of rock mechanics* (3rd ed.).London, UK: Chapman & Hall.
52. King, R. C., Hillis, R. R., & Reynolds, S. D. (2008). In situ stresses and natural fractures in the Northern Perth Basin, Australia. *Australian Journal of Earth Science, 55*, 685–701. doi:10.1080/08120090801982843
53. King, R. C., Holford, S. P., Hillis, R. R., Backe, G., Tingay, M., & Tuitt, A. (2012).Reassessing the in situ stress regimes of Australia's petroleum basins. *APPEA Journal, 52*, 415–426. doi:10.1029/2009JB006997
54. King, R. C., Neubauer, M., Hillis, R. R., & Reynolds, S. D. (2010). Variation of vertical stress in the Carnarvon Basin, NW Shelf, Australia. *Tectonophysics, 482*,73–81. Retrieved from http://dx.doi.org/10.1016/j.tecto.2009.06.018
55. Kirsch, V. (1898). Die Theorie der Elastizita¨ t und die Beddu¨ rfnisse der Festigkeitslehre. *Zeitschrift des Vereines Deutscher Ingenieure, 29*, 797–807.
56. Kuchuk, F. J. (1998). Interval pressure transient testing with MDT packer-probe module in horizontal wells. *SPE Reservoir Evaluation & Engineering, 1*, 509–518.

57. Maloney, S., & Kaiser, P. K. (1989). *Results of borehole breakout simulation tests.* ISRM International Symposium International Society for Rock Mechanics.
58. Mardia, K. V. (1975). Statistics of directional data. *Journal of the Royal Statistical Society, Series B (Methodological), 37,* 349–393
59. Mastin, L. (1988). Effect of borehole deviation on breakout orientations. *Journal of Geophysical Research: Solid Earth, 93*(B8), 9187–9195.
60. Mildren, S. D., Clarke, R., Holford, S. P., & Titus, L. (2013). Characterisation of fracture permeability at Cow Lagoon-1, MacArthur Basin, Northern Territory, Australia and recommendations for development of the Greater Cow Lagoon Structure. SPE Unconventional Resources Conference and Exhibition-Asia Pacific Society of Petroleum Engineers, November 2013, SPE167087.
61. Mildren, S. D., Hillis, R. R., Fett, T., & Robinson, P. H. (1994). Contemporary stresses in the Timor Sea: Implications for fault-trap integrity. The sedimentary basins of Western Australia (pp. 291–300). Proceedings of Petroleum Exploration Society of Australia Symposium, Perth, WA: PESA.
62. Muller, R. D., & Dyksterhuis, S. (2005). Current and palaeo-stress models for central Australian basins. In T. J. Munson & G. J. Ambrose (Eds.), Proceedings of the Central Australian Basins Symposium (CABS), Alice Springs, Northern Territory,16–18 August, 2005 (pp. 1–9). Northern Territory Geological Survey, Special Publication 2.
63. Munson, T. J. (2014). *Petroleum geology and potential of the onshore Northern Territory, 2014.* Darwin, NT: Northern Territory Geological Survey, Report 22.
64. Nelson, E., Hillis, R. R., & Mildren, S. D. (2006). Stress partitioning and wellbore failure in the West Tuna Area, Gippsland Basin. *Exploration Geophysics, 37,* 215–221.
65. Peška, P., & Zoback, M. D. (1995). Compressive and tensile failure of inclined well bores and determination of in situ stress and rock strength. *Journal of Geophysical Research: Solid Earth, 100*(B7), 12791–12811.
66. PetroFrontier (2012). *PetroFrontier Corp.* Operational Update October 12, 2012.Calgary, Alberta. Retrieved from www.petrofrontier.com
67. PetroFrontier (2014). *PetroFrontier Corp.* Operational Update July 22, 2014.Calgary, Alberta. Retrieved from www.petrofrontier.com
68. Plumb, R. A., & Hickman, S. H. (1985). Stress-induced borehole elongation; a comparison between the four-arm dipmeter and the borehole televiewer in the Auburn geothermal well. *Journal of Geophysical Research,* B90, 5513–5521.
69. Rajabi, M., Tingay, M., King, R., & Heidbach, O. (2016a). Present-day stress orientation in the Clarence-Moreton Basin of New South Wales, Australia: A new high density dataset reveals local stress rotations. *Basin Research,* doi:10.1111/bre.12175.
70. Rajabi, M., Tingay, M., & Heidbach, O. (2016b). The present-day stress field of New South Wales, Australia. *Australian Journal of Earth Sciences, 63,* 1–21.
71. Reynolds, S. D., Coblentz, D. D., & Hillis, R. R. (2002). Tectonic forces controlling the regional intraplate stress field in continental Australia: Results from new finite element modeling. *Journal of Geophysical Research: Solid Earth, 107*(B7),ETG-1. doi:10.1029/2001JB000408
72. Reynolds, S. D., & Hillis, R. R. (2000). The in situ stress field of the Perth Basin, Australia. *Geophysical Research Letters, 27,* 3421–3424.
73. Reynolds, S. D., Mildren, S. D., Hillis, R. R., & Meyer, J. J. (2006). Constraining stress magnitudes using petroleum exploration data in the Cooper–Eromanga Basins, Australia. *Tectonophysics, 415,* 123–140.
74. Sibson, R. H. (1974). Frictional constraints on thrust, wrench and normal faults.*Nature, 249,* 542–544.
75. Sibson, R. H. (1977). Fault rocks and fault mechanics. *Journal of the Geological Society of London, 133,* 191–213.
76. Sibson, R. H. (1990). Conditions for fault-valve behaviour. *Geological Society, London, Special Publications, 54*(1), 15–28.
77. Smith, M. G., & Bustin, R. M. (2000). Late Devonian and Early Mississippian Bakken and Exshaw black shale source rocks, Western Canada Sedimentary Basin: A sequence stratigraphic interpretation. *AAPG Bulletin, 84,* 940–960.

78. Smith, T. E., Kelman, A., Laurie, J. R., Nicoll, R. S., Carr, L. K., Hall, L., & Edwards, D.(2013). An updated stratigraphic framework for the Georgina Basin, Northern Territory and Queensland. *APPEA Journal, 53*, 1–6.
79. Summons, R. E., Powell, T. G., & Boreham, C. J. (1988). Petroleum geology and geochemistry of the Middle Proterozoic McArthur Basin, Northern Australia: III. Composition of extractable hydrocarbons. *Geochimica et Cosmochimica Acta, 52*,1747–1763.
80. Thiercelin, M. J., & Plumb, R. A. (1994). A core-based prediction of lithologic stress contrasts in east Texas formations. *SPE Formation Evaluation, 9*(04), 251–258.
81. Tiem, V. T. A., Horsfield, B., & di Primo, R. (2011). Gas shale potential of the Amadeus and Georgina Basins, Australia: Preliminary insights. Geoscience Australia, Record 2011/10.
82. Tingay, M. R. P., Hillis, R. R., Morley, C. K., Swarbrick, R. E., & Okpere, E. C. (2003).Variation in vertical stress in the Baram Basin, Brunei: Tectonic and geomechanical implications. *Marine and Petroleum Geology, 20*, 1201–1212. Retrieved from http:// dx.doi.org/ 10.1016/j.marpetgeo.2003.10.003
83. Vernik, L., & Zoback, M. D. (1992). Estimation of maximum horizontal principal stress magnitude from stress-induced well bore breakouts in the Cajon Pass scientific research borehole. *Journal of Geophysical Research, 97*(B4), 5109–5119.
84. Walter, M. R., Veevers, J. J., Calver, C. R., & Grey, K. (1995). Neoproterozoic stratigraphy of the Centralian Superbasin, Australia. *Precambrian Research, 73*,173–195. doi:10.1016/0301-9268(94)00077-5
85. White, A. J., Traugott, M. O., & Swarbrick, R. E. (2002). The use of leak-off tests as means of predicting minimum in-situ stress. *Petroleum Geoscience, 8*, 189–193.
86. Yew, C. H., & Weng, X. (2014). *Mechanics of hydraulic fracturing*. Oxford UK: Gulf Professional Publishing.
87. Zheng, Z., Kemeny, J., & Cook, N. G. (1989). Analysis of borehole breakouts.*Journal of Geophysical Research: Solid Earth, 94*(B6), 7171–7182.
88. Zhou, S., Hillis, R., & Sandiford, M. (1996). On the mechanical stability of ilnclined wellbores. *SPE Drilling & Completion, 11*(02), 67–73.
89. Zoback, M. D. (2007). *Reservoir geomechanics*. New York, NY: Cambridge University Press, ISBN: 9780521770699
90. Zoback, M. D., & Healy, J. H. (1984). Friction, faulting and in situ stress. *Annals of Geophysics, 2*, 689–698.
91. Zoback, M. D., Moos, D., Mastin, L., & Anderson, R. N. (1985). Well bore breakouts and in situ stress. *Journal of Geophysical Research, 90*, 5523–5530.
92. Zoback, M. L., & Zoback, M. (1980). State of stress in the conterminous United States. *Journal of Geophysical Research, 85*(B11), 6113–6156. doi:10.1029/JB085iB11p06113
93. Zoback, M. L., Zoback, M. D., Adams, J., Assumpcao, M., Bell, S., Bergman, E. A., &Fuchs, K. (1989). Global patterns of tectonic stress. *Nature, 341*(6240), 291–298.

CHAPTER 5

Effects of subsurface structures of source regions on long-period ground motions observed in the Tokyo Bay area, Japan

Tomiichi Uetake*

TEPCO Research Institute, Tokyo Electric Power Company Holdings, Inc., 4-1 Egasaki-cho, Tsurumi-ku, Yokohama, Kanagawa 230-8510, Japan

ABSTRACT

We compared the long-period ground motion observed in the Tokyo Bay area during two shallow M6.7 earthquakes that occurred in northern Nagano Prefecture, Japan, on March 12, 2011, and November 22, 2014. The magnitudes, focal depths, and source mechanisms of these events were almost identical, but their seismograms were quite different. Significant long-period later arrivals with a predominant period of 5 s were recognized in the velocity traces of the 2011 event, but there were no such remarkable later arrivals in the 2014 event traces. The ground motions at stations located outside the basin area were studied as incident waves to the Kanto Basin. A large wave packet with a predominant period of 5 s was recognized in the velocity traces of the 2011 event, but there was no significant wave packet in the 2014 event traces. Based on particle motion, this wave packet was hypothesized to be a Rayleigh wave. The source regions of the two events have quite different subsurface structures. The different characteristics in long-period ground motion in the Tokyo Bay area during the two events were due to different Rayleigh wave excitations in the source regions.

Keywords: Long-period ground motion Rayleigh wave Velocity structure Source region Kanto Basin

1. INTRODUCTION

A shallow M6.7 earthquake occurred in northern Nagano Prefecture, Japan, on November 22, 2014. This event was the first M7 class event to occur in this area after a shallow M6.7 earthquake occurred at the Niigata–Nagano prefectural border on March 12, 2011. The epicenters of these events are shown in Fig. 1. These two events had the same magnitude, according to the Japan Meteorological Agency (JMA) earthquake catalog (Japan Meteorological Agency 2016) and similar source mechanisms, according to the F-net moment tensor solution (Fukuyama et al. 1998; Okada et al. 2004). Both epicenters were located northwest of Tokyo Bay, and the distances from Tokyo Bay were comparable. Usually, ground motion characteristics are estimated based on source parameters, path distance, and site conditions (e.g., Kanno et al. 2006; Yokota et al. 2011; Yuzawa and Nagumo 2012). Considering this, although the ground motions observed in the Tokyo Bay area during these two events should have shown similar characteristics, their seismograms showed some differences. Such differences are usually considered to be due to deviations in ground motion in an empirical evaluation. In this paper, we describe the waveforms around the Tokyo Bay area during the two events and discuss the causes of their different characteristics.

1.1 Events and observation stations used in this study

We studied observation data from two shallow earthquakes that occurred in northern Nagano Prefecture, Japan. One event occurred at the Niigata–Nagano prefectural border in 2011; the other occurred in the northern part of Nagano Prefecture in 2014. The source parameters of these two events are shown in Table 1. According to the JMA, the magnitudes of both the events are the same, but the focal depth of the 2011 event is deeper than the focal depth of the 2014 event (Japan Meteorological Agency 2016). According to F-net (Fukuyama et al. 1998; Okada et al. 2004), the moment magnitude of the 2011 event is slightly smaller than that of the 2014 event, but the focal depths are the same; furthermore, both source mechanisms are almost identical. The epicenters and source mechanisms of these events are shown in Figs. 1 and 2. The locations of the two events are different, but the source parameters are almost identical. The observation stations used in this study are shown in Fig. 2. The distance and back azimuth of the 2011 event from the SNG station on the west shore of Tokyo Bay are about 184 km and N325°E, respectively. Those of the 2014 event are about 206 km and N305°E, respectively. The differences in distance and back azimuth from the Tokyo Bay area are about 20 km and 20°, respectively.

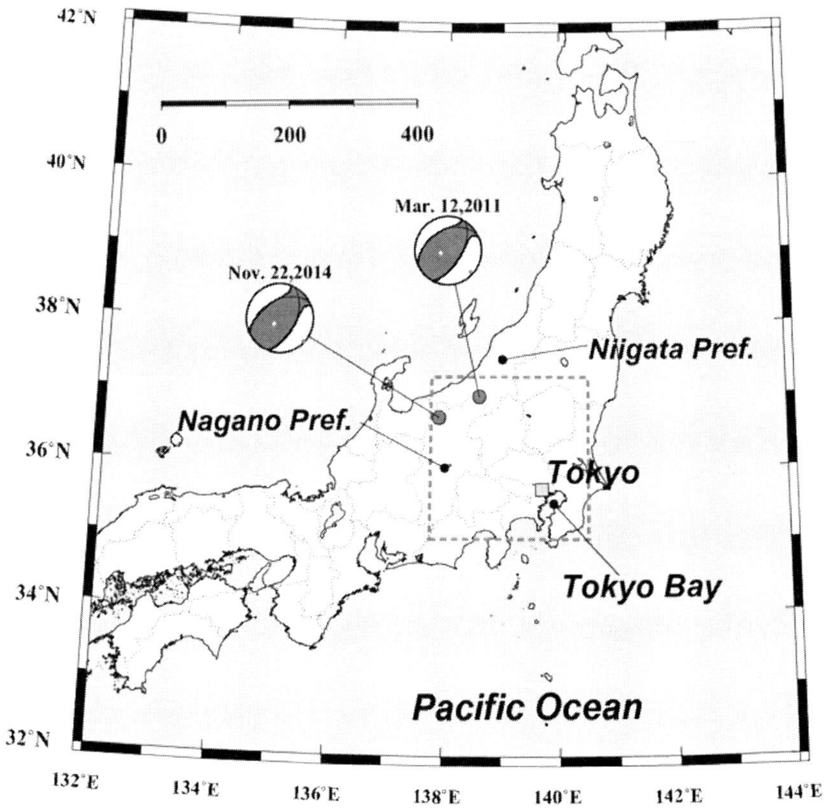

Figure. 1 Epicenters of the events used in this study. The area in the *red square* is enlarged in Fig. 2. *Dashed lines* show the prefectural borders in Japan. The event source mechanisms are the moment tensor solutions from F-net (Okada et al. 2004)

Table 1 Event source parameters

Date	Time (JST)	Epicenter		JMA		F-net	
		Latitude	Longitude	*M*	Depth (km)	Mw	Depth (km)
March 12, 2011	03:59	36.986N	138.598E	6.7	8.4	6.2	5
November 22, 2014	22:08	36.693N	137.891E	6.7	4.6	6.3	5

In this study, we consider data from two networks. One is the broadband velocity seismometer network on the shore of Tokyo Bay, operated by Tokyo Electric Power Company Holdings (TEPCO). The green squares in Fig. 2 indicate the TEPCO stations. Servo velocity seismometers (Tokyo Sokushin VSE-355G3) are installed in TEPCO's network. The full scale of the sensor is 2 m/s, and the sensor response is flat in the frequency range between 0.008 and 70 Hz. The data are recorded with 100-Hz sampling and 24-bit resolution. The other data are from the strong motion seismograph networks (K-NET and KiK-net) operated by the National Research Institute for Earth Science and Disaster Resilience, Japan (NIED) (Okada et al. 2004). The red and blue circles in Fig. 2 show the NIED stations used in this study. Feedback-type high-resolution accelerometers are installed at the K-NET and KiK-net stations. Because we primarily discuss the characteristics of velocity waveforms, the acceleration data were integrated into the velocity traces in the frequency domain.

The gray contours in Fig. 2 indicate the depth of seismic bedrock from the "Japan Integrated Velocity Structure Model Version 1" (Koketsu et al. 2012). The upper boundary of the seismic bedrock around Tokyo Bay is very deep, about 3000–4000 m. The contour also suggests that the velocity structures in the source regions of the two events are different. The depth of seismic bedrock in the source region of the 2011 event is deeper than the source region of the 2014 event.

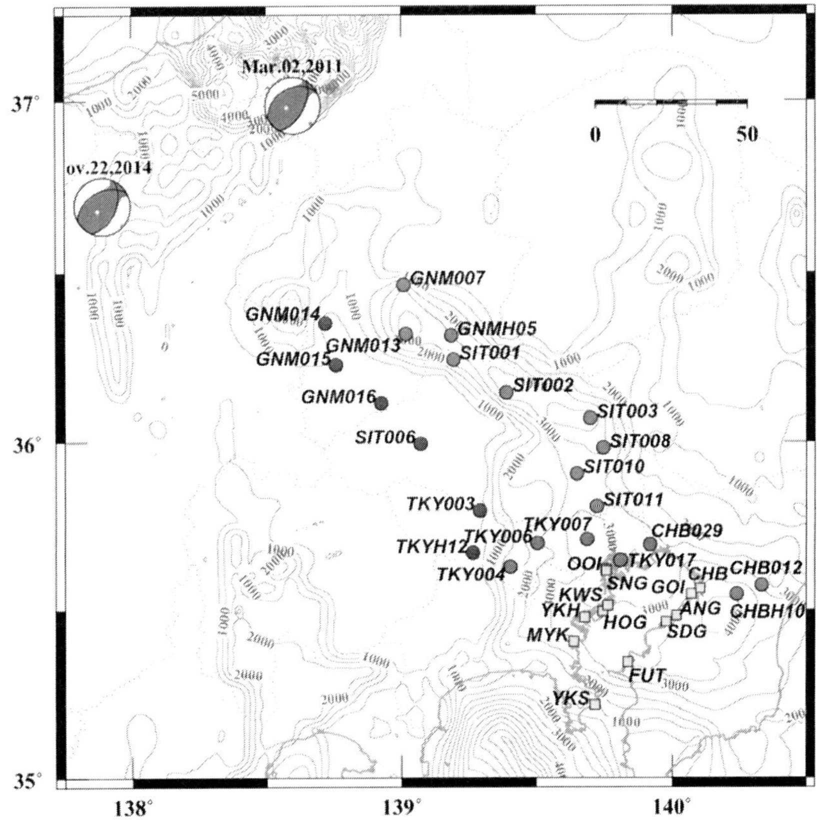

Figure. 2 Locations of the strong motion observation stations used in this study. *Green squares* indicate the Tokyo Electric Power Company Holdings stations. *Circles* indicate K-NET and KiK-net stations (Okada et al. 2004). *Red circles* show the stations located in the basin area, and *blue circles* show the stations located outside the basin. Characters next to *each mark* indicate the station code. *Dashed purple lines* show the prefectural borders in Japan. *Gray contour lines* indicate the depth of seismic bedrock from the "Japan Integrated Velocity Model version 1" (Koketsu et al. 2012). Event source mechanisms are the moment tensor solutions from F-net (Okada et al. 2004)

1.2 Velocity seismograms observed around the Tokyo Bay area

The velocity seismograms of both events observed at the TEPCO stations are shown in Fig. 3. There are some differences between the velocity waveforms of the 2011 and 2014 events. Remarkable wave packets are recognized in the waveforms of the 2011 event for all components; these remarkable later arrivals have a predominant period of about 5 s. The arrival times of these phases at the stations on the east shore of Tokyo Bay are later than those at the stations on the west shore, and there are no significant phases at the YKS, MYK, and FUT stations. There are no significant wave packets in the waveforms of the 2014 event, but the dominant period of the S-wave part in the horizontal components is longer than that of the 2011 event. In the waveforms of the NS component at GOI and CHB stations, the later arrivals with predominant periods of 6–8 s are recognized at around 200 s for both events. However, the predominant period of these wave packets and arrival time difference between GOI and CHB vary between the two events.

The velocity response spectra with 5% damping are compared in Fig. 4. The response spectra of the 2011 event have remarkable peaks at around 5 s in the horizontal and vertical components at all stations except for YKS. In the response spectra from CHB, the peak period in the NS component is longer than the peak period for other stations. The response spectra of the 2014 event have no significant 5-s peaks, and the peak period is longer than that of the 2011 event. There are significant peaks with periods of 7–8 s in the SNG spectrum for only the horizontal component and in the CHB spectrum for all components.

The arrival times of the significant later phases at the stations located on the west side are faster than those stations located on the east side of Tokyo Bay, as shown in Fig. 3. This suggests that these later phases propagated from west to east. To confirm this hypothesis, we provide the velocity waveforms at stations on west–east direction observation line north of Tokyo Bay in Fig. 5. The top two traces are velocity seismograms at the station located outside of the basin, and the other eight traces are velocity seismograms at the station in Kanto Basin. These traces show that the long-period wave packets propagated from west to east in the basin area.

1.3 Ground motions at the stations located outside of the basin

The ground motions outside the Kanto Basin were studied to examine the properties of the incident wave. The velocity waveforms at stations outside the basin are shown in Fig. 6. The station locations are shown in Fig. 2 with blue circles. A remarkable later phase is confirmed in the NS and UD components of the 2011 event. This phase seems to propagate from north to south. In contrast, no special later arrival was recognized in the velocity seismograms from the 2014 event.

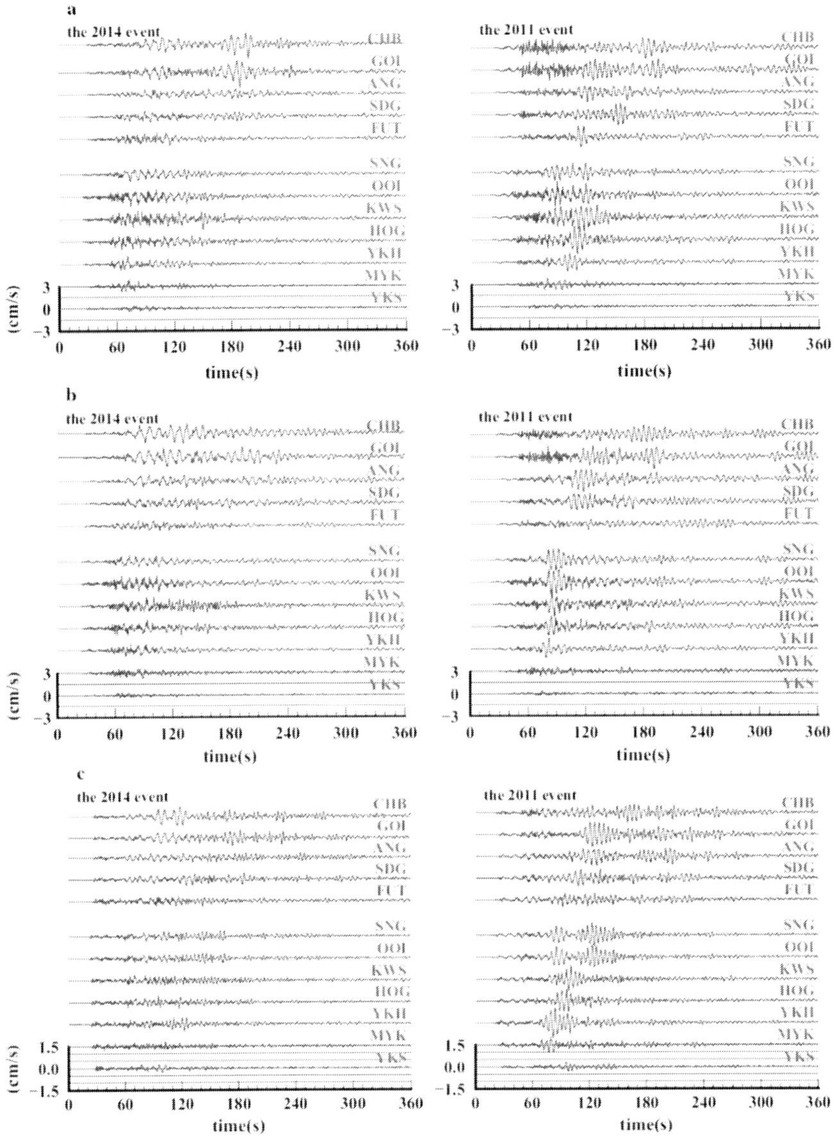

Figure. 3 Velocity seismograms observed at TEPCO stations in the Tokyo Bay area, **A** NS component, **B** EW component, and **C** UD component. The *left side* shows the records for the 2014 event, and the *right side* shows the records for the 2011 event. *Red characters associated with each trace* indicate the station code. The location of each station is shown in Fig. 2

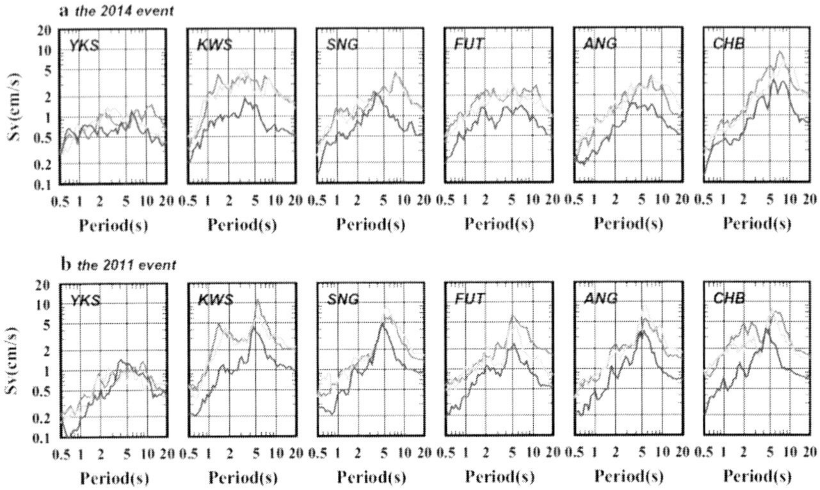

Figure. 4 Velocity response spectra with 5% damping for the Tokyo Bay area stations. **B** Show the spectra from the 2011 event. **A** Show those from the 2014 event. *Red, green,* and *blue lines* indicate the spectra for the NS, EW, and UD components, respectively. *Characters in each frame* indicate the station code. The location of each station is shown in Fig. 2

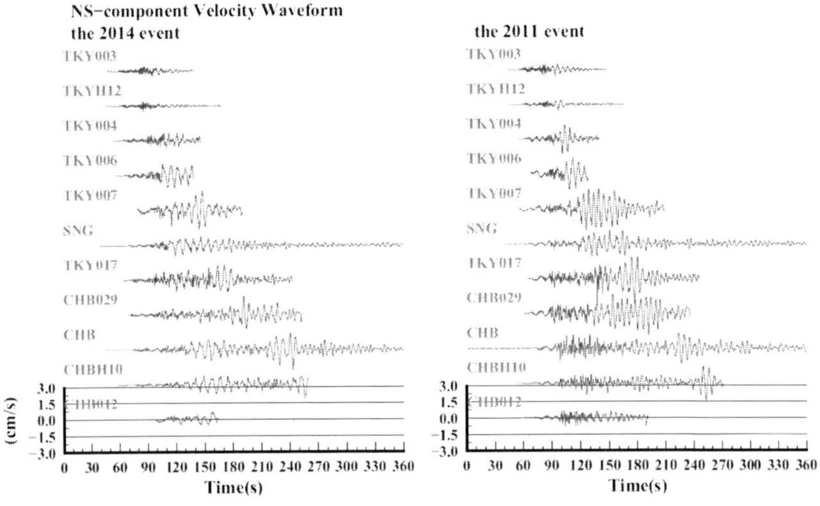

Figure. 5 Comparison of NS component velocity seismograms between the 2014 and 2011 events at stations on a west–east observation line in northern Tokyo Bay. *Characters associated with each trace* indicate the station code. The location of each station is shown in Fig. 2

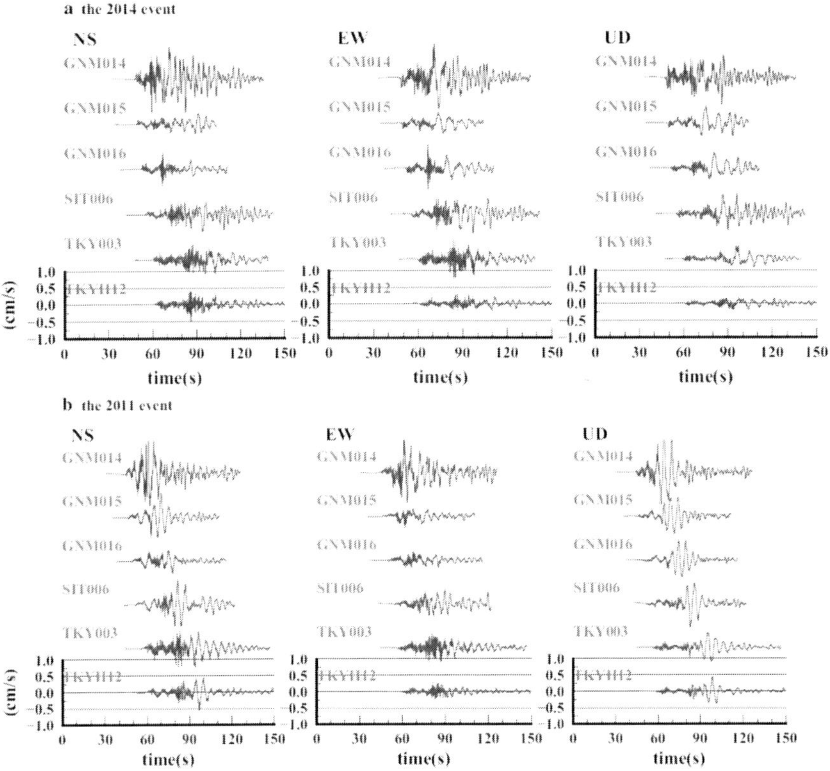

Figure. 6 Velocity waveforms at the stations located in the outside the basin area. **B** Show the 2011 event waveforms; **A** show those of the 2014 event. *Columns from left to right* show the NS, EW, and UD components, respectively. *Characters associated with each trace* indicate the station code. The location of each station is shown in Fig. 2

The velocity response spectra with 5% damping are shown in Fig. 7. A significant peak, with a 5-s period, is apparent in the velocity response spectra of the 2011 event. This peak is coincident with the predominant period of the significant later phases in the basin area. However, this significant later phase was not recognized in the waveforms of the 2014 event, and there is no significant peak in the 2014 event spectra. The difference between the 2011 and 2014 incident waves to the basin caused the variation in the observed ground motions in the Tokyo Bay area.

To evaluate the wave type of the later phase, particle motions were produced using the velocity waveforms at station TKY003 for the 2011 event, as shown in Fig. 8. The NS and the EW components were converted to the radial and transverse components using the back azimuth to the epicenter. The significant later phase is in the time window between 45 and 60 s. The vibrations of the later phase are dominantly in the radial direction, and orbital motion in the radial UD plane shows retrograde rotation to the radial direction. From this evidence, the significant later phase is interpreted as Rayleigh waves.

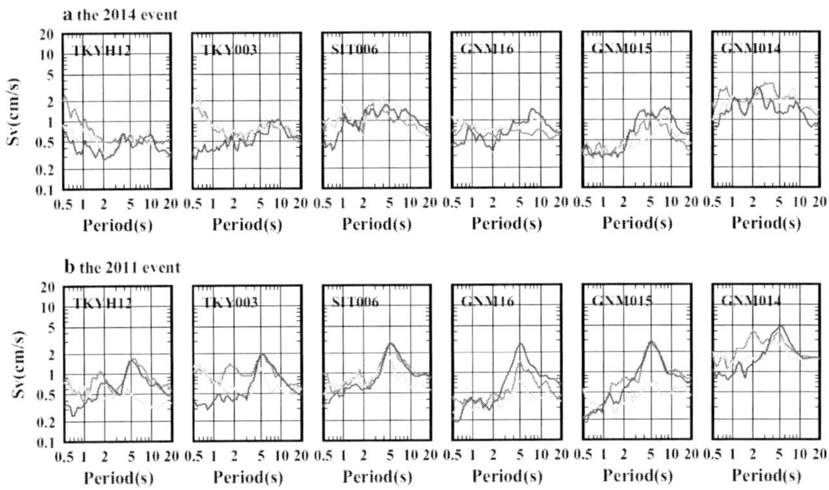

Figure. 7 Velocity response spectra at stations located in the outside the basin area. **B** Show the 2011 event spectra; **A** show those of the 2014 event. *Red, green,* and *blue lines* indicate the spectra of the NS, EW, and UD components, respectively. *Characters in each frame* indicate the station code. The location of each station is shown in Fig. 2

2. DISCUSSION

2.1 Effects of source region velocity structure

The Rayleigh waves observed at the stations located outside the basin were likely excited in the source region. The contouring of the seismic bedrock depth shown in Fig. 2 suggests that the velocity structures in the two source regions are different. The structure of the outside the basin area surrounding Kanto Basin is also different from the structures of the source regions. We studied the Rayleigh wave propagation characteristics using 2-D velocity structure models as shown in Fig. 9. These models were made by connecting the source region with the outside the basin area based on the

"Japan Integrated Velocity Structure Model version 1" (Koketsu et al. 2012). The seismic bedrock (layer no. 8) is deep in the source region but very shallow in the outside the basin area.

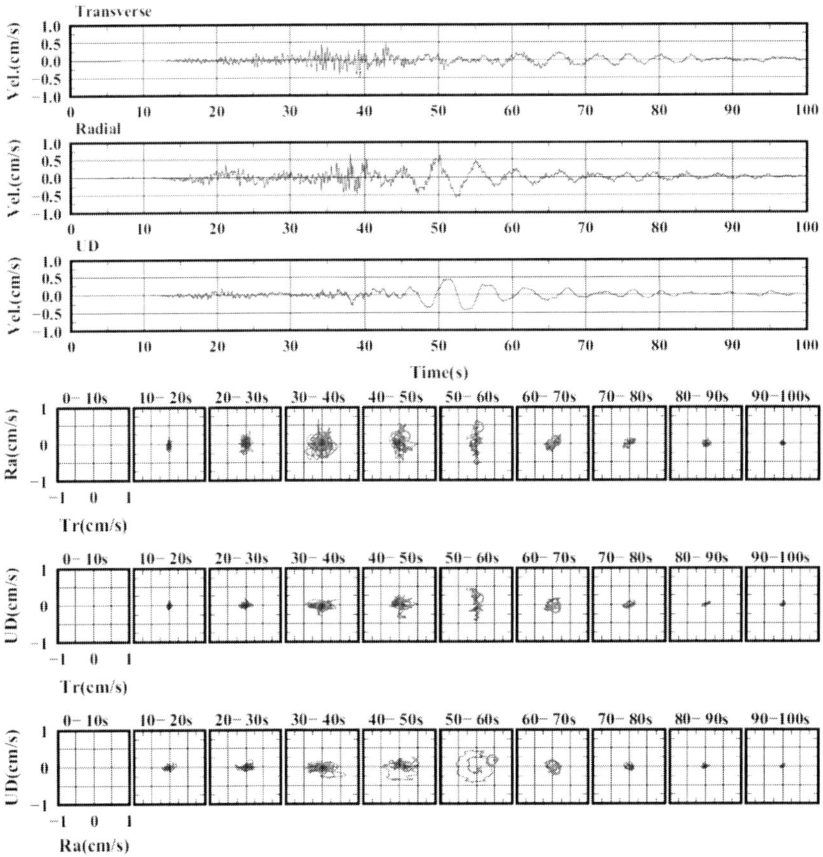

Figure. 8 Particle motion from velocity seismograms at station TKY003 during the 2011 event. The *three bottom rows* indicate the particle motions in the radial *vertical* (UD) plane, transverse *vertical* (UD) plane, and transverse radial *horizontal plane*. The particle motions are displayed every 10 s. *Red circles* and *crosses* indicate the start and end points, respectively. The *three top figures* are velocity traces in the transverse, radial, and UD components, respectively

A fourth-order staggered grid FDM was used for the P-SV problem. The model size was 60 km in the horizontal direction and 30 km in the depth direction with a 50-m grid size. The bottom, left, and right sides of the model were set with absorbing boundary conditions (Cerjan et al. 1985). The Q value was taken into account according to Graves (1996), and the reference frequency was 0.2 Hz. The time step of calculation was 0.002 s. We set the same moment source based on a reverse fault with a dip of 45° at 6 km depth in both simulations. The Herrmann function (Herrmann 1979) with a 5-s period was used as the source function.

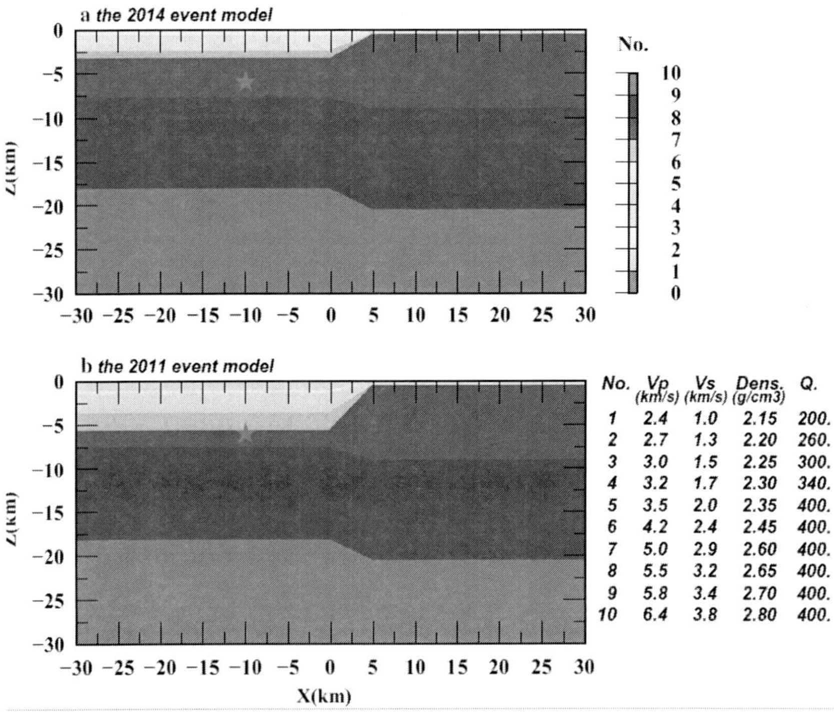

No.	Vp (km/s)	Vs (km/s)	Dens. (g/cm3)	Q.
1	2.4	1.0	2.15	200.
2	2.7	1.3	2.20	260.
3	3.0	1.5	2.25	300.
4	3.2	1.7	2.30	340.
5	3.5	2.0	2.35	400.
6	4.2	2.4	2.45	400.
7	5.0	2.9	2.60	400.
8	5.5	3.2	2.65	400.
9	5.8	3.4	2.70	400.
10	6.4	3.8	2.80	400.

Figure. 9 2-D velocity structure models using a comparison of wave propagation characteristics. *Lower* and *upper* models correspond to the source regions for the 2014 and 2011 events, respectively. *Red stars* indicate the source points. *Color contours* show the layer numbers. The table at the *lower right* provides the layer properties

The results from the two models are compared in Fig. 10. The later arrivals in the traces of the 2011 event model are more pronounced than those in the traces of the 2014 event model, especially in the deep bedrock area ($X < 0.0$). In the shallow bedrock area ($X > 5.0$), the later arrivals of the 2011 event model are also delayed and larger than those of the 2014 event model. However, the difference in amplitude between the two models is not as large as in the observation records. We hypothesize that the differences between the observations and model results are due to the simplified simulation; the actual ground motions contain the combined effects of 3-D subsurface structures and detailed fault ruptures. Zama (1996) made the pioneering proposal to consider the combination of focal region and target region at the time long-period ground motion is evaluated. When evaluating earthquake ground motion, we should consider not only the seismic parameters but also the velocity structure of the focal region.

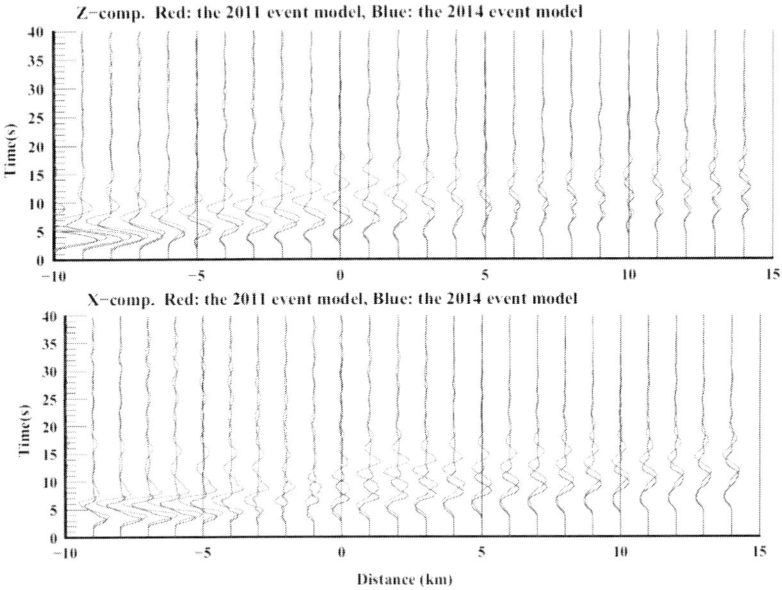

Figure. 10 Comparison of ground motion traces using different velocity structures in the source regions. *Red* and *blue lines*, respectively, show the ground motion traces using the 2014 and 2011 event models shown in Fig. 9

2.2 Effects of basin features on surface wave propagation

The deep bedrock area of the Kanto Basin extends in the northwest direction similar to a trench, as shown in Fig. 2, and this structure guides the propagation of surface waves. Koyama et al. (1988) reported that this trench guided the propagation of Love waves from events that occurred in western Nagano Prefecture. Furumura and Hayakawa (2007) reported that long-period ground motions observed in the Tokyo area during the 2004 Niigata-ken Chuetsu M6.8 earthquake were propagated thorough this trench. The epicenter of this earthquake was located about 40 km north of the epicenter of the 2011 event. Therefore, the effect of this trench on the later arrivals during the 2011 event was studied from a consideration of travel time. The velocity waveforms in the NS and UD components are shown in Fig. 11. The first onset of S waves propagated from north to south with almost the same velocity outside the basin area as in the trench-like basin area. However, long-period later arrivals in the trench-like basin area propagated more slowly than those in the outside the basin area. The significant later arrivals in the trench-like basin area and significant later arrivals in the Tokyo Bay area were not temporally continuous. The significant later arrivals in the Tokyo Bay area likely propagated from around TKYH12, as shown in Fig. 5. The later arrivals with a predominant period of 5 s were likely excited by the incidence of Rayleigh waves from outside the basin area to the basin. The excitation and propagation of significant later arrivals with a predominant period of 5 s during the 2011 event were not affected by the trench-like structure of the Kanto Basin.

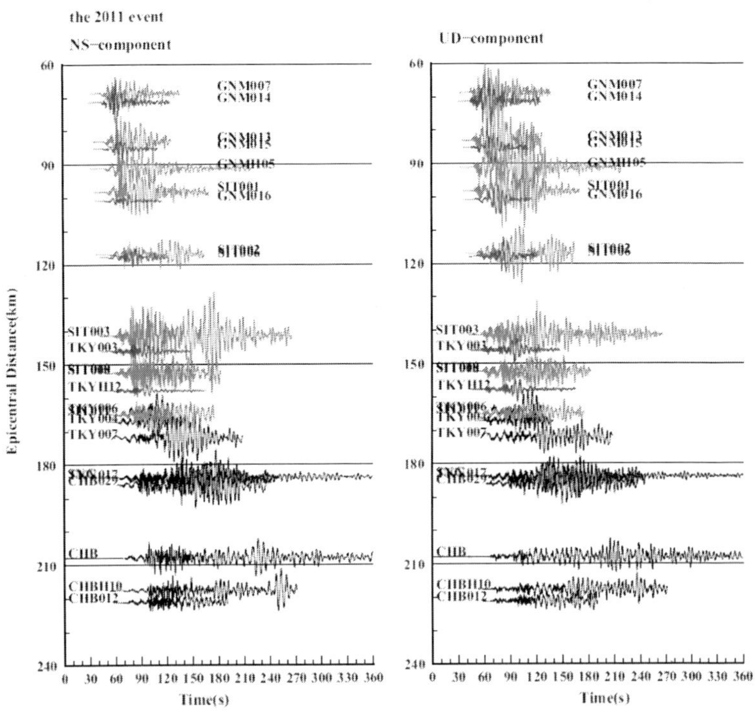

Figure. 11 Sections of velocity seismograms observed during the 2011 event. The *left-side figure* shows the traces in the NS component. The *right-side figure* shows the traces in the UD component. *Red traces* are observed at stations on the deep trench of the Kanto Basin, as shown in Fig. 2. *Blue traces* indicate the ground motions at stations located outside the basin, as shown in Fig. 2. *Black traces* indicate the stations used in Fig. 5

Yuzawa and Nagumo (2012) studied the shakability of long-period ground motion in Kanto Basin and pointed out that its variability is due to source region. They interpreted the amplitude variation as due to the distance that the surface wave travels through sedimentary layers. Our results indicate that the variation in shakability is not only affected by propagation distance in the basin but also by input wave characteristics.

3. CONCLUSIONS

A shallow M6.7 earthquake occurred in northern Nagano Prefecture, Japan, on November 22, 2014. The magnitude, focal depth, and source mechanism of this event were almost identical to an event that occurred near the border of Nagano and Niigata prefectures on March 12, 2011. However, the seismograms of these events observed from the Tokyo Bay area were quite different. Significant long-period later arrivals with a predominant period of 5 s were recognized in the traces of the 2011 event, but were not recognized in the traces of the 2014 event. Because the incident wave to the Kanto Basin is a controlling factor on long-period ground motion in the Tokyo Bay area, we examined the ground motion at an outside the basin site. A large wave packet with a predominant period of 5 s was clearly recognized in the velocity traces of the 2011 event. This wave packet was likely a Rayleigh wave. From a numerical simulation, we confirmed that the subsurface structure of the focal region affected the excitation of the Rayleigh waves. Therefore, the difference in long-period ground motion between the two events in the Tokyo Bay area was affected by excitation of Rayleigh waves in the focal regions. This suggests that it is important to consider the effects of both the local and source regions.

ACKNOWLEDGEMENTS

The source parameters for the earthquakes used in this study were provided by the Japan Meteorological Agency. The CMT solution for the F-net and strong motion data from K-NET and KiK-net were provided by the National Research Institute for Earth Science and Disaster Resilience, Japan. Generic Mapping Tools (Wessel and Smith 1998) was used to draw the figures. Editor Takuto Maeda, reviewer Yadab P. Dhakal, and an anonymous reviewer helped to significantly improve the manuscript.

REFERENCES

1. Cerjan C, Kosloff D, Kosloff R, Reshelf M (1985) A nonreflecting boundary condition for discrete acoustic and elastic wave equations. Geophysics 50:705–708
2. Fukuyama E, Ishida M, Dreger DS, Kawai H (1998) Automated seismic moment tensor determination by using on-line broadband seismic waveforms. J Seismol Soc Jpn (Zisin 2) 51:149–156 **(in Japanese with English abstract)**
3. Furumura T, Hayakawa T (2007) Anomalous propagation of long-period ground motions recorded in Tokyo during the 23 October 2004 Niigata-ken Chuetsu (Mw6.6) earthquake, Japan. Bull Seismol Soc Am 97(3):863–880
4. Graves RW (1996) Simulating seismic wave propagation in 3D elastic media using staggered-grid finite differences. Bull Seismol Soc Am 86:1091–1106
5. Herrmann RB (1979) SH-wave generation by dislocation source—a numerical study. Bull Seismol Soc Am 69:1–15
6. Japan Meteorological Agency (JMA) (2016) The Seismological Bulletin of Japan. http://www.data.jma.go.jp/svd/eqev/data/bulletin/hypo_e.html. Accessed 18 Dec 2016
7. Kanno T, Narita A, Morikawa N, Fujiwara H, Fukushima Y (2006) A new attenuation relation for strong ground motion in Japan based on recorded data. Bull Seismol Soc Am 96(3):879–897. doi:10.1785/0120050138
8. Koketsu K, Miyake H, Suzuki H (2012) Japan integrated velocity structure model version 1. In: Proceedings of the 15th world conference on earthquake engineering. Paper no. 1773
9. Koyama S, Seo K, Samano T (1988) On the significant later phase of seismograms at Kumagaya, Japan. In: Proceedings of 9th world conference on earthquake engineering, vol 2, pp 585–590
10. Okada Y, Kasahara K, Hori S, Obara K, Sekiguchi S, Fujiwara H, Yamamoto A (2004) Recent progress of seismic observation networks in Japan—hinet, K-NET and KiK-net. Earth Planets Space 56:xv–xxviii. doi:10.1186/ BF03353076
11. Wessel P, Smith W (1998) New, improved version of the generic mapping tools released. EOS Trans AGU 79:579
12. Yokota T, Ikeuchi K, Yahagi T, Kaida Y, Suzuki H (2011) Attenuation and amplification of long-period component of ground motion. J Jpn Assoc Earthq Eng 11(1):1_81–1_101. doi:10.5610/jaee.11.1_81 **(in Japanese with English abstract)**
13. Yuzawa Y, Nagumo H (2012) Factors of variability and measures for the shakability of long-period ground motion—Kanto Basin as an example. J Jpn Assoc Earthq Eng 12(2):2_41–2_59 **(in Japanese with English abstract)**
14. Zama S (1996) Regionality of long-period ground motion using JMA strong motion displacement records. In: Proceedings of 11th world conference on earthquake engineering. Paper 1891

CHAPTER 6

Seismic imaging of slab metamorphism and genesis of intermediate-depth intraslab earthquakes

Akira Hasegawa[1]*Junichi Nakajima[2]

[1]*Research Center for Prediction of Earthquakes and Volcanic Eruptions, Graduate School of Science, Tohoku University, Aramaki, Aoba-ku, Sendai 980-8578, Japan.*

[2]*Department of Earth and Planetary Sciences, School of Science, Tokyo Institute of Technology, 2-12-1 Ookayama, Meguro-ku, Tokyo 152-8551, Japan.*

AUTHORS' CONTRIBUTIONS

AH proposed the topic, wrote the manuscript, and prepared the figures. JN collaborated with the corresponding author in the construction of the manuscript and in the preparation of the figures. Both authors read and approved the final manuscript.

ABSTRACT

We review studies of intermediate-depth seismicity and seismic imaging of the interior of subducting slabs in relation to slab metamorphism and their implications for the genesis of intermediate-depth earthquakes. Intermediate-depth events form a double seismic zone in the depth range of c. 40–180km, which occur only at locations where hydrous minerals are present, and are particularly concentrated along dehydration reaction boundaries. Recent studies have revealed detailed spatial distributions of these events and a close relationship with slab metamorphism. Pressure–temperature paths of the crust for cold slabs encounter facies boundaries with large H_2O production rates and positive total volume change, which are expected to cause highly active seismicity near the facies boundaries. A belt of upper-plane seismicity in the crust nearly parallel to 80–90km depth contours of the slab surface has been detected in the cold Pacific slab beneath eastern Japan, and is probably caused by slab crust dehydration with a large H_2O production rate. A seismic low-velocity layer in the slab crust persists down to the depth of this upper-plane seismic belt, which provides evidence for phase transformation of dehydration at this depth. Similar low-velocity subducting crust closely related with intraslab seismicity has been detected in several other subduction zones. Seismic tomography studies in NE Japan and northern Chile also revealed the presence of a P-wave low-velocity layer along the lower plane of a double seismic zone. However, in contrast to predictions based on the serpentinized mantle, S-wave velocity along this layer is not low. Seismic anisotropy and pore aspect ratio may play a role in generating this unique structure. Although further validation is required, observations of these distinct low P-wave velocities along the lower seismic plane suggest the presence of hydrated rocks or fluids within that layer. These observations support the hypothesis that dehydration-derived H_2O causes intermediate-depth intraslab earthquakes. However, it is possible that dual mechanisms generate these earthquakes; the initiation of earthquake rupture may be caused by local excess pore pressure from H_2O, and subsequent ruptures may propagate through thermal shear instability. In either case, slab-derived H_2O plays an important role in generating intermediate-depth events.

Keywords: Internal slab structure, Slab metamorphism, Dehydration-related embrittlement, Intermediate-depth earthquake, Intraslab earthquake, Earthquake generation, mechanism Subduction zone

1. INTRODUCTION

Subduction of oceanic lithosphere causes high seismic activity in subduction zones. The shallow portion of the plate interface between a subducting oceanic plate and an overriding continental plate is locked by friction. To release the stress that accumulates from this interplate locking, the locked portion of the plate interface occasionally slips dynamically and causes interplate earthquakes. Plate convergence at subduction zones also causes stress accumulation within both the overriding continental lithosphere and the subducting oceanic lithosphere. To release this accumulated stress, earthquakes take place both in the overriding and subducting plates. Earthquakes that occur in a subducting plate are called intraslab earthquakes because the subducting oceanic lithosphere is referred to as the slab. Intraslab earthquakes can occur as deeply as the bottom of the upper mantle. With the exception of shallow events, all earthquakes occur within the subducting slab. Therefore, the maximum depth of earthquakes on Earth is c. 660 km, or potentially slightly deeper. The global distribution of the focal depths of seismicity is bimodal (Fig. 1). The first, and largest, peak of earthquake frequency with depth occurs at very shallow depths between 0 and 30 km. The frequency decreases almost exponentially to a depth of c. 300km, where a minimum emerges between c. 300 and 450km in depth. The frequency then increases to a depth of c. 550km, where it reaches a local maximum of 550–600 km (Frohlich,2006; Houston,2007). Generally, intraslab earthquakes are divided into two groups based on focal depth. Events shallower than 300 km are defined as intermediate-depth earthquakes, and those deeper than 300 km are considered deep-focus earthquakes.

At the depths of intraslab earthquakes, lithostatic pressure, and therefore frictional strength, becomes too great to allow dynamic slips, i.e., earthquakes, along fault planes. In order to generate earthquakes at such great depths, shear stress must exceed that very large frictional strength. However, it is difficult to envision the application of such large shear stress on a fault where intermediate-depth and deep-focus events take place. Nevertheless, intermediate-depth and deep-focus earthquakes do occur in the subducting slab, and therefore the frictional strength must not be as high as expected. Therefore, some special mechanism must operate to lower the strength of friction and generate intermediate-depth and deep-focus intraslab earthquakes (e.g., Frohlich,1994,2006).

Although several plausible mechanisms have been proposed, the exact mechanism that generates intermediate-depth and deep-focus earthquakes remains unresolved. Some early studies explored the possibility that implosive phase change of mantle materials could cause intraslab earthquakes (e.g., Leith and Sharpe,1936). However, subsequent studies have revealed that both intermediate-depth and deep-focus earthquakes generally do not have the large isotropic components that would be expected if this mechanism occurred (e.g., Hara et al.1996; Kawakatsu, 1991, 1996; Kuge and Kawakatsu, 1993; Okal, 1996; Russakoff et al.1997; Estabrook, 1999). Therefore, some other mechanism must exist to generate intraslab events. To date, leading hypotheses proposed to explain the occurrence of earthquakes at such great depths include thermal shear instability of plastic deformation (e.g., Ogawa,1987; Hobbs and Ord,1988; Karato et al.2001; Kelemen and Hirth,2007), transformational faulting and anticrack formation (e.g., Kirby,1987; Green and Burnley,1989; Burnley et al.1991; Green and Houston, 1995; Kirby et al.1996), and dehydration embrittlement (e.g., Griggs and Handin, 1960; Raleigh and Paterso1965; Raleigh,1967; Meade and Jeanloz,1991; Nishiyama, 1992; Green and Houston,1995; Kirby,1995; Kirby et al.1996; Seno and Yamanaka,1996).

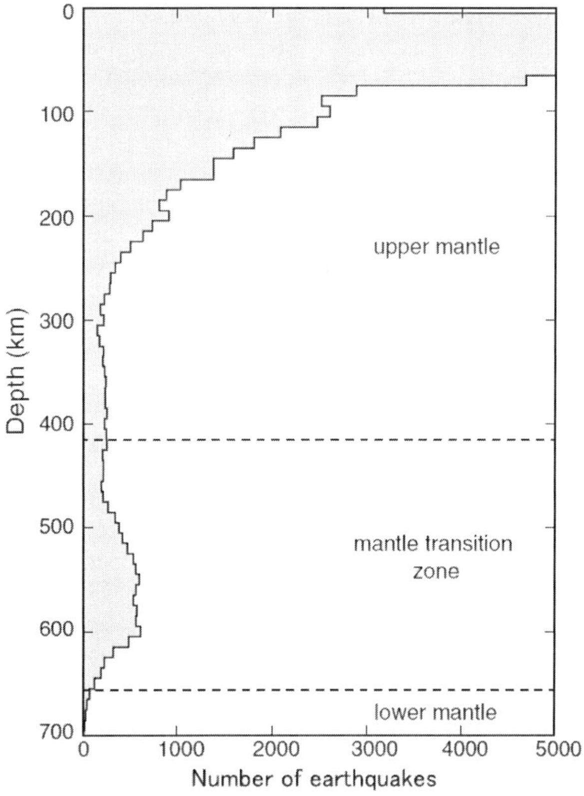

Figure. 1 Focal depth distribution of global seismicity for the period of 1960–2007. The shallower peak is truncated to show the deeper peak in detail

The thermal shear instability model assumes that intraslab earthquakes occur as follows. If plastic deformation progresses at a faster rate than the rate at which the generated heat dissipates, a positive feedback is produced that further promotes plastic deformation. Consequently, plastic deformation proceeds rapidly until materials in that portion of the slab start to melt, which results in sudden slips, i.e., earthquakes. The transformational faulting hypothesis assumes the following mechanism. As the slab subducts into the mantle transition zone, the phase transformation from olivine to spinel proceeds in the slab. Initially, small lens-shaped spinel crystals (spinel lenses) are formed in this phase transformation. Under the deviatoric stress field within the slab, spinel lenses grow analogously to the cracks that develop at shallow depths. These spinel lenses align along a plane that obliquely intersects with the axis of maximum compressive stress, and eventually, the crystals coalesce in that plane, which results in abrupt slips and therefore earthquakes. The dehydration embrittlement hypothesis posits that hydrous minerals that were present in the lithosphere before subduction decompose as the slab subducts and that the released H_2O lowers the effective normal stress, thereby enabling brittle fractures that cause earthquakes. However, all three of these hypotheses are based on extrapolation from the results of laboratory experiments or theoretical calculations to simulate the great depths and scale of natural intraslab earthquakes; therefore, the exact mechanism has not yet been conclusively determined.

The present manuscript is focused on the genesis of intermediate-depth intraslab earthquakes. The transformational faulting model applies only for events that occur at the great depths of the mantle transition zone; therefore, it is not applicable to intermediate-depth events. Dehydration-related embrittlement is thought to be a promising hypothesis as a mechanism of lowering frictional strength sufficiently to generate intermediate-depth intraslab events (Nishiyama,1992; Kirby,1995; Seno and Yamanaka,1996). According to this hypothesis, temperature and pressure increase within the slab with subduction, and as a result, hydrous minerals that have been fixed in the slab are decomposed. H_2O is liberated by this process then lowers the effective normal stress. This effect would allow brittle faulting or dynamic slip along a fault even at depths with high lithostatic pressure. The traditional dehydration embrittlement hypothesis refers only to the embrittlement of rock that occurs during dehydration reactions. In a broader sense, however, dehydration-related embrittlement may also include the embrittlement caused by fluid released through dehydration reactions at any distance from the fault and at any time (e.g., Barcheck et al.2012). We use the term in the latter, broader sense. The thermal shear instability model is another plausible mechanism for causing not only deep-focus earthquakes but also intermediate-depth intraslab seismicity. In some recent studies, the results of laboratory experiments and numerical simulations have supported this hypothesis (e.g., Kelemen and Hirth, 2007; Chernak and Hirth,2011; Proctor and Hirth2015).

In the following sections, we review recent research findings about intermediate-depth intraslab seismicity and the internal structure of subducting slabs in relation to slab metamorphism and discuss the generation mechanisms of intermediate-depth intraslab earthquakes.

2. REVIEW

2.1 The double seismic zone and dehydration of metamorphosed crust and serpentinized mantle

Seismic studies of precise intermediate-depth seismicity and the detailed seismic structures of subducting slabs have provided growing evidence that supports the dehydration-related embrittlement hypothesis for earthquakes nucleation.

A non-negligible amount of H_2O is contained within subducting lithosphere in the form of hydrous minerals. Hydration of oceanic lithosphere can occur in a number of tectonic settings. (1) When new lithosphere is created at a mid-ocean ridge, hydrous minerals are formed in the shallow portion of the crust by hydrothermal circulation and reactions with seawater (e.g., Dixon et al.2002; Staudigel et al. 1995). (2) As oceanic lithosphere approaches a trench, it begins to sink and deforms under the bending force, which produces the outer rise. This bending force causes normal faulting in the trench–outer rise, and seawater enters along faults, in which hydrous minerals are then formed (Fig. 2) (e.g., Kirby,1995; Peacock,2001; Ranero et al.2003; Faccenda et al.2009; Iyer et al.2012). (3) Seawater enters along transform faults near mid-ocean ridges because of fault movement, and hydrated minerals form along these faults as well (e.g., Bonatti and Honnorez,1976; Gregg et al.2007). (4) In addition, Seno and Yamanaka (1996) described a scenario whereby H_2O is fixed to the oceanic plate in the form of hydrous minerals as the plate passes directly over mantle plumes or superplumes during its horizontal movement along the earth's surface. Magma of the plumes or superplumes intrudes into or accretes to the overlying oceanic lithosphere, and H_2O contained in the magma is added and eventually fixed in the oceanic plate.

H_2O is thus incorporated into oceanic lithosphere and fixed in the form of hydrous minerals through several different processes. It is therefore estimated that oceanic lithosphere contains a considerable amount of H_2O prior to its subduction (e.g., Rüpke et al. 2004; Faccenda, 2014). When lithosphere reaches a trench, it then sinks beneath the continental plate, and its temperature and pressure rise with increasing depth. Hydrous minerals that were fixed in the oceanic lithosphere then decompose and release H_2O. It is proposed in the dehydration-related embrittlement hypothesis that this liberated H_2O causes intraslab earthquakes, especially intermediate-depth events.

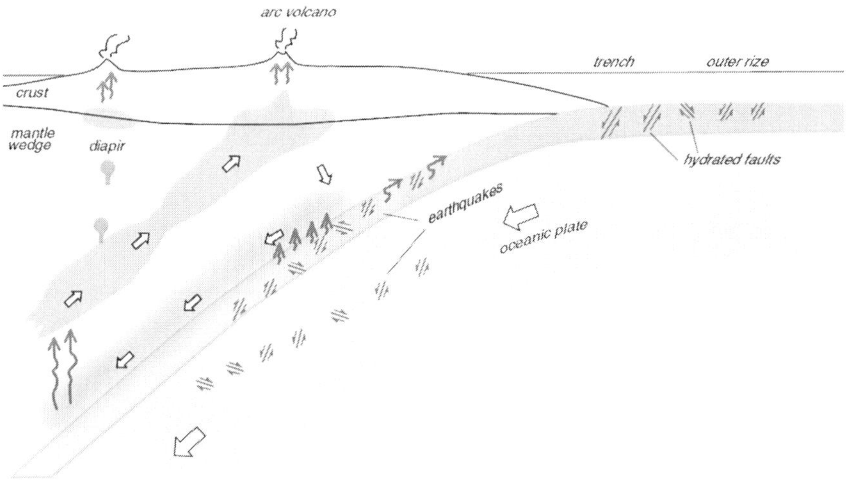

Figure. 2 Schematic diagram showing normal faults in the trench–outer rise, intermediate-depth intraslab earthquakes, and arc magmatism. Estimated transportation paths of H_2O from the subducting slab to the arc crust are indicated by *blue arrows* (Hasegawa and Nakajima,2004)

If hydrous minerals are present along normal faults in the trench–outer rise region prior to plate subduction, those hydrous minerals decompose and expel H_2O along the faults as the plate subducts to depth. It has been proposed that intraslab earthquakes occur along these preexisting planes of weakness that originated as normal faults prior to subduction (e.g., Kirby, 1995). Jiao et al. (2000) tested this hypothesis by examining fault plane solutions of intraslab events in Tonga. The results show that for outer rise events, features of the geometry of the fault system prior to subduction persist down to c. 450 km depth in the subducted slab, which suggests that intraslab events down to this depth are caused by the reactivation of preexisting faults containing hydrous minerals that formed prior to subduction. Similar results were obtained for intermediate-depth intraslab events in the Middle America and Chile subduction zones (Ranero et al. 2005). These observations support the dehydration-related embrittlement hypothesis for the genesis of intermediate-depth intraslab events, especially those that occur within the slab crust. However, subsequent studies by Warren et al. (2007, 2008) have shown that this issue may be more complex. By examining the fault plane orientations of intraslab events in Tonga and Middle America, they identified many subhorizontal faults, which suggest the formation of new faults. Therefore, the outer rise faults are not always reactivated. Further investigations are required to confirm the hypothesized reactivation of preexisting faults.

If intermediate-depth intraslab events are caused by dehydration-derived H_2O, these events would not occur everywhere throughout a slab; they would occur only in parts of the slab where hydrous minerals or H_2O exist. In particular, it is hypothesized that many of these events occur at the phase boundaries (facies boundaries) of dehydration reactions across which H_2O content changes. As described in detail below, recent studies have demonstrated that intermediate-depth intraslab events do in fact show such distribution. To verify whether fluid-related embrittlement is the cause of intermediate-depth events, it is necessary to determine where hydrous minerals are present within the slab and where they decompose and expel H_2O. Based on experimentally and theoretically derived phase diagrams of the rocks that compose the oceanic crust and mantle (Fig. 3), Hacker et al. (2003a, b) estimated the existence range of hydrous minerals and the locations of the facies boundaries of dehydration reactions within the slabs beneath the subduction zones of Cascadia, Costa Rica and NE and SW Japan. As predicted, comparing the phase diagrams with hypocenter distributions of intermediate-depth intraslab earthquakes in these subduction zones shows that the earthquakes appear to occur only within the existence range of hydrous minerals and not outside this range. Examples from NE Japan and Costa Rica are shown in Fig. 4.

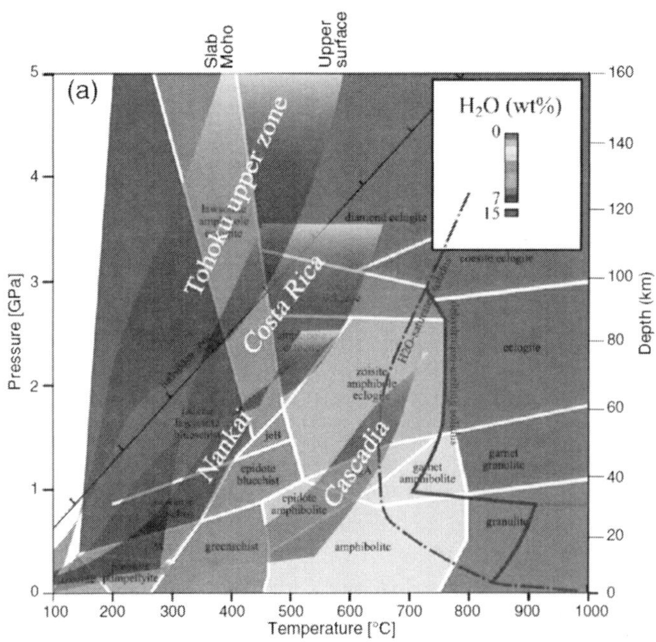

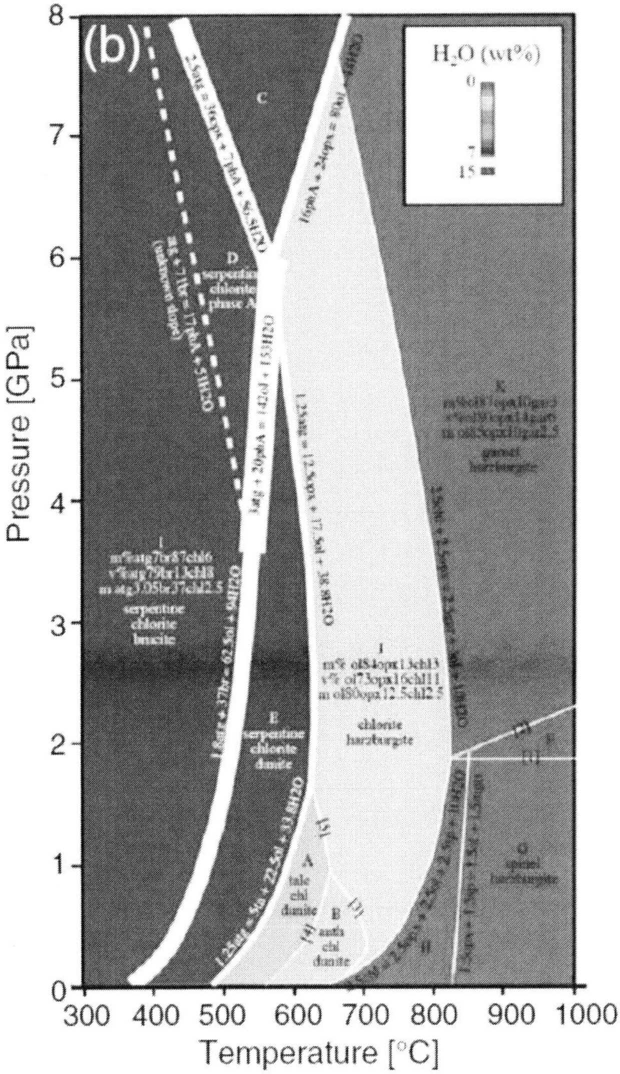

Figure. 3 Metamorphic facies and maximum H_2O contents. a Mid-ocean ridge basalt (MORB) and b harzburgite. H_2O contents in wt% are shown by the color scale. Estimated pressure–temperature paths for the slab crust under Tohoku, Costa Rica, Nankai, and Cascadia are indicated as *shaded areas* in a *PA*, prehnite–actinolite, *eA* epidote amphibolite, *eB* epidote blueschist, and *jeB* jadeite–epidote blueschist. (After Hacker et al. 2003b)

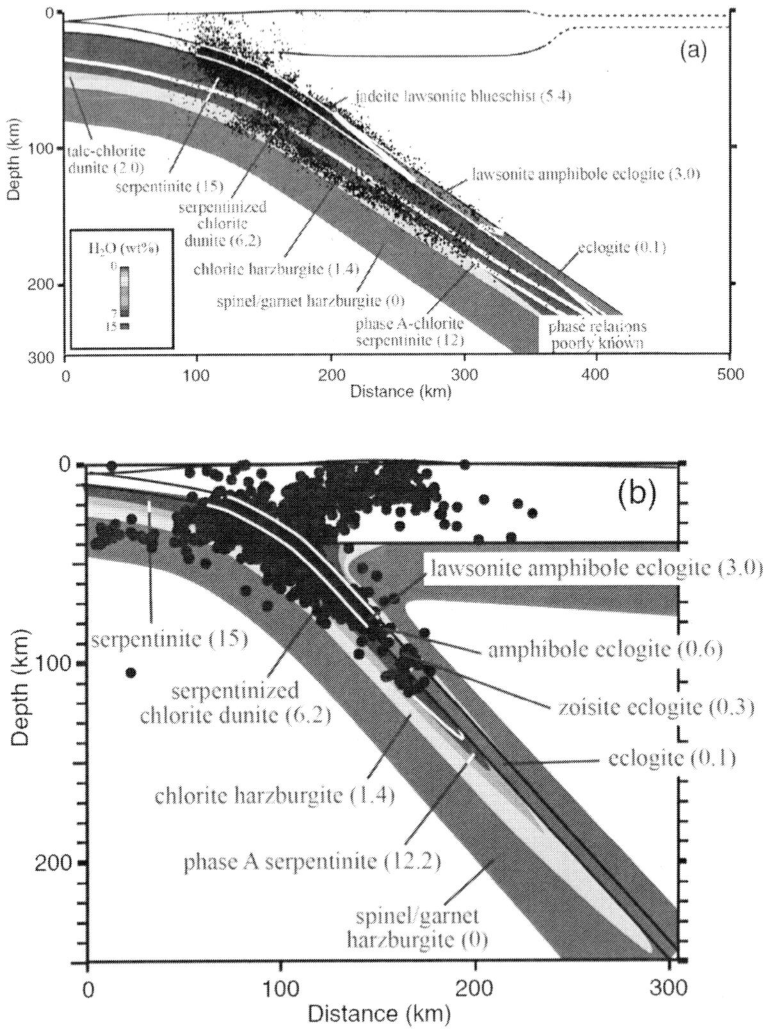

Figure. 4 Across-arc vertical cross-sections of intraslab seismicity and existence range of hydrous minerals in the slab. a Tohoku and b Costa Rica subduction zones. Maximum H_2O contents are shown by the *color scale. Dots* show earthquakes. (After Hacker et al. 2003)

Moreover, Yamasaki and Seno (2003) found that intermediate-depth intraslab events tend to concentrate along the dehydration reaction boundaries within slabs. Based on experimentally derived phase diagrams, they estimated the locations of dehydration decomposition in metamorphosed crust and serpentinized mantle within the slabs of subduction zones beneath NE and SW Japan, Taiwan, northern Chile, Cape Mendocino, and the eastern Aleutians, and compared these estimated locations with the hypocenter distributions of intermediate-depth earthquakes. Results of this analysis for the Pacific slabs beneath NE Japan and the eastern Aleutians are shown in Fig. 5. In NE Japan, intermediate-depth earthquakes occur in a double-planed deep seismic zone 70–150 km in depth, with subparallel upper and lower planes 30–40 km apart from each other (Tsumura, 1973; Umino and Hasegawa, 1975: Hasegawa et al. 1978a, b). Figure 5a shows that the estimated locations of the dehydration boundary correspond approximately to the upper and lower planes of this double seismic zone. Under the eastern Aleutians, the Wadati–Benioff zone also comprises the double seismic zone, with a slightly smaller separation between the upper and lower planes than that of the NE Japan slab. Figure 5b again shows that the estimated dehydration reaction boundaries roughly correspond to the upper and lower seismic planes. Similar results were also obtained for the remaining four subduction zones included in the study. This correspondence further explains why the lower-plane events of the double seismic zone are generated in the middle of the lithospheric mantle, as previously reported by Peacock (2001).

Since the discovery of an unambiguous double seismic zone in Tohoku (Hasegawa et al. 1978a, b), numerous subduction zones across the world have been similarly investigated, and double seismic zones have been detected with certainty in several subduction zones (e.g., Frohlich, 2006), such as in slabs subducted below Hokkaido (Suzuki et al. 1983), Kuril–Kamchatka (Kao and Chen, 1994), Alaska (Abers, 1992, 1996), Cascadia (Cassidy and Waldhauser, 2003), Mendocino (Smith et al. 1993), northern Chile (Comte and Suarez, 1994; Comte et al. 1999), New Zealand (Reyners et al. 1997), Tonga (Kawakatsu, 1985, 1986), and northern Taiwan (Kao and Rau, 1999). However, double seismic zones are not apparent in all subduction zones worldwide. Furthermore, even when a double subduction zone is present, its depth may range from c. 40 to 180 km, and the separation distance between the upper and lower planes may range between c. 10 and 40 km.

Hypocenters of earthquakes must be accurately determined, especially their focal depths, to evaluate whether a double seismic zone is present in a given subduction zone. Dense seismic observation network data are required for accurate determination of earthquake hypocenters, but few subduction zones are installed with such dense networks. Without relying on data from such local seismic networks, Brudzinski et al. (2007) investigated a number of subduction zones worldwide to evaluate whether double seismic zones were present. They used global seismic network data and selected events with well-constrained focal depths, such as those determined based on depth phases. Using hypocenter data from the EHB catalog (Engdahl et al. 1998; Engdahl and Villaseñor, 2002), they verified the presence of the double seismic zone by investigating whether two peaks appear in the frequency distribution of intermediate-depth intraslab earthquakes with respect to the distance from the slab upper surface.

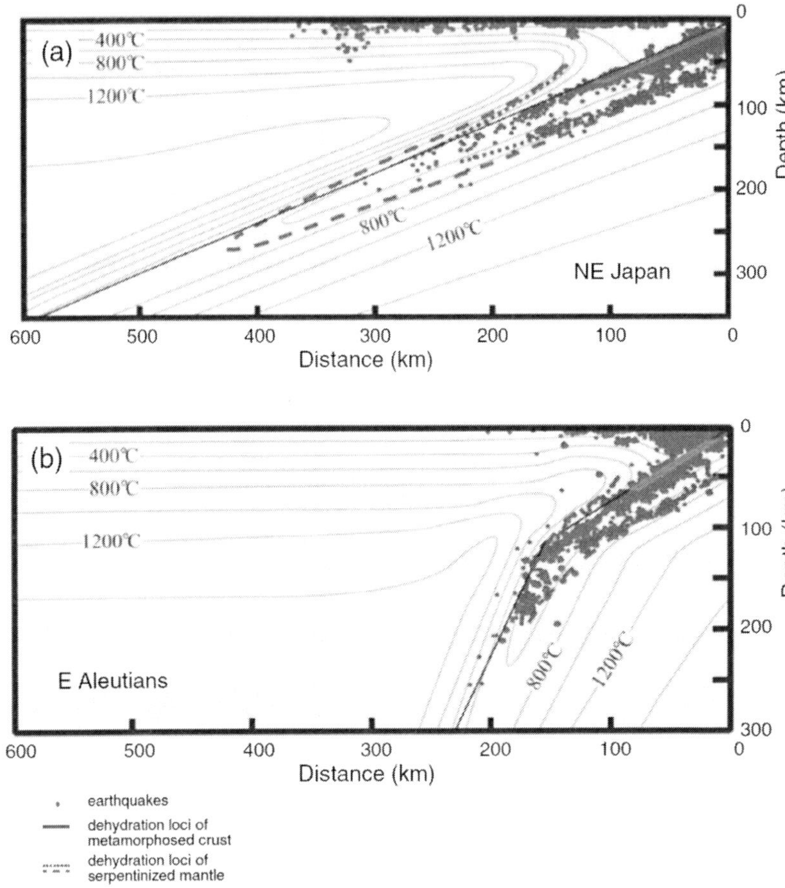

Figure. 5 Across-arc vertical cross-sections of intraslab earthquakes and dehydration loci in the slabs. a Tohoku and b eastern Aleutian subduction zones. Dehydration loci of metamorphosed crust and serpentinized mantle are indicated by the *green* and *red lines*, respectively. *Red broken lines* and *red dotted lines* are those based on the antigorite reaction boundaries of Ulmer and Trommsdorff (1995) and Wunder and Schreyer (1997), respectively. *Blue dots* are earthquakes. Thermal structure is shown as isothermal contours. (After Yamasaki and Seno, 2003)

The two peaks in the frequency distribution are interpreted to correspond to the upper and lower planes of the double seismic zone. It thus became clear that (1) the two peaks were apparent for all 30 of the investigated subduction zones, and (2) the distance between the two peaks, i.e., the separation distance between the upper and lower planes, varies by subduction zone, and is proportional to the age of the subducting plate (Fig. 6). These findings are consistent with the predictions of the dehydration-related embrittlement hypothesis. In fact, the lowerplane events correspond to the expected positions of dehydration decomposition in the serpentinized mantle in the slab, as shown in Fig. 6.

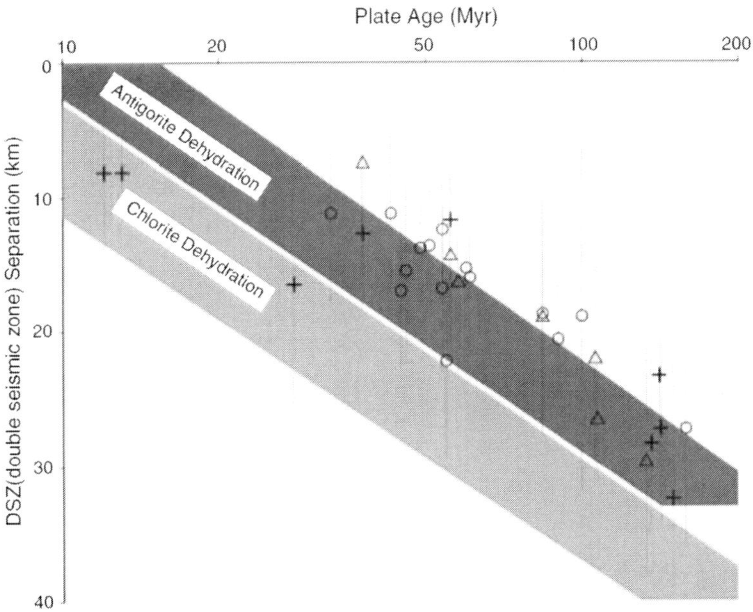

Figure. 6 Separation of the double seismic zone plotted against the age of the subducting slab. Distances between the upper and lower seismic planes are shown by three different symbols: *triangles* show typical down-dip compression overlying down-dip extension, *circles* indicate a different pattern, and *crosses* denote unknown. *Vertical lines* attached to the symbols are one-standard deviation uncertainties from Gaussian fits. Predicted separations of the double seismic zones based on the lower seismic planes resulting from the dehydration of antigorite and chlorite based on Hacker et al. (2003) are shown as *blue* and *green shaded areas*, respectively. (After Brudzinski et al. 2007)

Detailed analyses of intermediate-depth intraslab earthquakes in NE Japan have further revealed a characteristic structure of the Wadati–Benioff zone. Igarashi et al. (2001) reported that normal fault type events also occur at the top of the upper plane of the double seismic zone in the depth range of c. 70–100 km. In terms of focal mechanisms, the Wadati–Benioff zone beneath NE Japan is triple-planed; from top to bottom, these three planes are characterized by normal fault, down-dip compressional and down-dip extensional events. Wang (2002) argued that plate unbending combined with dehydration-related embrittlement may account for the morphology and stresses of this triple-planed seismic zone. Metamorphic reactions densify materials and therefore induce tensile stress in the uppermost transformed crust. This tensile stress, as well as high pore fluid pressure, in the uppermost slab crust thus generates the normal fault type seismicity that occurs there. A recent study by Nakajima et al. (2013) revealed that even in a small seismic cluster, normal fault type events occur just 1 km above the down-dip compressional events in the lower crust of the subducting Pacific Plate beneath Niigata, NE Japan. These observations provide evidence for the role of densification-related stresses caused by metamorphic reactions and fluid-related embrittlement in earthquake nucleation within the slab crust.

3. SEISMIC WAVE VELOCITY STRUCTURE AND UPPER-PLANE SEISMICITY IN THE SLAB CRUST

3.1 A belt of intraslab seismicity and phase transition in the slab crust beneath eastern Japan

As described in the previous section, the locations of upper-plane events in the Pacific slab crust beneath NE Japan roughly correspond to the dehydration reaction boundaries of the metamorphosed crust (Fig. 5a). Recent studies have revealed a more detailed spatial distribution of this upper-plane seismicity, as well as a clear relationship between that distribution and the seismic wave velocity structure of the slab crust, which is likely associated with dehydration reactions. This relationship may be well explained if dehydration-related embrittlement is the cause of intermediate-depth intraslab events.

The hypocenter distribution of events that occurred in the crust of the Pacific slab under eastern Japan is shown in Fig. 7a (Kita et al. 2010). As shown with arrows in this figure, events in the slab crust in Tohoku and eastern Hokkaido form a significant belt-like seismic zone nearly parallel to the 70–90 km depth contours of the upper surface of the slab. The frequency distributions of events in the slab crust as a function of depth along lines (b) through (l) in Fig. 7a are shown in Fig. 7b–l, respectively. The distribution for each of these lines shows a clear peak between the depths of 70 and 90 km, as indicated in the figure by arrows. Figure 8a shows an across-arc vertical cross-section of intraslab seismicity under central Tohoku. The belt-like concentrated earthquake activity shown in the plan view of Fig. 7a is visible at c. 70–90 km in depth in the vertical cross-sectional view in Fig. 8a. Kita et al. (2006) named this seismically active band-shaped region of the slab crust the "upper-plane seismic belt."

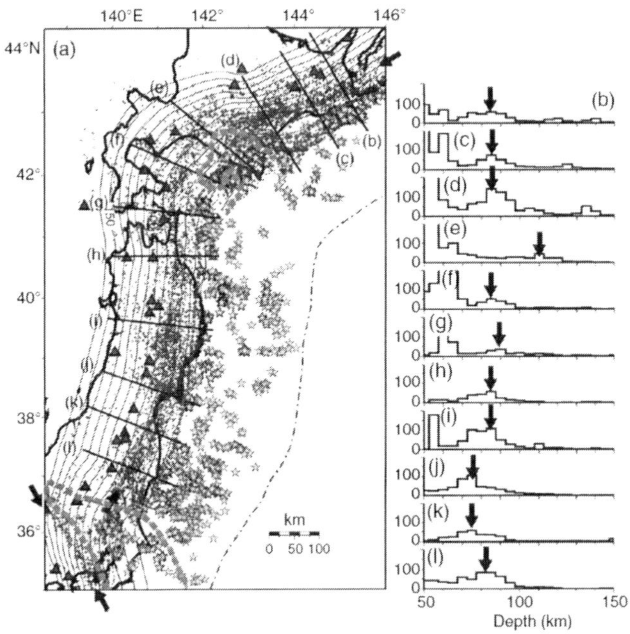

Figure. 7 A belt of upper-plane seismicity in the slab crust. a Hypocenter distribution of earthquakes in the crust of the Pacific Plate subducting under eastern Japan. Events 0–10 km below the upper plate surface are shown by *red dots. Bold arrows* denote estimated locations of the upper-plane seismic belt. The contact zone between the Philippine Sea and Pacific Plate beneath Kanto and that between the subducted crust and the Pacific Plate beneath western Hokkaido are indicated by *blue thick broken curves. Solid curves, red triangles*, and *gray stars* denote iso-depth contours of the upper plate surface, active volcanoes, and small repeating earthquakes, respectively. b–l Frequency distributions of upper-plane events along the *lines b* through *l* in Fig. 7a plotted against depth. Frequency is plotted for intervals of 10 km. The *black arrows* show the peak that corresponds to the upper-plane seismic belt. (After Kita et al. 2010)

If intermediate-depth intraslab events are caused by dehydration-related embrittlement, this upper-plane seismic belt should be located in the existence range of hydrous minerals in the slab crust, particularly around the facies boundaries of the dehydration reaction.

Beneath Kanto, the upper-plane seismic belt is not parallel to the 70–90 -km depth contours of the upper surface of the slab; it becomes deeper locally and is oblique to the 70–90 -km depth contours, as shown in Fig. 7a. This difference is caused by the shielding effect of the Philippine Sea slab that directly overlies the Pacific slab (Hasegawa et al. 2007, 2009, 2016). High-quality data obtained in Japan with a nationwide seismic network have enabled precise estimation of the positions and shapes of both the Pacific and the Philippine Sea Plates subducting under the Japanese archipelago, as well as the contact of the two plates over a wide area under Kanto (Nakajima and Hasegawa, 2007, 2010; Nakajima et al. 2009b; Hirose et al. 2008; Kita et al. 2010a; Uchida et al. 2009, 2010). As shown in Fig. 9a, b, the Pacific slab under Kanto is directly overlain by, and in contact with, the Philippine Sea slab. The extent of the slab–slab contact area, which is enclosed by two thick, broken green lines in Fig. 7a, corresponds approximately to the extent of the Kanto Plain. Figure 7a shows that the down-dip end of the contact area is parallel to the upper-plane seismic belt, adjacent to its shallow side, and oblique to the 70–90 -km depth contours of the upper surface of the slab. This clear spatial correspondence can be understood if we consider the following scenario. (1) The Pacific slab subducting under Kanto is capped by the overlying Philippine Sea slab. (2) This cap prevents temperature rise in the Pacific slab due to heating from the hot mantle wedge above. (3) Therefore, the phase transition associated with the dehydration reaction is significantly delayed in the Pacific slab crust. (4) Consequently, the upper-plane seismic belt in the Pacific slab deepens locally and is oblique to the 70–90 km depth contours of the upper surface of the slab, as shown in Fig. 7a.

A similar local deepening of the upper-plane seismic belt was detected below SW Hokkaido, associated with the junction of the NE Japan and Kuril arcs (Kita et al. 2010). As indicated by the arrows in Fig. 7a, e, the upper-plane seismic belt becomes deeper locally in the area beneath this arc–arc junction and reaches the vicinity of the 110 -km depth contour of the slab's upper surface. Beneath Hokkaido, on the southwestern edge of the Kuril arc, the Pacific Plate is subducting obliquely to the trench axis. This oblique subduction causes the forearc sliver of the Kuril arc to collide with the forearc of the NE Japan arc (e.g., Kimura, 1986). A seismic tomography study by Kita et al. (2010) clearly imaged the crust portion of the colliding forearc, which has subducted into the mantle wedge and reached the upper surface of the Pacific slab immediately below (Fig. 9c, d).

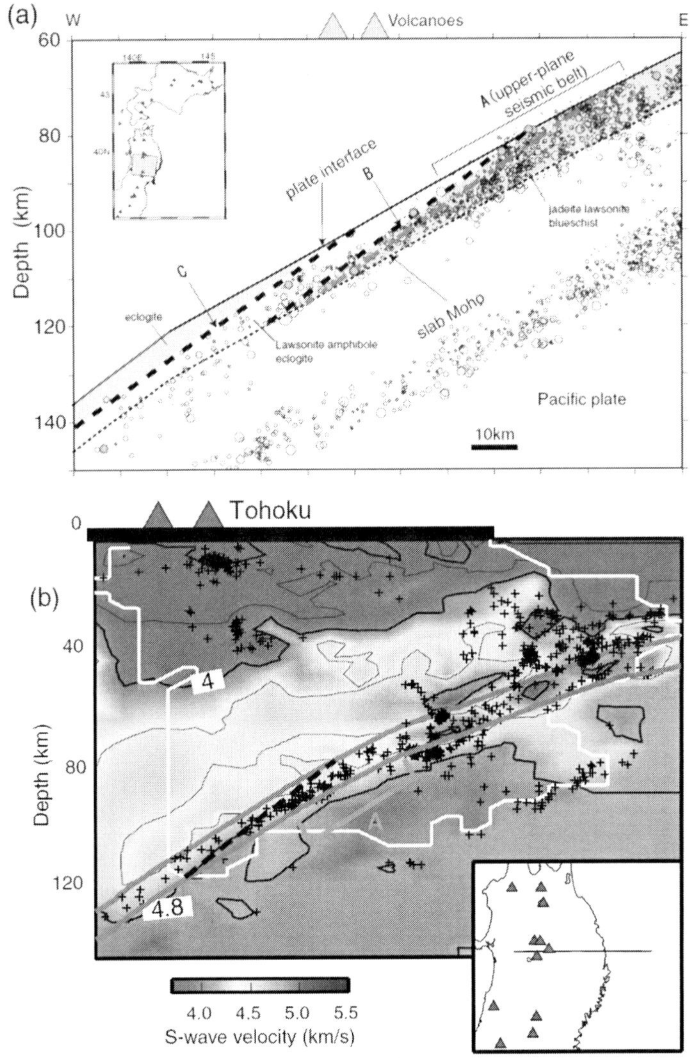

Figure. 8 Seismicity and facies boundaries in the slab crust and low S-wave velocity layer. a Across-arc vertical cross-section of intraslab earthquakes and facies boundaries in the crust of the Pacific slab under central Tohoku (modified from Kita et al.(2006)). Facies boundaries are indicated by *broken black lines* and a *long and short dashed red line. Open circles* are earthquakes. *Solid* and *thin dashed lines* are the upper plate interface and the Moho within the slab, respectively. b Across-arc vertical cross-section of S-wave velocity in central Tohoku (Tsuji et al. 2008). S-wave velocity is indicated by the color scale at the *bottom. Solid line* is the upper plate interface, and *A* denotes the upper-plane seismic belt

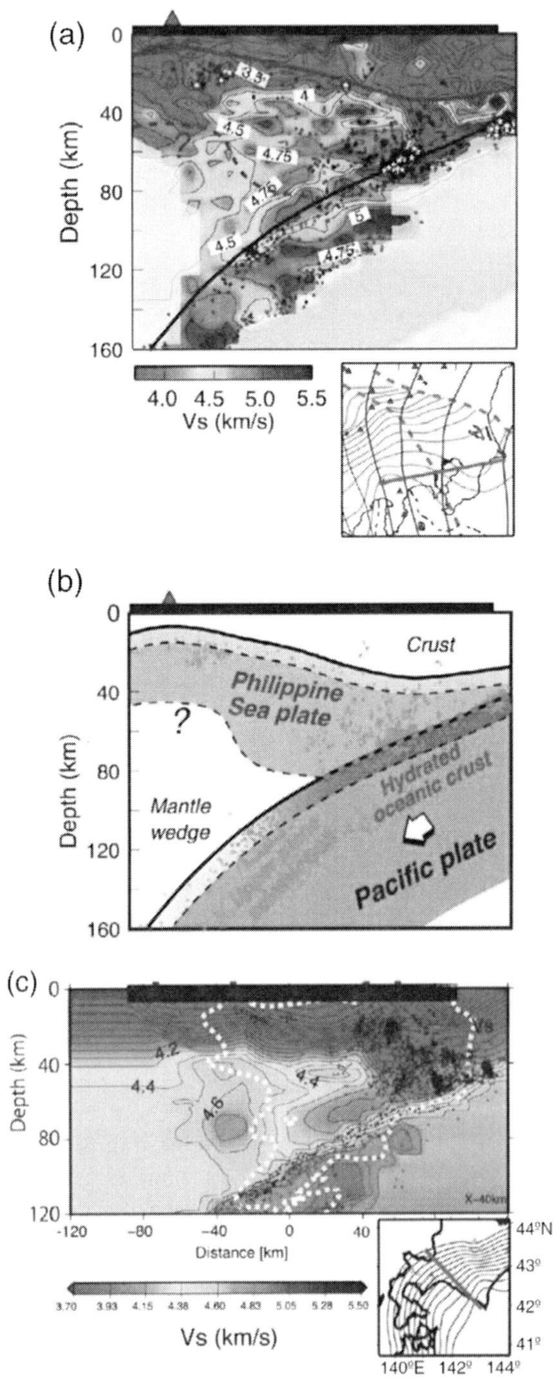
(a)
Depth (km)
0
40
80
120
160
3.3
4
4.5
4.75
4.75
5
4.5
4.75
4.0 4.5 5.0 5.5
Vs (km/s)
(b)
Depth (km)
0
40
80
120
160
Crust
Philippine Sea plate
?
Mantle wedge
Hydrated oceanic crust
Pacific plate
(c)
Depth (km)
0
40
80
120
4.2
4.4
4.6
4.4
Vs
X-40km
-120 -80 -40 0 40
Distance [km]
3.70 3.93 4.15 4.38 4.60 4.83 5.05 5.28 5.50
Vs (km/s)
44°N
43°
42°
41°
140°E 142° 144°

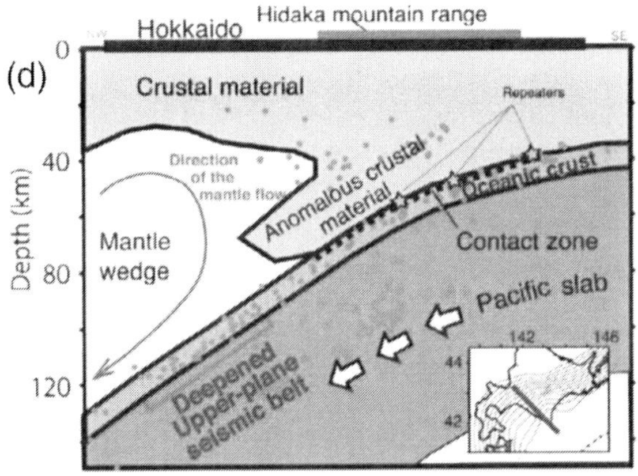

Figure. 9 The contact of the Pacific slab with the overlying Philippine Sea slab or forearc crust. a Vertical cross-section of S-wave velocity along a line under Kanto in the *inset map* and b schematic figure showing the contact between the Philippine Sea and Pacific slabs (Nakajima et al. 2009b). c Across-arc vertical cross-section of S-wave velocity under western Hokkaido along a line in the *inset map* and d schematic figure showing the contact between the subducted forearc crust and the Pacific slab (Kita et al. 2010a). S-wave velocity is indicated by the *color scale* at the *bottom*

The estimated contact area between the subducted crust thus imaged, and the Pacific slab is enclosed by a thick, broken green line in Fig. 7a. This figure shows that the contact area corresponds closely with the local deepening of the upper-plane seismic belt. Therefore, similarly to the case in Kanto, the local deepening of the upper-plane seismic belt in SW Hokkaido can be understood if we consider the following. (1) The Pacific slab under SW Hokkaido is capped by subducted crust that directly overlies it (2) his capping prevents the Pacific slab from being heated by the mantle wedge immediately above it (3) herefore, the phase transition and dehydration is delayed in the slab crust (4) his delay causes local deepening of the upper-plane seismic belt in this region, as well (Kita et al. 2010).

If the phase transition of the dehydration reaction occurs in the slab crust, seismic wave velocity is expected to become faster across the phase boundary. Therefore, it can be verified based on the distribution of seismic wave velocities obtained through seismic tomography whether the phase transition of the dehydration reaction causes the formation of the upper-plane seismic belt. Figure 8b shows an across-arc vertical cross-section of S-wave velocity in central Tohoku obtained by Tsuji et al. (2008) using the double-difference (DD) tomography method of Zhang and Thurber (2003). A low-velocity layer that corresponds to the slab crust extends down to the depth of the upper-plane seismic belt (A in the figure), and no further. Figure 10a shows the distribution of S-wave velocity in the Pacific slab crust estimated through DD tomography (Nakajima et al. 2009). As also shown in Fig. 8b, the low-velocity layer under Tohoku extends to a depth of approximately 80 km, the depth of the upper-plane seismic belt, and no further. In contrast, the low-velocity layer under Kanto becomes locally deeper in the area that corresponds to contact with the overlying Philippine Sea slab; this area is enclosed by two thick, broken green lines in Fig. 10a. Figure 10a shows that the areal extent of the low-velocity layer is reasonably consistent with the slab–slab contact area. Comparison of Fig. 10a with the hypocenter distribution in the slab crust (Fig. 10b) shows that the location of the locally deepened upper-plane seismic belt is in excellent agreement with the down-dip end of the low-velocity layer in the slab crust. Consequently, the obtained distribution of S-wave velocity in the slab crust further supports the prediction that the phase transition is locally delayed in this region.

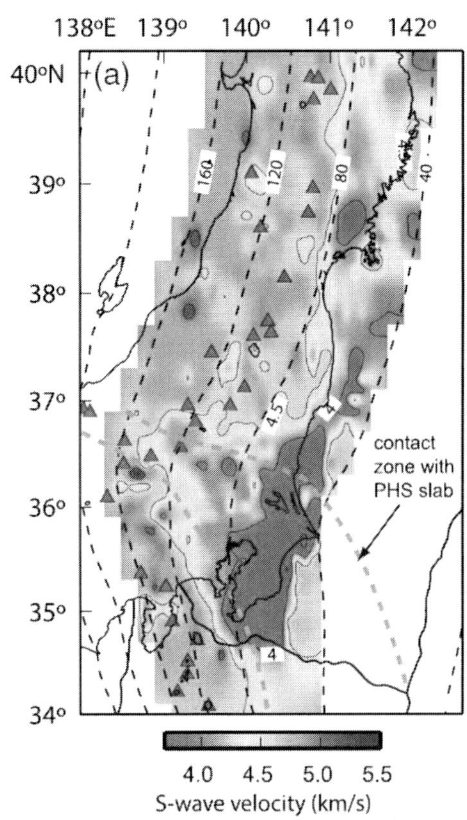

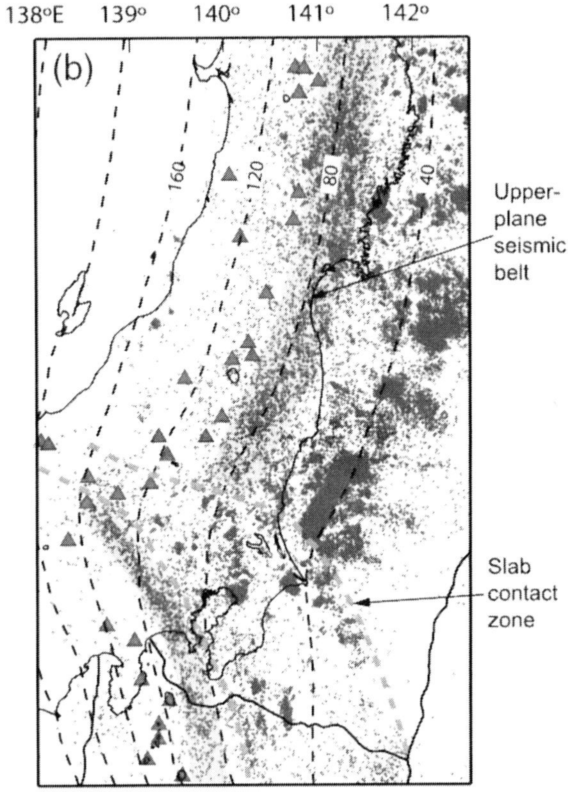

Figure. 10 Seismicity and S-wave velocity in the crust of the Pacific slab. a S-wave velocity distribution in the crust of the Pacific slab under NE Japan on a curved plane 5 km below the upper plate surface. b Hypocenters of earthquakes within the crust of the Pacific slab. Events 0–10 km below the upper plate surface are shown as *blue dots*, and the upper-plane seismic belt is shown in *pink. Broken black curves* denote iso-depth contours of the upper plate surface. The slab–slab contact zone is enclosed by *two green broken curves. Red triangles* are active volcanoes. (After Nakajima et al. 2009b)

Mode-converted seismic waves at the upper plate interface from intermediate-depth intraslab events also provide evidence that the seismic low-velocity layer corresponds to the subducting slab crust (e.g., Matsuzawa et al. 1986, 1990; Zhao et al. 1997; Abers, 2005). P-to-S mode-converted waves at the upper surface of the plate propagate through the slab crust for longer distances than do direct P- and S-waves; therefore, their arrival-time data are convenient for estimating the seismic velocity structure within the slab crust. A recent analysis of P-to-S converted waves under central Tohoku by Shiina et al. (2013) reported that the seismic low-velocity layer persists down to the depth of the upper-plane seismic belt in the subducted oceanic crust. Estimated P-wave velocity at the depth of the upper-plane seismic belt in the low-velocity crust is c. 10% lower than theoretically predicted values for metamorphosed mid-ocean ridge basalt (MORB). Shiina et al. (2013) proposed the coexistence of fluids and hydrous minerals to explain this substantial decrease in P-wave velocity, which provides additional seismic evidence that earthquakes in the slab crust are facilitated by fluid-related embrittlement.

Thus far, we have discussed the phase transition based on the seismic observations that the belt-like concentration of seismic activity is detectable in the slab crust and that the seismic wave low-velocity layer extends down to the depth of this upper-plane seismic belt. Here, we also compare the locations of the upper-plane seismic belt and the down-dip end of the seismic low-velocity layer with thermal–petrological estimates of the positions of the dehydration reaction boundaries in the slab crust, although there is high uncertainty in their positions because of error in the estimation of temperature distribution in the subducting slab. The phase boundaries of jadeite lawsonite blueschist (hereafter JLB) → lawsonite amphibole eclogite (hereafter LAE) and of LAE → eclogite in the slab crust beneath central Tohoku are indicated by the broken black lines B and C, respectively, in Fig. 8a. These boundaries are estimated based on the MORB phase diagram of Hacker et al. (2003b) (Fig. 3a) and the temperature distribution of the slab by Peacock and Wang (1999) (the P–T trajectory of the oceanic crust of Tohoku shown in Fig. 3a). The upper-plane seismic belt indicated as A in the figure is located near or on the shallower side of the phase boundary of JLB with a H_2O content of 5.4% → LAE with a H_2O content of 3.0%, as indicated by B in the figure. These findings are roughly consistent with the predictions based on the dehydration-related embrittlement hypothesis. Furthermore, earthquakes occur only in the lower crust on the deeper side of the phase boundary of JLB → LAE, and no longer occur after exceeding the phase boundary of LAE → eclogite, as indicated by C in the figure. This observation is also consistent with the predictions drawn from the dehydration-related embrittlement hypothesis. For another example, the estimated dehydration reaction boundary based on the MORB phase diagram by Omori et al. (2009) and the temperature distribution of the slab given by van Keken et al. (2002) is indicated by a thick, one-dot-chain red line in Fig. 8a. In this case, the phase boundary also roughly coincides with the position of the upper-plane seismic belt as predicted based on the dehydration-related embrittlement hypothesis.

The production rate of dehydration-derived H_2O appears to clearly explain the formation of the belt-like concentration of upper-plane seismicity at depths of c. 80–120 km in the oceanic crust of the Pacific slab subducting under eastern Japan. Abers et al. (2013) calculated the accumulated amount of dehydration-derived H_2O, the H_2O production rate, and the overall system volume change from the dehydration reaction, as a function of depth for two representative slabs: the Pacific slab beneath eastern Hokkaido as a cold slab and the Philippine Sea slab beneath Kii as a warm slab (Fig. 11b–d). Their calculations were based on estimated P–T (pressure-temperature) trajectories for the subducting slab crust on the P–T diagram of maximum H_2O contents for a hydrated MORB metabasalt, as shown in Fig. 11a. The frequency of earthquakes within the slab crust as a function of depth beneath eastern Hokkaido (Kita et al. 2010) is shown in Fig. 11e for comparison. These figures show that the H_2O production rate, the overall volume change from dehydration, and the frequency of earthquakes in the slab crust below eastern Hokkaido (blue lines in Fig. 11c–e) all have peaks at a depth of c. 80–90 km. Although Fig. 11c, d is based on calculations for the top of the slab, the strong correlations of both the H_2O production rate and the total volume change with seismicity in the slab crust strongly support the interpretation that embrittlement caused by increased pore pressure from dehydration-derived H_2O is the reason for the upper-plane seismic belt.

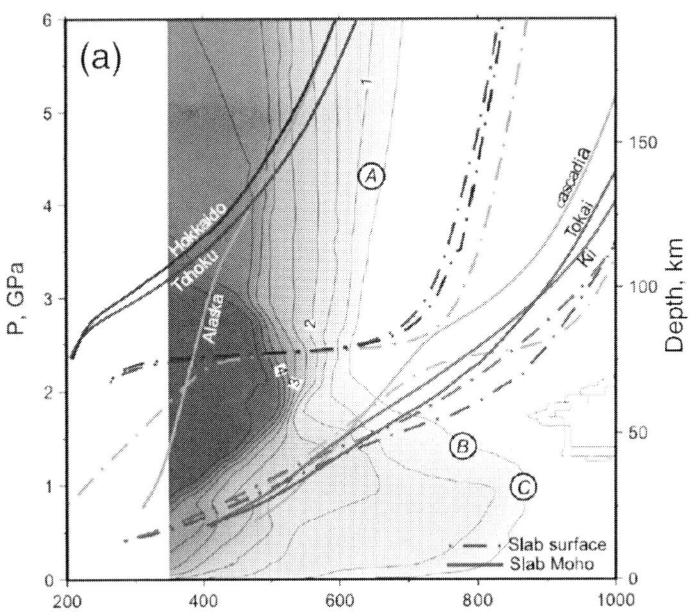

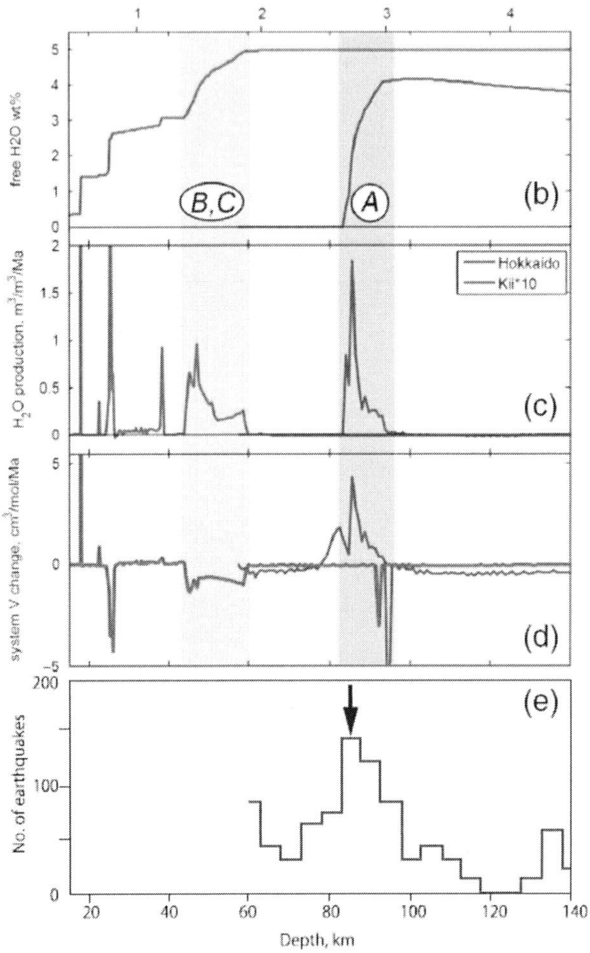

Figure. 11 P–T path, H_2O production, system volume change, and frequency of earthquakes in the crust. a Pressure–temperature (P–T) paths of the subducting oceanic crust for six subduction zone segments plotted on a diagram of maximum H_2O contents (Abers et al. 2013). P–T paths for the upper slab surface and slab Moho are shown as *colored solid* and *dot dashed lines*, respectively. Shading shows maximum H_2O contents in wt% for a hydrated metabasalt of MORB bulk composition (Hacker, 2008). Major dehydration reactions are indicated by *A* (lawsonite breakdown), *B* (zoisite breakdown), and *C* (amphibole breakdown). b–d Evolution of pore fluid pressure and H_2O content for cold and warm slabs (Abers et al. 2013). Hokkaido is indicated by *blue lines* as a representative cold slab and Kii by *red* lines as a representative warm slab. b Accumulated dehydration-derived H_2O content in wt% as dehydration reactions progress. c Rate of

H_2O production, with Kii scaled by a factor of 10 relative to Hokkaido. d Total system (solid + H_2O) volume. Shaded in light *blue* and *yellow* denote depth intervals of major dehydration for Hokkaido and Kii, respectively. *A*, *B*, and C denote specific reactions shown in Fig. 11a. e Frequency of upper-plane earthquakes beneath central Hokkaido (Fig. 7d) as a function of depth (Kita et al. 2010)

At present, however, there are large uncertainties in estimating the temperature distribution within subducting slab crust; therefore, further verification of the thermal–petrological estimates of the position of the dehydration reaction boundary is necessary. Nevertheless, the following seismic observations strongly support the dehydration-related embrittlement hypothesis: (1) that there is a belt of upper-plane seismicity nearly parallel to the 80–90 km depth contours of the upper slab surface, (2) that the seismic low-velocity layer corresponding to the slab crust persists down to the upper-plane seismic belt and, no further, (3) that both the upper-plane seismic belt and the down-dip end of the seismic low-velocity layer deepen locally under Kanto and SW Hokkaido, with these local deepenings coinciding spatially with the areas of contact with the Philippine Sea slab and the overlying subducted forearc crust, and (4) that both the H_2O production rate and the net volume change caused by dehydration in the slab crust have peak depths at c. 80–90 km, which is approximately consistent with the location of the upper-plane seismic belt.

3.2 The upper-plane seismic belt and seismic low-velocity layer in slab crust in other subduction zones

Does a similar upper-plane seismic belt form in the slab crust in other subduction zones? The formation of upper-plane seismic belts should be common to all subduction zones if they are caused by fluids expelled by the dehydration of slab crust materials.

Abers et al. (2013) systematically investigated the relationship between trends in earthquake locations and thermal–petrological conditions within the crust of subducting slabs by comparing several well-characterized subduction zones from different parts of the world. They found that in young (and therefore warm) subduction zones, earthquakes preferentially lie within the slab mantle or very close to the slab Moho. In contrast, in old (and therefore cold) subduction zones, earthquakes occur throughout the oceanic slab crust as well as the slab mantle. They proposed that separation into these two classes is caused by differences in the P–T paths that the slab crust follows as it descends into the mantle. The subducting slab P–T trajectories of young and warm subduction zones and those of old and cold subduction zones encounter different facies boundaries, which produce differences in the way that H_2O is released via dehydration. Estimated temperatures along the upper surface of the slab and the slab Moho for each of the six subduction zone segments are plotted on a pressure–temperature diagram of H_2O content for hydrated MORB in Fig. 11a. The figure shows that the P–T paths for the cold and warm slabs are characterized by considerably different conditions. At high pressures, the main dehydration reaction is the breakdown of lawsonite (A in Fig. 11a), and the crusts of cold slabs, such as those beneath Hokkaido, Tohoku, and central Alaska, experience this dehydration reaction to eclogite. At lower pressures, the breakdown of zoisite and amphibole (B and C in Fig. 11a) takes place and the crusts of warm slabs, such as those beneath Cascadia, Tokai, and Kii, undergo this dehydration reaction.

The above two dehydration reaction boundaries produce different H_2O production rates, as shown in Fig. 11b, c. These figures show a gradual release of H_2O in the depth range of c. 20–60 km for the warm slab beneath Kii, but a considerably abrupt release in the depth range of c. 80–95 km for the cold slab beneath Hokkaido. The production rate of H_2O in cold slabs is an order of magnitude higher than that for warm slabs (Fig. 11c). Moreover, because of the positive Clapeyron slope, the overall volume change from the dehydration reaction is positive for the cold slab (Fig. 11d), which causes effective increase in pore fluid pressure and thus favors overpressuring. The large H_2O production rate and the increase in total volume through the breakdown of blueschist are likely to promote earthquake occurrence in cold slabs. In contrast, the slow H_2O production rate and decrease in total volume through the breakdown of zoisite and amphibole probably inhibit, or at least do not promote, earthquake occurrence in the crusts of warm slabs. Abers et al. (2013) showed that the boundary between these two cases can be expressed by the thermal parameter (Kirby, 1995; (plate age) × (vertical descent rate of plate)) of c. 500–1000 km.

The results given by Abers et al. (2013) predict that the formation of the upper-plane seismic belt in the slab crust does not occur only in the subducting Pacific slab beneath eastern Japan but occurs also commonly in cold subduction zones with thermal parameters larger than c. 500–1000 km. In contrast, in the crust of warm slabs with thermal parameters less than c. 500–1000 km, earthquakes occur preferentially within the slab mantle, and therefore, the upper-plane seismic belt does not form. Dense local seismic network data and intense seismic activity in the slab crust are required to ascertain the presence of an upper-plane seismic belt. Few such subduction zones exist, and consequently, clear evidence of the presence of upper-plane seismic belts in other subduction zones has not yet been reported.

The nationwide dense seismic network of Japan also covers the SW Japan subduction zone where the Philippine Sea slab is subducting beneath it. The age of the subducting Philippine Sea Plate at the trench axis varies in the along-arc direction. It is as young as 15–25 Ma between western Shikoku and Tokai, whereas it is older in Kanto (>40 Ma) and Kyushu (> 50 Ma) (Seno and Maruyama, 1984). Therefore, it may be possible to ascertain the presence or absence of the upper-plane seismic belt. Figure 12 shows the hypocenter distribution of events in the crust of the Philippine Sea slab. This figure shows that events in the slab crust under the region from western Shikoku to Tokai are distributed to a maximum depth of c. 40, except for some outliers. In contrast, the down-dip end of the seismicity in the slab crust is deeper under regions where older lithosphere of the Philippine Sea Plate is subducting; it is as deep as 80–90 km in Kanto and is c. 100 km deep in Kyushu. These findings are consistent with the expectation that the colder Philippine Sea slab under Kanto and Kyushu is able to generate earthquakes at greater depths.

Another prominent feature of Fig. 12 is a belt-like concentration of seismicity, as indicated by the arrow, that is nearly parallel to the c. 60 km depth contour of the upper surface of the slab under southern Kyushu (strictly speaking, under the region to the south of the NW extension of the Kyushu–Palau ridge (thick broken line)). This feature is similar to the upper-plane seismic belt detected in the crust of the Pacific slab under eastern Japan. Although a similar belt-like concentration of seismic activity seems to be detectable in the depth range of 40–50 km beneath Kanto and may reflect the same activity, it is not discussed further here because it is unclear. In contrast, such a significant belt-like concentrated earthquake zone is not apparent in the much larger area from western Shikoku to Tokai. This along-arc variation in crustal seismicity is also inferred to be the result of the difference in age (or temperature) of the Philippine Sea slab (Hasegawa et al. 2009). The thermal parameters calculated by Syracuse et al. (2010) indicate that the slab beneath southern Kyushu at 31.22°N, 130.57°E has a value of 1560 km, which means that it belongs to the cold slab group. In contrast, the slab beneath Nankai at 35.3°N, 137.3°E has a value of 450 km, which means it belongs to the warm slab group. Consequently, the belt-like concentrated seismicity in the crust of the older Philippine Sea slab beneath southern Kyushu, which occurs at depths approximately 20 km shallower than the belt of the Pacific slab beneath Hokkaido and Tohoku, is probably caused by the same dehydration reactions of slab crust materials. The absence of such a clear belt-like concentrated seismicity in the crust of the younger portion of the Philippine Sea slab beneath the area from Tokai to western Shikoku (or northern Kyushu) is attributed to the warmer conditions there.

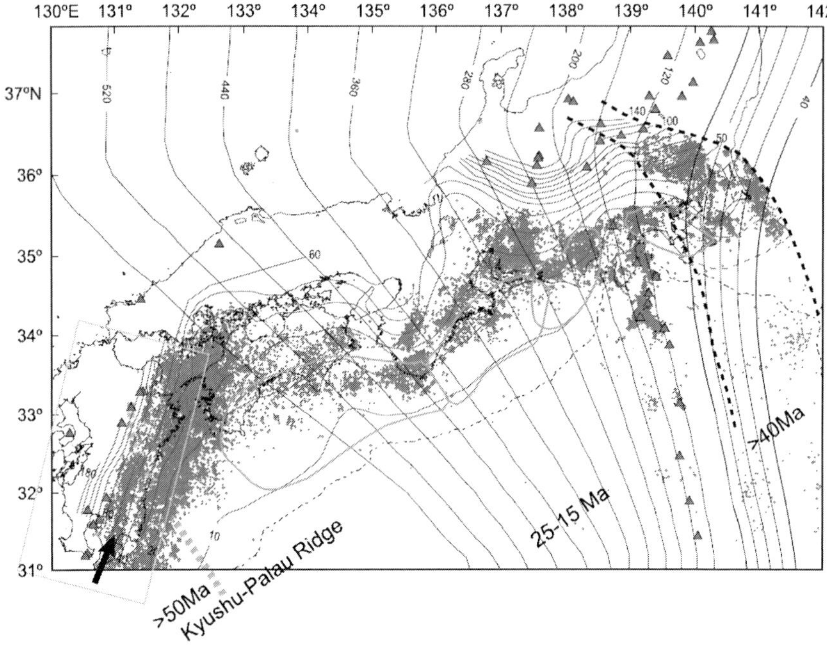

Figure. 12 Hypocenter distribution of earthquakes in the crust of the Philippine Sea plate under western Japan. Events 0–10 km below the upper plate surface are shown as *red dots. Bold arrows* denote estimated locations of the upper-plane seismic belt. The contact zone between the Philippine Sea and Pacific plates beneath Kanto is indicated by *thick broken curves. Solid curves, red triangles,* and *orange curves* denote iso-depth contours of the upper plate surface, active volcanoes, and source areas of large interplate earthquakes, respectively. *Thick, broken gray line* shows the Kyushu–Palau Ridge and its northeastern prolongation on top of the Philippine Sea slab. The northward bend near the northern end appears because of the projection of the slab's top surface with an increasing subduction angle relative to Earth's surface. (After Hasegawa et al. 2010)

As shown in Fig. 12, the estimated upper-plane seismic belt in southern Kyushu abruptly disappears north of the NW prolongation of the Kyushu–Palau ridge (thick, broken gray line in Fig. 12). Figure 13 illustrates an along-arc vertical cross-section of earthquakes that occurred within the rectangle shown in Fig. 12. This figure also shows that the upper-plane seismic belt, which is visible at depths of c. 50–70 km, and abruptly disappears to the north of the NW prolongation of the Kyushu–Palau ridge, as shown by the thick, broken red line. The sudden disappearance of the upper-plane seismic belt strongly suggests that the formation of the belt is controlled by slab age or temperature, as shown in Fig. 11, because the slab age abruptly changes between the north and south of the Kyushu–Palau ridge.

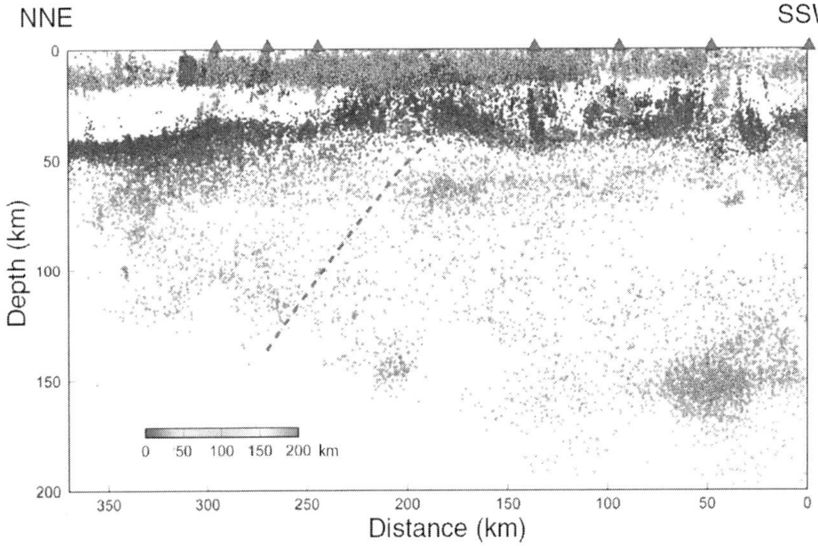

Figure. 13 Along-arc vertical cross-section of earthquakes beneath Kyushu. Earthquakes that occurred in the rectangular region shown in Fig. 12 are shown as *circles.* The distance of each event from the east-southeastern edge of the rectangular region is shown as the *color scale.* The estimated northeastern prolongation of the Kyushu–Palau ridge from the *top* of the slab is shown as the *thick, broken red line. Red triangles* at the *top* denote volcanoes

Studies of several subduction zones have revealed the presence of seismic low-velocity zones that correspond to the subducting crust, although the upper-plane seismic belt has not yet been detected at other sites (e.g., Fukao et al. 1983; Hori et al. 1985; Hori, 1990; Helffrich, 1996; Reyners et al. 1999; Abers, 2000, 2005; Bostok et al. 2002; Kodaira et al. 2000, 2004; Eberhart-Phillips et al. 2005; Hirose et al. 2008a, b; Syracuse et al. 2008; Rondenay et al. 2008, 2010; Abers et al. 2009, 2013; Eberhart-Phillips and Bannister, 2015). These seismic low-velocity layers in subducting crust have been interpreted as evidence for the presence of hydrated rocks or fluids liberated by dehydration reactions. Typical examples of these layers are shown in Figs. 14, 15, and 16.

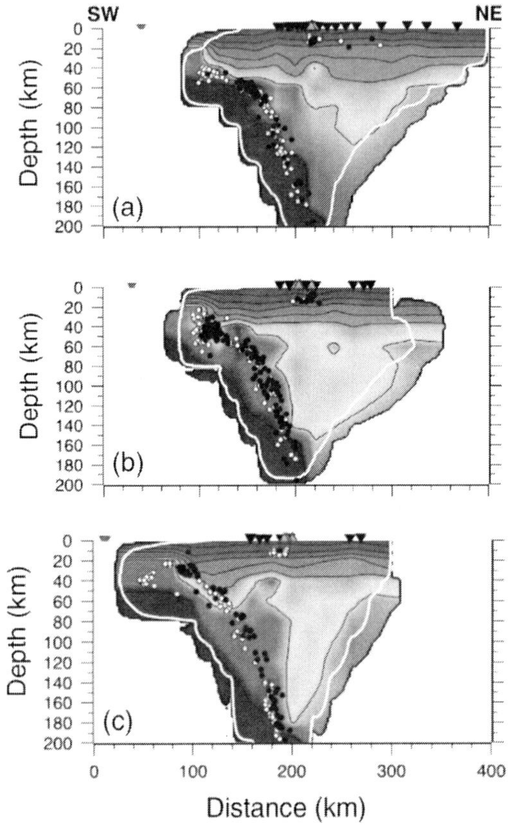
SW
NE
Depth (km)
(a)
(b)
(c)
Distance (km)
0
100
200
300
400

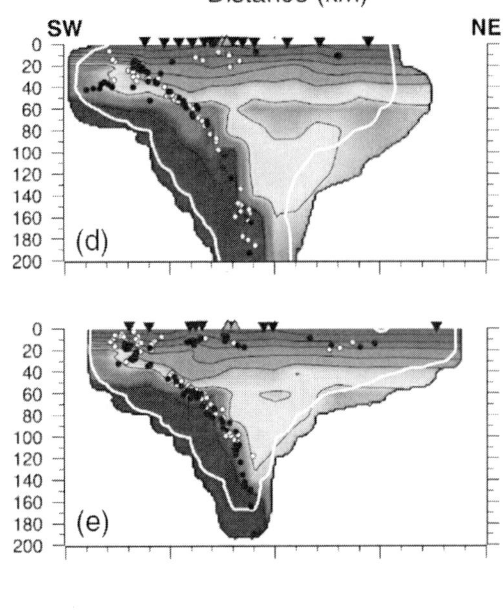
SW
NE
(d)
(e)

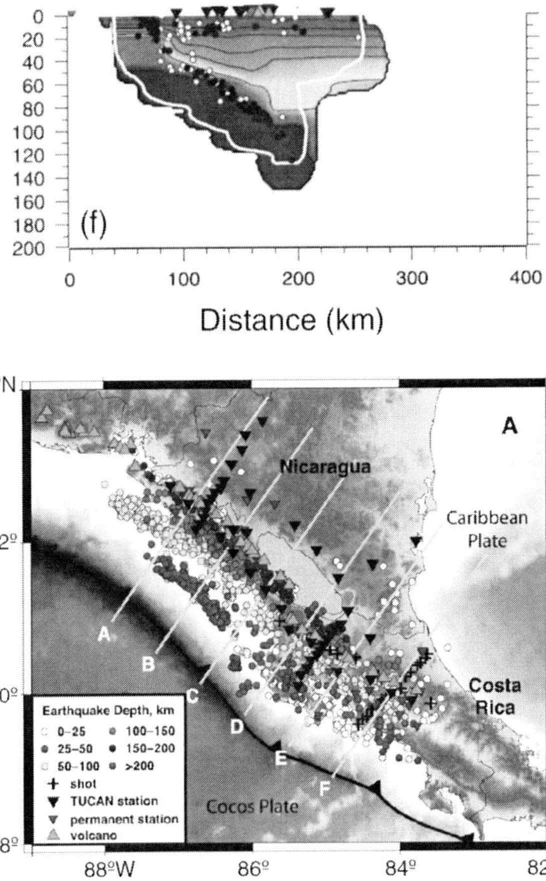

Figure. 14 P-wave velocity structure in Nicaragua and Costa Rica (Syracuse et al. 2008). P-wave velocity is shown on across-arc vertical cross-sections along *lines* a through f in the *inset map* by the *color scale* at the *bottom*. Regions surrounded by *white contours* contain 50 or more seismic rays. Earthquakes within 40 km to the northwest of each cross-section line are shown in *white* and those within 40 km to the southeast are in *black*. *Red inverted triangles* show the trench axis

Figure 14 shows across-arc vertical cross-sections of the P-wave velocity in the slab below Nicaragua and Costa Rica estimated based on inversions of P- and S-wave arrival times of local earthquakes (Syracuse et al. 2008). Nicaragua and Costa Rica lie on the Caribbean Plate, under which the young Cocos Plate is subducting northeastward at a rate of c. 88 mm/year. Inversion results show the presence of P-wave low-velocity zones in this area that probably correspond to the crust of the subducted Cocos Plate. This observation is particularly clear in Fig. 14c, which shows a P-wave low-velocity zone that extends as far as 70 km in depth. The down-dip end may be produced by the phase transformation of materials that compose the slab crust. Moreover, P-wave low-velocity zones were also imaged at the locations of the slab mantle c. 20–30 km below Wadati–Benioff zone seismicity, depths of c. 140 km beneath Nicaragua, and c. 60 km beneath Costa Rica. Syracuse et al. (2008) interpreted these low-velocity zones beneath Nicaragua as 10–20% serpentinized mantle of the subducted Cocos Plate, which may introduce relatively high amounts of H_2O into the mantle wedge beneath Nicaragua.

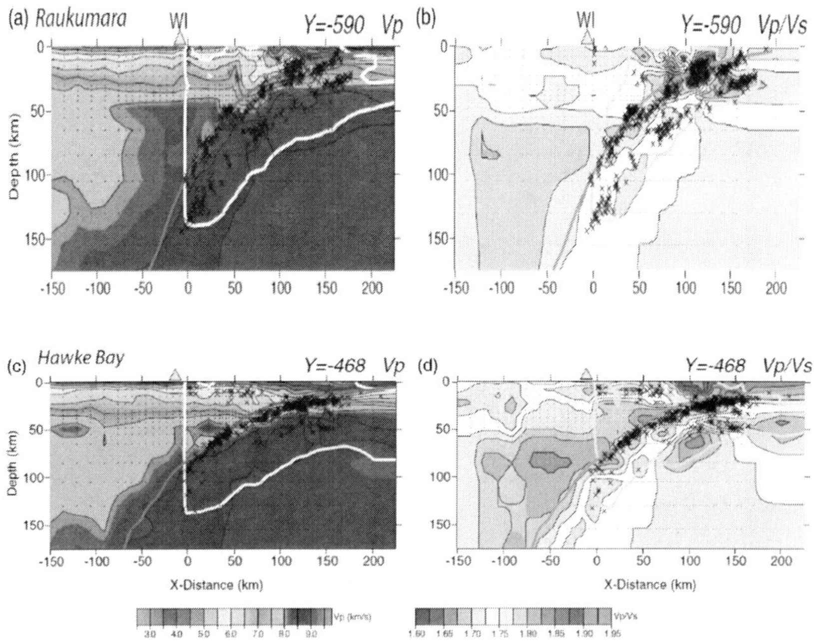

Figure. 15 P-wave velocity and Vp/Vs ratio structures in northern Hikurangi, New Zealand (Eberhart-Phillips and Bannister, 2015). P-wave velocity and Vp/Vs values are shown on across-arc vertical cross-sections by the *color scales* at the *bottom* in a and c and b and d, respectively. a–b Raukumara Peninsula and c–d Hawke Bay. *White* and *yellow lines* denote the limit of resolution for the spread function = 3.5. Earthquakes and volcanoes are shown as *crosses* and *triangles*, respectively. *Magenta* and *green lines* show the upper surface of the slab

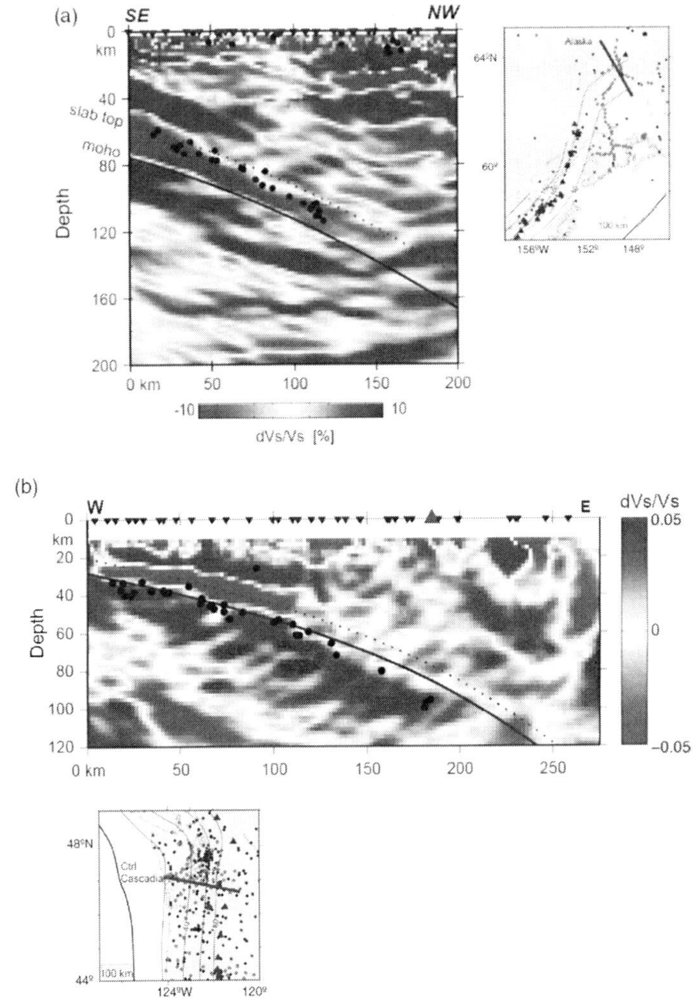

Figure. 16 S-wave velocity structures in central Alaska and central Cascadia. Across-arc seismic sections are shown along lines in the *inset maps* under a central Alaska and b central Cascadia (Abers et al. 2013, originally from Rondenay et al. 2008, 2010 and Abers et al. 2009). *Colored images* show migrated teleseismic P coda as dVs/Vs (S-wave velocity perturbation). dVs/Vs values are shown by the *color scale* at the *bottom*; *red* and *blue* correspond to slow and fast velocities, respectively. *Black circles* denote earthquakes. *Dotted* and *solid lines* indicate the upper slab surface and the slab Moho, respectively. *Inverted black triangles* show seismic stations used, and the *red triangle* in b shows a volcano

Figure 15 shows across-arc vertical cross-sections of P-wave velocity and Vp/Vs (the ratio of P-wave velocity to S-wave velocity) along the northern Hikurangi subduction zone of New Zealand. These data were obtained by inverting arrival-time data of local earthquakes (Eberhart-Phillips and Bannister, 2015). Onshore–offshore marine-seismic data were also incorporated in these inversions, which improved constraints on the shallow portion of the target area. In this area, the Pacific Plate is subducting obliquely under the Australian Plate, on which the North Island of New Zealand lies, at a rate of 42–47 mm/year. The obtained seismic velocity images shown in Fig. 15 clearly indicate the presence of a P-wave low-velocity zone and a high Vp/Vs zone to depths of 80–100 km associated with the oceanic crust of the subducted Pacific slab. This zone of low Vp and high Vp/Vs is probably caused by hydrated rocks or fluids liberated by dehydration reactions.

Figure 16a shows an across-arc seismic section beneath central Alaska obtained through seismic receiver-function analyses (Rondenay et al. 2008, 2010). In this region, the Pacific Plate has an age of 35–45 Ma at the trench axis and converges with the North American Plate at a rate of c. 55 mm/year, which produces a thermal structure of the slab that falls in the middle of the range globally (Kirby et al. 1996; Rondenay et al. 2008). Along the profile shown in Fig. 16a, the Yakutat terrane, which has a thickness from 15 to 30 km and can be described as an oceanic plateau, is subducting. This subduction was clearly imaged as a dipping low-velocity layer to a depth of c. 130 km. Intraslab earthquakes occur near the top of the subducting slab crust at depths shallower than c. 80 km and then gradually deepen toward the oceanic Moho with increasing depth. Both the intraslab seismicity and the low-velocity slab crust terminate at c. 120–130 km in depth, which suggests that the eclogite-forming phase transition operates in this depth range (Abers et al. 2013).

Figure 16b shows an across-arc seismic migration image below central Cascadia obtained via seismic receiver-function analyses (Abers et al. 2009). In this region, the Juan de Fuca Plate, with an age of 6–9 Ma at the trench axis, is subducting obliquely at a rate of 35–45 mm/year. This slow subduction rate in the vertical direction and the young age of the plate make it one of the hottest slabs in the world (Wada and Wang, 2009; Syracuse et al. 2010). The subducting slab crust was clearly imaged as a dipping low-velocity layer with a thickness of 7–9 km to a depth of c. 40 km. Almost all intraslab earthquakes in this region are located at or just below the slab Moho, and very little seismicity is produced within the subducting crust. The shallow termination depths of the low-velocity slab crust and the very low level of seismicity in the slab crust are consistent with the high-temperature conditions of this slab (Abers et al. 2013). Based on estimations of the thermal structure of the slab and on the phase diagram of crust materials, Rondenay et al. (2008) compared two subduction zones, central Alaska and central Cascadia, and claimed that the termination depths of the low-velocity subducting crust correspond to eclogitization of the metamorphosed crust through dehydration of hydrous minerals, which depends largely on the thermal structure of the subducting slab crust. Overall, seismicity in the slab crust appears to be closely related to the dehydration reactions, although the upper-plane seismic belt is not apparent even in the case of the cold slab described here, which can most likely be attributed to the very small number of earthquake data.

All of these observations seem to support the dehydration-related embrittlement hypothesis as the mechanism behind the genesis of intermediate-depth events within the slab crust.

3.3 Seismic wave velocity structure and lower-plane seismicity in the slab mantle: seismic low-velocity zones and serpentinized slab mantle

Until recently, seismic imaging of the subducting slab was performed not to image detailed, inhomogeneous structures inside the slab, but to determine the shape of the subducting slab as a seismic high-velocity zone separate from the surrounding mantle with normal seismic wave velocity. Very high spatial resolution is required to detect the inhomogeneous structure inside the slab, and in the past, this resolution could not be achieved. A high-density seismic network is required for such high-resolution seismic imaging, and recently, this resolution has become possible through dense observation network data. The nationwide dense seismic network has been deployed in Japan, a typical subduction zone with a very high seismic activity, and this development has enabled us to image even the internal structure of the subducted slab below Japan. The seismic low-velocity layer that persists to the depth of the upper-plane seismic belt in the slab crust, as described in the "The double seismic zone and dehydration of metamorphosed crust and 149 serpentinized mantle" section, is one example of such internal structure. Recent seismic tomography studies based on nationwide dense network data have also revealed the detailed structure of seismic velocity inside even the slab mantle.

If hydrous minerals or the H_2O released through dehydration reactions are present in some regions of the slab mantle, those regions are expected to have low seismic wave velocities. Precisely, imaging of heterogeneous structures inside the slab mantle with seismic tomography can be challenging, however. The slab mantle exists at greater depths than the slab crust, and consequently, seismic rays arriving at seismic stations above earthquake hypocenters do not sufficiently crisscross the slab mantle, which results in poor resolution in seismic velocity imaging. Therefore, research into the heterogeneous seismic velocity structure of the slab mantle remains very limited.

If many densely distributed events occur in a particular area, that area can be imaged very precisely via seismic tomography because a large number of ray paths will crisscross the area. The occurrence of numerous aftershocks associated with a large intraslab earthquake provides such a rare opportunity. Based on the tomography method of Zhao et al. (1992, 1993), Mishra and Zhao (2004) determined the detailed seismic velocity structure in and around the hypocenter of the 2003 M 7.1 Miyagi–Oki earthquake, which occurred at a depth of c. 70 km in the Pacific slab near the oceanic Moho beneath NE Japan. In their tomographic inversions, they used many data from aftershocks in addition to the background earthquake data, which enabled precise imaging of the source area within the slab. P-wave velocity structures revealed through this method are shown in Fig. 17 in N–S and E–W vertical cross-sections that pass through the mainshock hypocenter. This figure shows an anomalously low-velocity zone across the crust and mantle in the Pacific slab and that the source area of the M 7.1 intermediate-depth intraslab event is located in this zone characterized by a velocity anomaly lower than its surroundings in the slab. Mishra and Zhao (2004) concluded that this low-velocity anomaly and the occurrence of earthquakes within it could be attributed to the process of fluid-related embrittlement caused by the dehydration of hydrous minerals in the slab.

Nakajima et al. (2011) examined the seismic wave velocity structure of the source area of the 2011 M 7.1 Miyagi–Oki earthquake that took place at a depth of c. 65 km in the slab mantle about 1 month after the 2011 Mw 9.0 Tohoku–Oki earthquake. As shown in Fig. 18, the seismic wave velocity structure obtained by using numerous aftershock data in conjunction with the background seismicity data revealed the existence of a low-velocity zone around the fault plane of this M 7.1 intermediate-depth intraslab event. The lateral extent of the low-velocity zone is comparable to the extent of the fault plane of the M 7.1 event. The angle between the mainshock fault plane and the upper slab surface is c. 60°, which is consistent with the dip of the oceanward-dipping normal faults in the outer rise region. Based on these observations, Nakajima et al. (2011) suggested that the M 7.1 intraslab earthquake was generated by reactivation of a buried hydrated fault that had formed in the plate prior to its subduction.

Zhang et al. (2004) were the first to determine the detailed internal structure of the slab mantle to the depth of the lower seismic plane of the double seismic zone. Their tomographic imaging of the slab beneath central Tohoku with DD tomography revealed the existence of a P-wave low-velocity layer along the lower seismic plane, as shown in Fig. 19. The similar P-wave low-velocity layer along the lower seismic plane of the slab mantle was identified in subsequent seismic tomography studies by Shelly et al. (2006) and Nakajima et al. (2009c) under southern Tohoku and eastern Hokkaido, respectively. Figure 20 shows across-arc vertical cross-sections of P- and S-wave velocities under eastern Hokkaido as determined by Nakajima et al. (2009c). A clear P-wave low-velocity layer is visible at the lower plane of the double seismic zone.

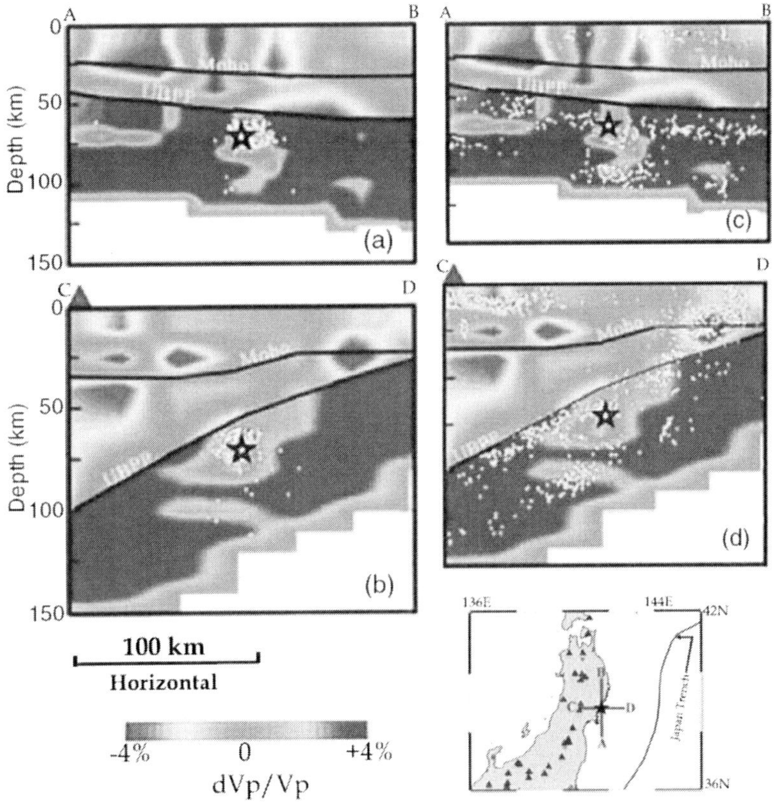

Figure. 17 P-wave velocity structure in and around the source area of the 2003 M 7.3 Miyagi–Oki intraslab earthquake. P-wave velocity perturbations are plotted based on the *color scale* at the *bottom* for a, c N–S and b, d E–W vertical cross-sections along *lines* shown in the *inset map. White dots* in a and b are aftershocks of the M 7.3 event, and those in c and d are background seismicity. Hypocenter of the M 7.3 event is indicated by a *star. Thick curves* are the Moho of the upper plate and the top surface of the subducting Pacific slab, in order from the *top*. (After Mishra and Zhao, 2004)

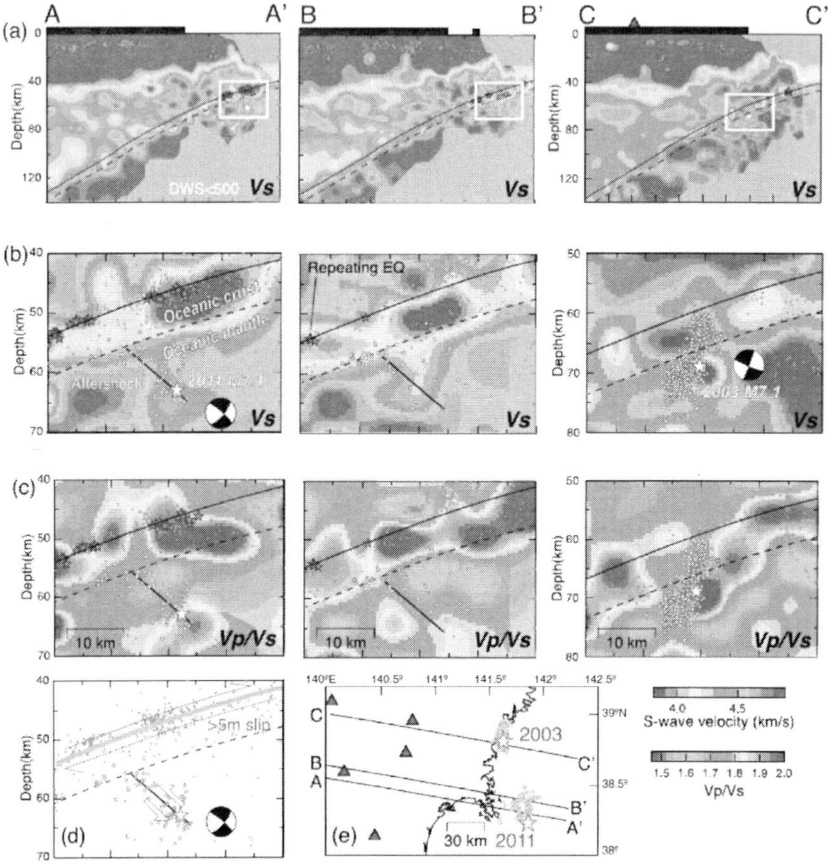

Figure. 18 Seismic velocity structure in and around the source area of the 2011 M 7.1 Miyagi–Oki intraslab earthquake. a Across-arc vertical cross-sections of S-wave velocity under central Tohoku along *three lines*: A–A', B–B', and C–C' in the *inset map e. White stars* in *line* A–A' and *line* C–C' denote the hypocenter of the 2011 M 7.1 event and the 2003 M 7.1 event, respectively. *Solid* and *dashed black curves* denote the upper surface of the Pacific slab and the Moho within it, respectively. *White rectangle* in a shows the area enlarged in b and c. b Across-arc vertical cross-sections of S-wave velocity and c Vp/Vs within the area outlined by the *white rectangle* in a. *Black lines* in A–A' and in B–B' show the fault plane of the 2011 M 7.1 event. d Schematic figure showing fault slips of the 2011 M 9.0 Tohoku–Oki earthquake (*purple arrows*) and the M 7.1 intraslab event (*white arrows*). e Map showing the aftershock distribution of the 2011 and 2003 intraslab events and the locations of the three lines, A–A', B–B', and C–C'. (After Nakajima et al. 2011)

However, both Figs. 19 and 20 indicate that S-wave velocity is not necessarily low for that layer. Therefore, the Vp/Vs ratio for this layer is not necessarily as high as is expected for the serpentinized mantle, but may be fairly low, as shown in Fig. 19c.

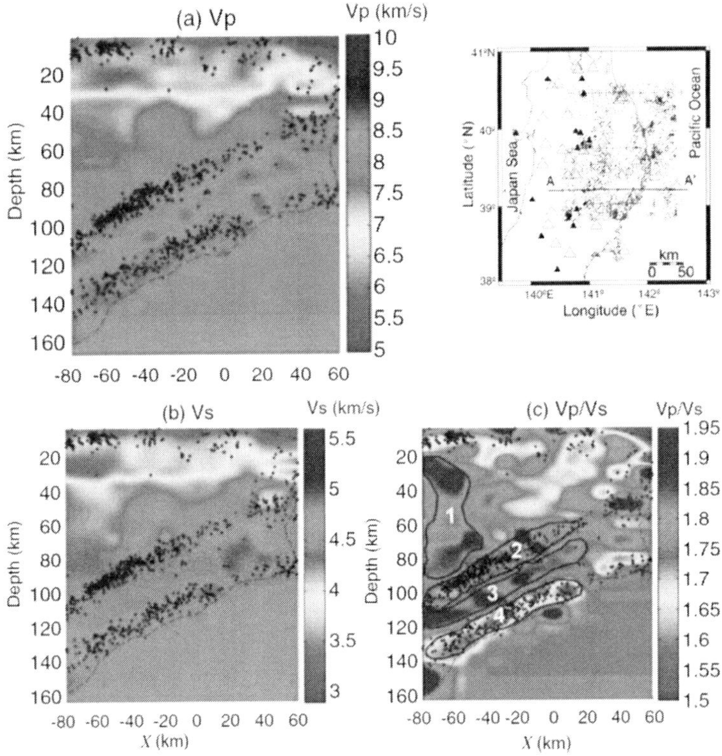

Figure. 19 Seismic velocity structure along the lower seismic plane under central Tohoku. Shown are across-arc vertical cross-sections of a P-wave velocity (Vp), b S-wave velocity (Vs), and c Vp/Vs ratio under central Tohoku. P-wave velocity, S-wave velocity, and Vp/Vs ratio along the line in the *inset map* are shown by the *color scale* at the *right* of each figure. *Black dots* are earthquakes. (After Zhang et al. 2004)

Reynard et al. (2010) investigated the cause of the low Vp and Vp/Vs values imaged via seismic tomography for this layer of the mid-slab mantle below eastern Hokkaido (Fig. 20). They claimed that this anomalous layer is the result of the anisotropy of anhydrous peridotite, which suggests that the same mineralogical composition occurs along the lower-plane seismicity as in the surrounding slab mantle.

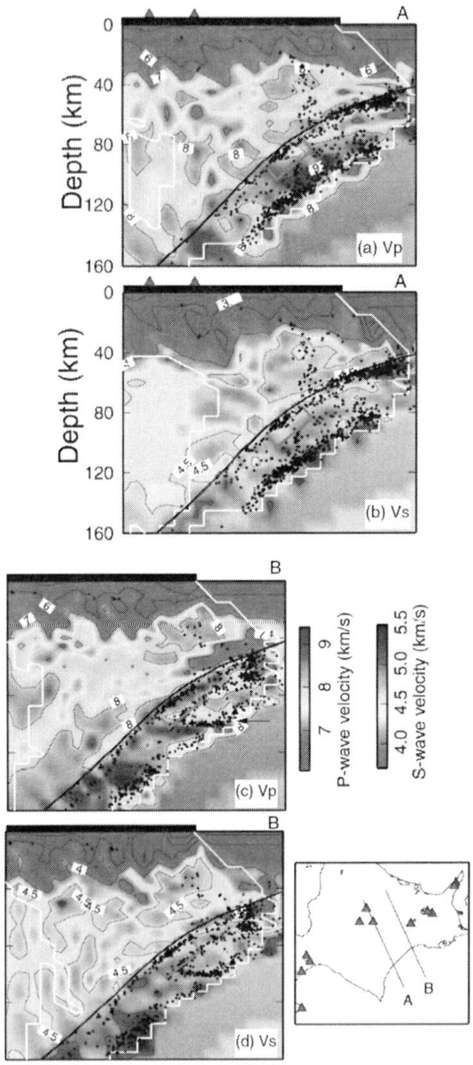

Figure. 20 Seismic velocity structure along the lower seismic plane under eastern Hokkaido. Shown are across-arc vertical cross-sections of a, c P-wave velocity and b, d S-wave velocity under eastern Hokkaido. P- and S-wave velocities along *two lines* in the *inset map* are shown by the *color scales* at the *right*. Aftershock activity of the 1993 M 7.5 Kushiro–Oki intraslab earthquake is shown by an *arrow* in c. *Solid lines* denote the upper surface of the subducting slab. *Solid lines* and *red triangles* at the *top* show the land area and active volcanoes, respectively. (After Nakajima et al. 2009c)

Based on this interpretation, they further argued that the shear heating instability of plastic deformation (Griggs and Handin,1960; Ogawa,1987; Hobbs and Ord, 1988) would therefore be the mechanism behind the lower-plane earthquakes.

Conventionally, the thermal shear instability model was thought to be unable to explain the observation that the lower plane of the double seismic zone formed along the isotherm of approximately 600–700 C in the slab mantle (see Figs. 5 and 6; Yamasaki and Seno,2003;. Brudzinski et al.2007). However, a recent investigation by Kelemen and Hirth (2007) based on numerical modeling of a fine-grained viscous shear zone in a coarse-grained elastic half-space revealed that the shear heating instability of peridotite occurs effectively within a narrow temperature range of 650–800 °C. If this is the case, the shear heating instability model of plastic deformation can also explain the observations that the lower-plane events occur along the isotherm of approximately 600–700 °C.

However, if we inspect the seismic tomography images (Fig.20) of Nakajima et al. (2009c) in detail, regions with P-wave velocity below 7 km/s are distributed entirely along the lower plane of the double seismic zone. It is difficult to explain the distribution of this low P-wave velocity anomaly along the lower seismic plane, but not in the surrounding mantle, based on the anisotropy of anhydrous peridotite alone. The low Vp values over the entire lower seismic plane imaged with seismic tomography suggest that materials with lower Vp were more abundant there than in the surrounding slab mantle, although the anisotropy effect of those materials may contribute in part to the lower Vp and Vp/Vs values detected via seismic tomography. The low Vp and Vp/Vs layer observed along the lower seismic plane is still not fully explained, and further examination and verification are required to determine the cause. Moreover, seismicity in the slab mantle is limited not only in the upper and lower planes of the double seismic zone but also between these planes, although the amount of seismic activity is lower (Kita et al. 2010b). Such interplane events are also shown in Fig. 20. In particular, the aftershock activity associated with the 1993 M 7.5 Kushiro–Oki earthquake at a depth of c. 100 km was distributed along a distinct P-wave low-velocity layer that extended horizontally from the lower seismic plane to the vicinity of the upper seismic plane (black arrow in Fig. 20c). This low P-wave velocity is also difficult to explain through the mechanism of peridotite anisotropy alone. The presence of interplane events (Kita et al. 2010b) indicates that intermediate-depth intraslab events occur even at temperatures much lower than 650–800 °C, which further contradicts predictions based on the estimate of Kelemen and Hirth (2007).

Similar P-wave low-velocity layers along the lower plane of the double seismic zone were detected beneath northern Chile using DD tomography (Dorbath et al. 2008), although these results were less clear than those for beneath Tohoku and Hokkaido (Zhang et al. 2004; Shelly et al. 2006; Nakajima et al. 2009c). Across-arc vertical cross-sections of the estimated P- and S-wave velocities are shown in Fig. 21. This figure shows the presence of P-wave low-velocity zones along the lower seismic plane in all three cross-sections. As in the Pacific slab below eastern Japan, S-wave velocities are not low, but normal along the lower plane, which suggests that the low P-wave velocity and normal S-wave velocity along the lower seismic plane is a common feature across subduction zones. Dorbath et al. (2008) inferred that these P-wave low-velocity zones form along the slab's flexural neutral plane where fluids released through dehydration of the surrounding hydrated minerals accumulate. Therefore, the P-wave velocities are low and trigger lower-plane seismicity.

For lower-plane events to be triggered by the aqueous fluids liberated from hydrous minerals via dehydration reactions, at least part of the slab mantle near the lower plane must be hydrated. However, the lower seismic plane is usually located at depths of several tens of kilometers, to as deep as c. 40 km, below the upper slab surface. In contrast, hydrothermal circulation and alteration at mid-ocean ridges are limited in many cases to shallow crustal levels. Peacock (2001) suggested that normal faulting in the trench–outer rise regions promotes infiltration of seawater several tens of kilometers into the oceanic plate, which facilitates the serpentinization of affected mantle materials. However, this hypothesis of the deep hydration of the oceanic mantle before the plate subduction has received criticism, and it is still controversial whether seawater infiltration extends to such great depths through bending-related normal faulting.

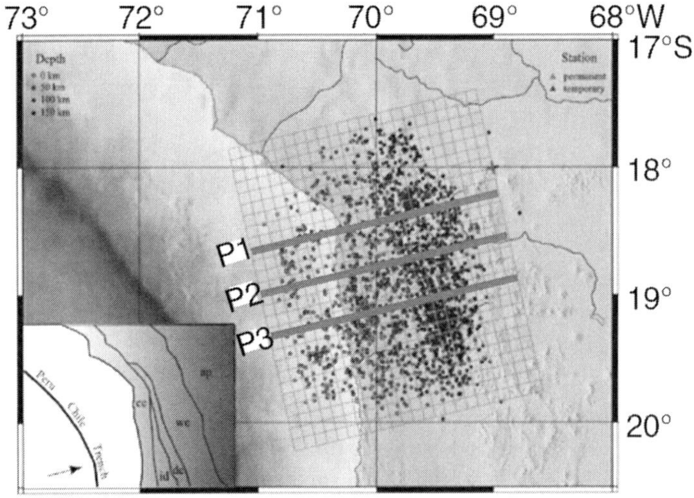

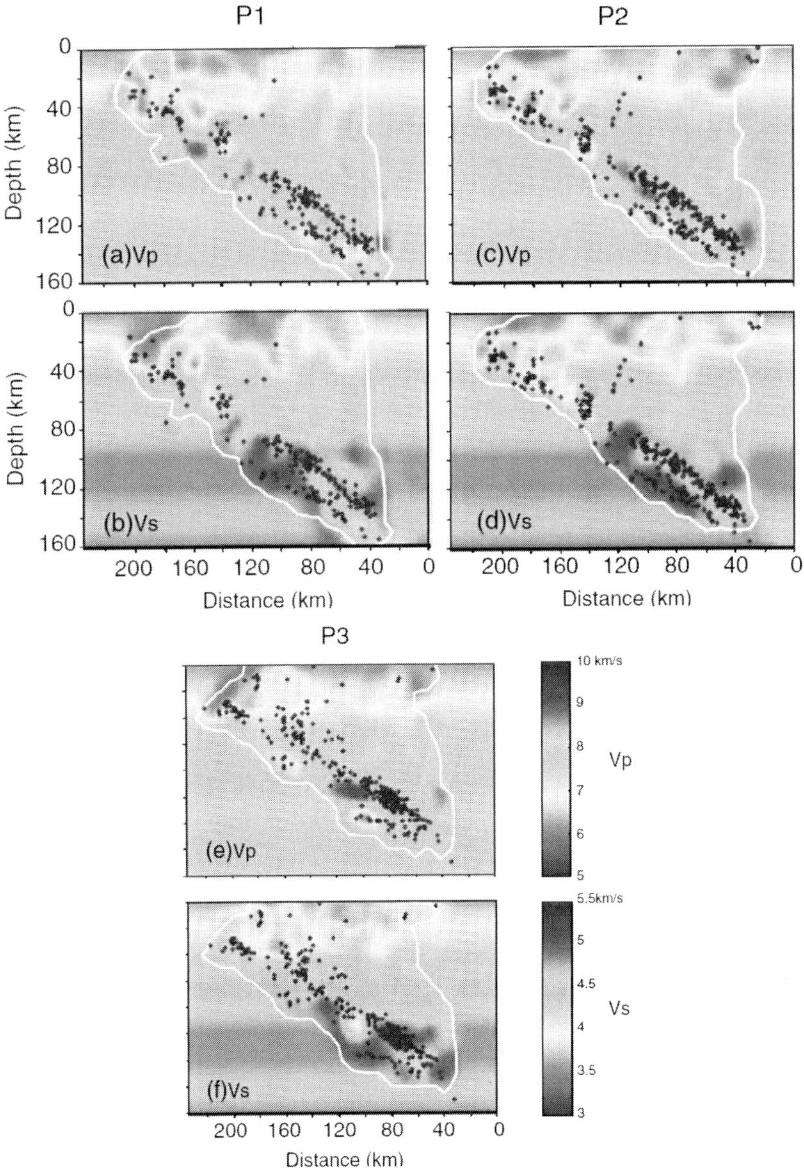

Figure. 21 Seismic velocity structure along the lower seismic plane under northern Chile. Shown are across-arc vertical cross-sections of a–c P-wave velocity and d–f S-wave velocity beneath northern Chile. P- and S-wave velocities along *three lines* in the *inset map* are indicated by the *color scales* to the right. Earthquakes within a 20 km-wide section are plotted as *dots. White lines* denote the limit of resolution where the derivative weight sum is more than 10. (After Dorbath et al. 2008)

Seismic reflection/refraction surveys conducted at trench–outer rise regions have revealed the presence of such normal faults that cut across the crust and penetrate into the mantle of the incoming oceanic plate (e.g., Ranero et al.2003; Ranero and Sallares,2004). Seismic reflection/refraction profiling experiments have provided further evidence of alteration of the crust and mantle of the incoming plate: anomalously low seismic wave velocities in the crust and mantle (Ivandic et al. 2008; Grevemeyer et al.2007; Contreras-Reyes et al.2007,2011; Fujie et al.2013). Seismic surveys have thus detected traces of the altered crust and mantle of the incoming oceanic plate, but these surveys are restricted to the uppermost mantle levels, with a maximum depth of c. 20 km. However, this limitation is mainly related to poor resolution at such great depths in the mid-mantle of oceanic plates associated with seismic surveys. In fact, a recent seismic reflection survey along a 233 km-long profile beneath the Wharton Basin in the Indian Ocean clearly imaged a fault that extends down to 40 km depth below the plate upper surface (Qin and Singh,2015). The negative polarity of the reflection indicates the presence of a low P-wave velocity zone along the fault, likely formed by serpentinization of mantle materials caused by infiltration of seawater. Although it is unclear whether the fault was formed by the bending stresses of the trench–outer rise region, this observation suggests the possibility of deep hydration of the oceanic mantle prior to plate subduction. Moreover, an estimation by Garth and Rietbrock (2014) based on waveform modeling of P-wave coda from intermediate-depth events, whereby a 40 km-thick layer is inferred to be 17–31% serpentinized, indicates bulk hydration of the oceanic mantle.

A possible alternative mechanism for hydration of the deep mantle of an incoming oceanic plate is the hypothesis proposed by Seno and Yamanaka (1996). These workers discussed the possibility of hydration of the middle-to-deep portion of the oceanic plate through H_2O released from solidified magma that was involved in plumes. Most of the magma solidifies and releases H_2O at the plume head immediately below the oceanic plate; this H_2O then accretes to the plate as the plate cools and thickens. In addition, magma injected into the oceanic plate releases H_2O as it solidifies. H_2O thus released reacts with the surrounding mantle materials, which results in the hydration of the mid-mantle. There are arguments against this hypothesis, as well; for example, Omori et al. (2009) argued that extensive hydration of the lithospheric mantle would not result from such a mechanism because the amount of H_2O released decreases at greater depths due to depth-based variation in the solubility of H_2O in the melt, and because magma reservoirs are present at shallow depths in the crust rather than such great depths in the mantle.

A dynamic thermomechanical–petrological modeling study of plate subduction by Faccenda et al. (2012) has presented another possibility for the penetration of H_2O into the deep slab mantle. The results of their simulation show that during slab subduction, fluids liberated through dehydration of hydrous minerals follow a bimodal migration pattern. Fluids released from the slab crust and the sub-Moho mantle flow upward, whereas those released from deeper levels of the mantle are driven into the cold core of the slab because of the pressure gradients generated by plate unbending stresses. Fluids thus driven downward into the cold core of the slab accumulate progressively and percolate updip along a layer nearly parallel to the slab surface (Fig. 22). The predicted location of this layer of accumulated fluids coincides with the lower plane of the double seismic zone. This lower layer moves deeper with increasing plate age, and the separation distance between this lower layer and the middle of the slab crust is roughly consistent with the separation of double seismic zones observed globally by Brudzinski et al. (2007), as shown in Fig. 23. Both free fluids and serpentinized peridotite are present in this lower layer in the middle of the slab mantle because the mantle rocks there are saturated with bound H_2O to the assumed upper limit of serpentinization. If this assumption is accurate and the slab dip is not sufficiently shallow, extremely deep hydration of the slab mantle prior to subduction is unnecessary to explain the lower-plane seismicity. Moreover, if the free fluids are intensively distributed along the lower seismic plane as demonstrated here and if the pores that hold these fluids have small aspect ratios, the lower seismic plane should have low Vp and low Vp/Vs values (e.g., Nakajima et al. 2001; Takei, 2002). As suggested by Dorbath et al. (2008), this possibility may partially explain the low Vp and low Vp/Vs along the lower plane, as is commonly observed in Tohoku, Hokkaido, and northern Chile (Figs. 19, 20, and 21).

The possibility of deep hydration of the slab mantle prior to subduction was also demonstrated by a recent 2-D reactive-flow model simulation of hydration processes at the trench–outer rise region (Iyer et al.2012). Iyer et al. (2012) developed a reaction-transport model that enables better quantification of the amount and depth at which hydration occurs in an incoming oceanic plate due to normal faulting caused by the bending force. The hydration patterns determined with this model show the formation of a highly serpentinized band in the slab mantle centered at the 270 °C isotherm. The depth of the 270 °C isotherm increases with increasing plate age and is roughly consistent with the separation distance of globally observed double seismic zones (Fig. 6; Brudzinski et al. 2007). If this mechanism actually operates in the mantle, it could explain the formation of double seismic zones. Iyer et al. (2012) suggested that the double hydrated zone at the depths of the slab mantle, through slab unbending (Fig. 22; Faccenda et al. 2012) and maximum serpentinization at c. 270 °C, may together cause the variations in the observed separation of the double seismic zone.

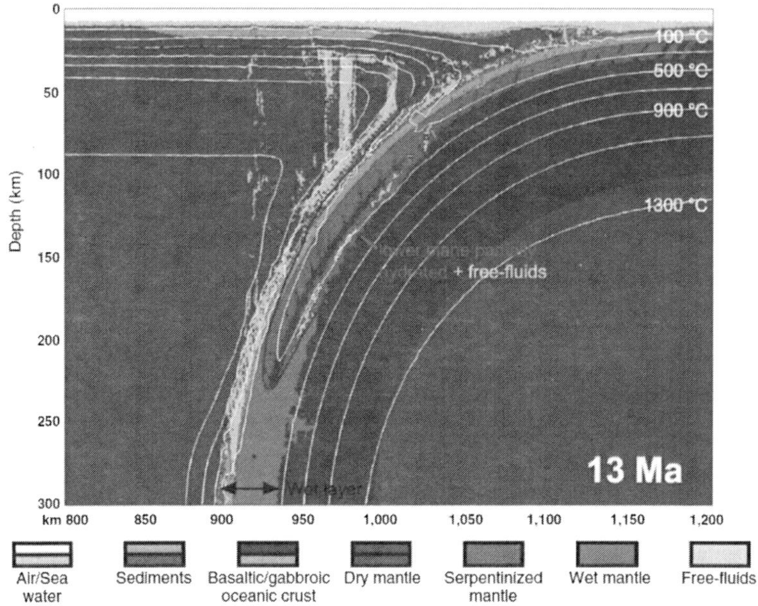

Figure. 22 Simulated hydration and dehydration of the subducting slab. Shown is an example of simulation results of a dynamic thermomechanical–petrological model that shows the hydration and dehydration of the slab caused by its subduction and bending/unbending stresses. A snapshot of the hydration and dehydration at 13 Ma after subduction is shown on an across-arc vertical cross-section of the case in which an oceanic plate with an age of 52 Ma is subducting. *White lines* are isotherms. (After Faccenda et al.2012)

Recently, Korenaga (2017) demonstrated another possible mechanism for deep hydration of the slab mantle prior to subduction. He proposed that mantle hydration first occurs through thermal cracking associated with the cooling of oceanic plates and is later enhanced by bending-related faults in the outer rise region. The amount of fluids required to generate the lower plane of the double seismic zones is not well quantified, but Korenaga estimated the lower bound of the amount of water contained in the slab mantle (0.03–0.07 wt%) and argued that this amount is sufficient to explain the lower-plane seismicity of double seismic zones.

Overall, observations of distinct low P-wave velocities along the lower plane of the double seismic zones under Tohoku, Hokkaido, and northern Chile, as shown in Figs.19,20, and21, strongly suggest the presence of hydrous minerals or fluids in the middle portion of the slab mantle, although further investigation is necessary. We expect this research to progress in the future.

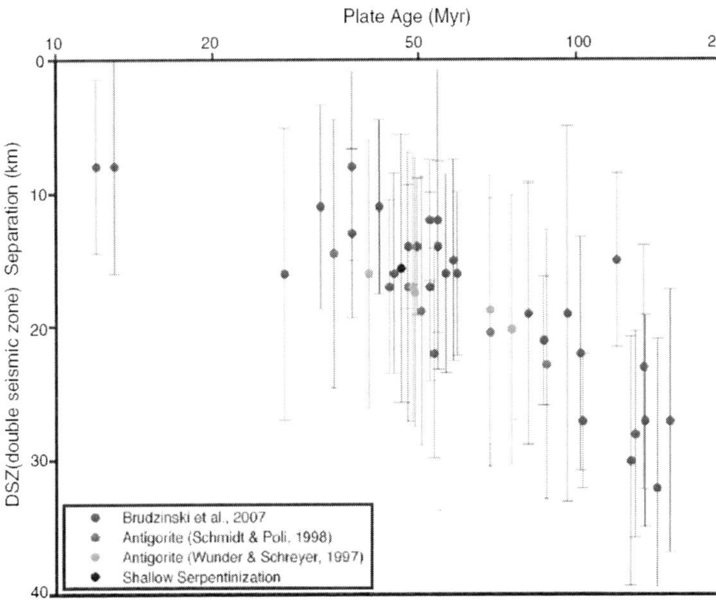

Figure. 23 Separation of the double seismic zone plotted against plate age. The separation of the double seismic zone obtained from a dynamic thermomechanical–petrological model is plotted against the age of the subducting slab. The distances from the middle of the oceanic crust to the level where fluids concentrate are shown by *red* and *green dots. Vertical lines* show *error bars. Red* and *green dots* are based on models of antigorite dehydration by Wunder and Schreyer (1997) and by Schmidt and Poli (1998), respectively. *Blue dots* are globally observed data reported by Brudzinski et al. (2007). (After Faccenda et al.2012)

4. GENESIS OF INTERMEDIATE-DEPTH INTRASLAB EARTHQUAKES

As described in the previous sections, spatial distribution of intraslab seismicity and the internal seismic structures of subducting slabs provide evidence that strongly supports the dehydration-related embrittlement hypothesis for nucleating intermediate-depth intraslab events. However, much of this evidence is circumstantial. The dehydration-related embrittlement hypothesis posits that H_2O expelled from hydrous minerals within the subducting slab generates earthquakes, but the details of the process that actually lead from a state of local excess pore fluid pressure to earthquake rupture are still subject to debate (e.g., Raleigh and Paterson,1965; Rutter and Brodie,1988; Ko et al.1997; Dobson et al.2002; Jung and Green,2004; Jung et al.2004,2006,2009; Kelemen and Hirth, 2007; Chernak and Hirth,2010,2011; Brantut and Sulem,2012; Sawai et al. 2013; Proctor and Hirth,2015; Auzende et al.2015).

Based on triaxial compression experiments on serpentinite samples at temperatures up to 700 °C and confining pressures up to 0.5 GPa in undrained conditions, Raleigh and Paterson (1965) reported a significant reduction in strength of the samples, and brittle fracturing occurred at temperatures higher than 375 °C, whereas plastic deformation occurred below 340 °C. They attributed the embrittlement and weakening over 375 °C to reduction in the effective normal stress caused by the release of the pore pressure of H_2O via serpentine dehydration. This mechanism of dehydration embrittlement would certainly apply at pressures below c. 0.5 GPa (corresponding to c. 15 km depth). However, it is not evident whether the same mechanism can be applied to intermediate-depth intraslab events that occur at much higher pressures. The effective stress law for sliding friction works at shallow depths, as seen in the experimental results of Raleigh and Paterson (1965), but this law does not work efficiently below the brittle–ductile transition depth (c. 15 km depth) (e.g., Hirth and Beeler, 2015). Another point, which is favorable to the dehydration-related embrittlement, in the experiments of Raleigh and Paterson (1965) is the positive Clapeyron slope of serpentine dehydration. The positive overall volume change associated with the dehydration in this case leads to an increase in pore pressure, which should operate to promote brittle fracture. However, at pressures higher than c. 2.2 GPa (which corresponds to c. 60 km depth), the Clapeyron slope of serpentine dehydration is negative (see Fig. 3b), and thus, the overall volume change that accompanies the dehydration becomes negative, which may suppress the occurrence of brittle instability.

In contrast, some laboratory experiments have shown that hydraulic embrittlement occurs even at this high-pressure range. For example, Dobson et al. (2002) detected strong acoustic emission signals during dehydration in serpentinite compression experiments that used a multianvil press at pressures up to 8.5 GPa and temperatures up to 900 °C. Recovered samples show characteristics of the brittle deformation associated with high pore fluid pressures, which suggests that dehydration embrittlement occurs even at pressures over 2.2 GPa. Based on triaxial compression tests of antigorite at pressures up to 6 GPa and temperatures up to 820 °C, Jung et al. (2004) found that antigorite dehydration results in faults delineated by ultrafine-grained solid reaction products independent of the sign of the Clapeyron slope. Jung et al. (2009) conducted in situ measurements of acoustic emissions under compression tests of serpentinite and showed that acoustic waves are generated by dehydration embrittlement even at pressures higher than 2.2 GPa. These experimental results suggest that dehydration embrittlement is a viable mechanism for generating earthquakes independent of depth.

On the contrary, the results of some recent laboratory experiments do not support the dehydration embrittlement hypothesis. For example, Chernak and Hirth (2011) performed deformation experiments on antigorite designed to maximize the possibility of dehydration instability. They showed that unstable fault behavior did not develop in the specimens, but rather that slow slip events took place. Compression experiments of antigorite-rich serpentinite by Gasc et al. (2011), who used a newly developed experimental setup that allows simultaneous collection of in situ X-ray diffraction data and recording of the full waveforms of acoustic emissions with a multi-anvil device, showed that the serpentinite samples did not produce detectable acoustic waves over the course of their dehydration. Proctor and Hirth (2015) conducted deformation tests with a new experimental method to vary the pore fluid pressure. Their results showed that dehydrating serpentinite causes deformation through semi-brittle flow with grain-scale ductile deformation, which suggests that earthquakes do not nucleate due to unstable frictional sliding.

Triaxial compression tests on antigorite conducted recently by Okazaki and Hirth (2016) using a Griggs-type deformation apparatus showed that antigorite samples are associated with slow slips during the dehydration reaction without producing acoustic waves. Their similar deformation tests on lawsonite, however, showed that unstable fault slip certainly takes place and that acoustic emissions are continuously generated in the dehydration of lawsonite, which suggests that lawsonite dehydration may be the cause of intermediate-depth events in the subducting slab crust. These results support the hypothesis that embrittlement through blueschist-facies dehydration reactions in the slab crust may lead to the formation of the belt of upper-plane seismicity observed in the crust of the cold slabs, such as that detected under eastern Japan.

As described above, the detailed process that leads to earthquake rupture is still under debate, and is left to future works. However, because intermediate-depth earthquakes have been found to occur at the dehydration boundaries within the subducting slab as described in the previous sections, a process analogous to hydraulic fracturing may be occurring due to local excess pore fluid pressure formed through dehydration and expulsion of H_2O from hydrous minerals. This phenomenon may cause intermediate-depth earthquakes within the slab.

Alternatively, intermediate-depth events may be caused by dual mechanisms, both fluid-related embrittlement and thermal shear instability, i.e., the local excess pore fluid pressure may initiate earthquake rupture via hydraulic fracturing, which may then trigger subsequent earthquake ruptures through thermal shear instability. Nevertheless, it is estimated that H_2O released by the dehydration of slab materials plays an important role in generating intermediate-depth intraslab earthquakes.

5. CONCLUSIONS

Intermediate-depth intraslab earthquakes occur within the existence range of hydrous minerals in the slab crust and slab mantle. In particular, intermediate-depth events tend to concentrate on the dehydration reaction boundaries in the slab, which forms the double seismic zone in the depth range of c. 40–180 km. Recent studies have revealed detailed spatial distribution of intermediate-depth seismicity that shows a close relationship with slab metamorphism. The pressure–temperature path of the crust for cold subducting slabs encounters facies boundaries with large rates of H_2O production and positive total volume change, which are likely to promote the occurrence of earthquakes near the facies boundaries in the crust due to increase in pore fluid pressure. Conversely, the path for warm slabs encounters boundaries with slow H_2O production rates and negative volumetric change, which may not effectively promote earthquake occurrence. Observations show that intermediate-depth earthquakes in warm slabs occur preferentially within the slab mantle or very close to the slab Moho, whereas those in cold slabs occur throughout the slab crust as well as in the slab mantle. Precise hypocenter locations of intermediate-depth events in NE Japan further demonstrate the presence of a belt of upper-plane seismicity in the crust of the cold Pacific slab that is nearly parallel to the 80–90 km depth contours of the upper slab surface. This concentration of seismic activity in the slab crust is probably caused by dehydration reactions with large H_2O production rates and positive net volume change. A seismic low-velocity layer detected in the slab crust persists down to the depth of this upper-plane seismic belt, but no further, which provides evidence that the phase transformation of the dehydration reaction occurs there. Both the belt of the upper-plane seismicity and the down-dip end of the seismic low-velocity layer deepen locally under Kanto and SW Hokkaido, and these deepenings coincide spatially with the areas of contact with the overlying Philippine Sea slab and subducted forearc crust, respectively. This local deepening is probably caused by delay of the phase transformation of the slab crust because of the thermal shielding effect of the overlying cold bodies. Although the belt of the upper-plane seismicity was also detected in the cold Philippine Sea slab beneath southern Kyushu, it has not been detected in cold slabs of other subduction zones of the world. This absence of evidence may simply reflect the shortage of high-quality data for intermediate-depth events precisely located in those subduction zones. However, similar low-velocity subducting crust with a close relationship with intraslab seismicity has been detected in several other subduction zones. Anomalously, low P- and S-wave velocity zones have been detected in the source area of M7-class large, intraslab events that occurred in the slab crust and the uppermost portion of the slab mantle in NE Japan, which suggests the presence of hydrous minerals or fluids. The situation for lower-plane events is somewhat more complicated. Seismic tomography studies in both eastern Japan and northern Chile again revealed the presence of a P-wave low-velocity layer along the lower plane of the double seismic zone. This P-wave low-velocity layer is not associated with low S-wave velocity, however, which is inconsistent with predictions based on the serpentinized mantle materials. Seismic anisotropy and pore aspect ratio may also play important roles in generating this unique structure. Although further validation is necessary for confirmation, the widespread distribution of a definite low-velocity P-wave layer along the entire lower seismic plane suggests the presence of hydrous minerals or fluids at that level, as well as the fluid-facilitated generation of lower-plane seismicity. All of these observations support the hypothesis that slab-derived H_2O causes intermediate-depth intraslab earthquakes. However, the possibility has not been ruled out that H_2O contributes only to earthquake nucleation and that seismic ruptures propagate through another mechanism. In other words, intermediate-depth events may be caused by both fluid-related embrittlement and thermal shear instability, i.e., local excess pore pressure may initiate earthquake rupture, and thus trigger subsequent earthquake ruptures through thermal shear instability. We anticipate significant progress in this area of research in the future to further verify this hypothesis for the genesis of intermediate-depth earthquakes, such as improving the accuracy of the temperature model, increasing the spatial resolution of seismic tomography imaging of the internal structures of subducting slabs, and deepening our understanding of the detailed processes that lead to earthquake rupture from a state of local excess pore fluid pressure based on rock deformation experiments or numerical simulations.

ACKNOWLEDGEMENTS

The authors thank T. Okada, N. Uchida, S. Kita, T. Matsuzawa, S. Kirby, S. Maruyama, and S. Omori for valuable discussions. The authors also wish to thank the editor Fenglin Niu and two anonymous reviewers for their constructive comments, which greatly improved the manuscript.

REFERENCES

1. Abers GA (1992) Relationship between shallow- and intermediate-depth seismicity in the eastern Aleutian subduction zone. Geophys Res Lett 20:2019–2022
2. Abers GA (1996) Plate structure and the origin of double seismic zones. In: Bebout GE, Scholl DW, Kirby SH, Platt JP (eds) Subduction, Top to Bottom, Geophys Mon 96. Amer Geophys Union, Washington, DC, pp 223–228
3. Abers GA (2000) Hydrated subducted crust at 100–250 km depth. Earth Planet Sci Lett 176:323–330
4. Abers GA (2005) Seismic low-velocity layer at the top of subducting slabs beneath volcanic arcs: observations, predictions, and systematic. Phys Earth Planet Int 149:7–29. doi:10.1016/jpepi200410002
5. Abers GA, MacKenzie LS, Rondenay S, Zhang Z, Wech AG, Creager KC (2009) Imaging the source region of Cascadia tremor and intermediate-depth earthquakes. Geology 37:1119–1122
6. Abers GA, Nakajima J, van Keken PE, Kita S, Hacker BR (2013) Thermal–petrological controls on the location of earthquakes within subducting plates. Earth Planet Sci Lett 369–370:178–187
7. Auzende AL, Escartin J, Walte NP, Guillot S, Hirth G, Frost DJ (2015) Deformation mechanisms of antigorite serpentinite at subduction zone conditions determined from experimentally and naturally deformed rocks. Earth Planet Sci Lett 411:229–240
8. Barcheck CG, Wiens DA, van Keken PE, Hacker BR (2012) The relationship of intermediate- and deep-focus seismicity to the hydration and dehydration of subducting slabs. Earth Planet Sci Lett 349–350:153–160
9. Bonatti E, Honnorez J (1976) Sections of the Earth's crust in the Equatorial Atlantic. J Geophys Res 81:4104–4116
10. Bostock MG, Hyndman RD, Rondenay S, Peacock SM (2002) An inverted continental Moho and serpentinization of the forearc mantle. Nature 417:536–538. doi:10.1038/417536a
11. Brantut N, Sulem J (2012) Strain localization and slip instability in a strain-rate hardening, chemically weakening material. J Appl Mechan 79:031004-1-10
12. Brudzinski MR, Thurber CH, Hacker BR, Engdahl ER (2007) Global prevalence of double Benioff zones. Science 316:1472–1474
13. Burnley PC, Green HW, Prior DJ (1991) Faulting associated with the olivine to spinel transformation in Mg_2GeO_4 and its implications for deep focus earthquakes. J Geophys Res 96:425–443
14. Cassidy JF, Waldhauser F (2003) Evidence for both crustal and mantle earthquakes in the subducting Juan de Fuca plate. Geophys Res Lett 30:1095. doi:10.1029/2002GL015511
15. Chernak LC, Hirth G (2010) Deformation of antigorite serpentinite at high temperature and pressure. Earth Planet Sci Lett 296:23–33
16. Chernak LC, Hirth G (2011) Syndeformational antigorite dehydration produces stable fault slip. Geology 39:847–850. doi:10.1130/G319191
17. Comte D, Suarez G (1994) An inverted double seismic zone in Chile: evidence of phase transformation in the subducted slab. Science 263:212–215. doi:10.1126/science2635144212
18. Comte DL, Dorbath L, Pardo M, Monfret T, Haessler H, Rivera L, Frogneux M, Glass B, Meneses C (1999) A double-layered seismic zone in Arica, northern Chile. Geophys Res Lett 26:1965–1968. doi:10.1029/1999GL900447
19. Contreras-Reyes E, Grevemeyer I, Flueh ER, Reichert C (2007) Upper lithospheric structure of the subduction zone offshore of southern Arauco peninsula Chile, at 38°S. J Geophys Res 113:B07303. doi:10.1029/2007JB005569
20. Contreras-Reyes E, Grevemeyer I, Watts AB, Flueh ER, Peirce C, Moeller S, Papenberg C (2011) Deep seismic structure of the Tonga subduction zone: implications for mantle hydration, tectonic erosion, and arc magmatism. J Geophys Res 116:B10103. doi:10.1029/2011JB008434

21. Dixon JE, Leist L, Langmuir C, Schilling J (2002) Recycled dehydrated lithosphere observed in plume-influenced mid-ocean-ridge basalt. Nature 420:385–389
22. Dobson DP, Meredith PG, Boon SA (2002) Simulation of subduction zone seismicity by dehydration of serpentine. Science 298:1407–1410
23. Dorbath C, Gerbault M, Carlier G, Guiraud M (2008) Double seismic zone of the Nazca plate in northern Chile: high resolution velocity structure, petrological implications, and thermomechanical modeling. Geochem Geophys Geosyst 9:Q07006. doi:10.1029/2008GC002020
24. Eberhart-Phillips D, Bannister S (2015) 3-D imaging of the northern Hikurangi subduction zone, New Zealand: variations in subducted sediment, slab fluids and slow slip. Geophys J Int 201:838–855
25. Eberhart-Phillips D, Reyners M, Chadwick M, Chiu J-M (2005) Crustal heterogeneity and subduction processes: 3-D Vp, Vp/Vs and Q in the southern North Island, New Zealand. Geophys J Int 162:270–288
26. Engdahl ER, Villaseñor A (2002) Global Seismicity: 1900–1999. In: Lee WHK, Kanamori H, Jennings PC, Kisslinger C (eds) International handbook of earthquake and engineering seismology, Part A, Chapter 41. Amsterdam: Academic Press; pp 665–690
27. Engdahl ER, van der Hilst R, Bauland R (1998) Global teleseismic earthquake relocation with improved travel times and procedures for depth determination. Bull Seismol Soc Amer 88:722–743
28. Estabrook CH (1999) Body wave inversion of the 1970 and 1963 South American large deep-focus earthquakes. J Geophys Res 104:28751–28767. doi:10.1029/1999JB900244
29. Faccenda M (2014) Water in the slab: A trilogy. Tectonophysics 614:1–30. doi:10.1016/jtecto201312020
30. Faccenda M, Gerya TV, Burlini L (2009) Deep slab hydration induced by bending-related variations in tectonic pressure. Nat Geosci 2(11):790–793. doi:10.1038/ngeo656
31. Faccenda M, Gerya TV, Mancktelow NS, Moresi L (2012) Fluid flow during slab unbending and dehydration: implications for intermediate-depth seismicity, slab weakening and deep water recycling. Geochem Geophys Geosyst 13:Q01010. doi:10.1029/2011GC003860
32. Frohlich C (1994) A break in the deep. Nature 368:101–101
33. Frohlich C (2006) Deep earthquakes. Cambridge University Press, New York
34. Fujie G, Kodaira S, Yamashita M, Sato T, Takahashi T, Takahashi N (2013) Systematic changes in the incoming plate structure at the Kuril trench. Geophys Res Lett 40:88–93. doi:10.1029/2012GL054340
35. Fukao Y, Hori S, Ukawa M (1983) A seismological constraint on the depth of the basalt–eclogite transition in a subducting oceanic crust. Nature 303:413–415
36. Garth T, Rietbrock A (2014) Order of magnitude increase in subducted H_2O due to hydrated normal faults within the Wadati–Benioff zone. Geology 42:207–210. doi:10.1130/G347301
37. Gasc J, Schbnel A, Brunet F, Guillon S, Mueller H-J, Lathe C (2011) Simultaneous acoustic emissions monitoring and synchrotron X-ray diffraction at high pressure and temperature: calibration and application to serpentinite dehydration. Phys Earth Planet Inter 189:121–133
38. Green HW, Burnley PC (1989) A new self-organizing mechanism for deep-focus earthquakes. Nature 341:733–737. doi:10.1038/341733a0
39. Green HW, Houston H (1995) The mechanics of deep earthquakes. Ann Rev Earth Planet Sci 23:169–213
40. Gregg PM, Lin J, Behn MD, Montesi LG (2007) Spreading rate dependence of gravity anomalies along oceanic transform faults. Nature 448:183–188
41. Grevemeyer I, Ranero CR, Flueh ER, Klaeschen D, Bialas J (2007) Passive and active seismological study of bending-related faulting and mantle serpentinization at the Middle America trench. Earth Planet Sci Lett 258:528–542. doi:10.1016/jepsl200704013
42. Griggs DT, Handin J (1960) Observations on fracture and a hypothesis of earthquakes. In: Griggs DT, Handin J (eds) Rock deformation. J Geolog Soc Amer Memoir, vol 79., pp 347–373

43. Hacker BR (2008) H2O subduction beyond arcs. Geochem Geophys Geodyn 9:Q03001. doi:10.1029/2007GC001707
44. Hacker BR, Abers GA, Peacock SM (2003a) Subduction factory 1 theoretical mineralogy, densities, seismic wave speeds, and H_2O contents. J Geophys Res 108(B1):2029. doi:10.1029/2001JB001127
45. Hacker BR, Peacock SM, Abers GA, Holloway SD (2003b) Subduction factory 2: are intermediate-depth earthquakes in subducting slabs linked to metamorphic dehydration reactions? J Geophys Res 108(B1):2030. doi:10.1029/2001JB001129
46. Hara T, Kuge K, Kawakatsu H (1996) Determination of the isotropic component of deep focus earthquakes by inversion of normal-mode data. Geophys J Int 127:515–528
47. Hasegawa A, Nakajima J (2004) Geophysical constraints on slab subduction and arc magmatism. In: The State of the Planet: Frontiers and Challenges in Geophysics, Geophys Monogr Ser 150. AGU, Washington, DC, pp 81–94
48. Hasegawa A, Umino N, Takagi A (1978a) Double-planed structure of the deep seismic zone in the northeastern Japan arc. Tectonophysics 47:43–58
49. Hasegawa A, Umino N, Takagi A (1978b) Double-planed deep seismic zone and upper-mantle structure in the northeastern Japan arc. Geophys J Int 54:281–296
50. Hasegawa A, Nakajima J, Kita S, Okada T, Matsuzawa T, Kirby S (2007) Anomalous deepening of a belt of intraslab earthquakes in the Pacific slab crust under Kanto, central Japan: possible anomalous thermal shielding, dehydration reactions, and seismicity caused by shallower cold slab material. Geophys Res Lett 34:L09305. doi:10.1029/2007GL029616
51. Hasegawa A, Nakajima J, Uchida N, Okada T, Zhao D, Matsuzawa T, Umino N (2009) Plate subduction, and generation of earthquakes and magmas in Japan as inferred from seismic observations: an overview. Gondwana Res 16:370–400
52. Hasegawa A, Nakajima J, Uchida N, Hirose F, Kita S, Matsuzawa T (2010) Slab structure beneath the Japanese Islands and earthquake generation. J Geogr 119:190–204 (in Japanese with English abstract)
53. Hasegawa A, Nakajima J, Zhao D (2016) Deep seismic structure. In: Geology of Japan. The Geological Society, London, pp 339–370
54. Helffrich GR (1996) Subducted lithospheric slab velocity structure: observations and mineralogical inferences. In: Bebout GE, Scholl D, Kirby SH, Platt JP (eds) Subduction top to bottom. Amer Geophys Union Geophys Mono, vol 96., pp 215–222
55. Hirose F, Nakajima J, Hasegawa A (2008a) Three-dimensional seismic velocity structure and configuration of the Philippine Sea slab in southwestern Japan estimated by double-difference tomography. J Geophys Res. 113. doi:10.1029/2007JB005274.
56. Hirose F, Nakajima J, Hasegawa A (2008a) Three-dimensional seismic velocity structure and configuration of the Philippine Sea slab beneath Kanto district, central Japan, estimated by double-difference tomography. J Seismol Soc Jpn 60:123–138 (in Japanese with English abstract)
57. Hirth G, Beeler NM (2015) The role of fluid pressure on frictional behavior at the base of the seismogenic zone. Geology 43:223–225
58. Hobbs BE, Ord A (1988) Plastic instabilities: implications for the origin of intermediate and deep focus earthquakes. J Geophys Res 93:10521–10540
59. Hori S (1990) Seismic waves guided by untransformed oceanic crust subducting into the mantle: the case of the Kanto district central Japan. Tectonophysics 176:355–376
60. Hori S, Inoue H, Fukao Y, Ukawa M (1985) Seismic detection of the untransformed 'basaltic' oceanic crust subducting into the mantle. Geophys J Int 83:169–197
61. Houston H (2007) Deep earthquakes. In: Schubert G (ed) Treatise on Geophysics, vol 4, Earthquake Seismology. Elsevier, Amsterdam, pp 321–350
62. Igarashi T, Matsuzawa T, Umino N, Hasegawa A (2001) Spatial distribution of focal mechanisms for inter- and intraplate earthquakes associated with the subducting Pacific plate beneath the northeastern Japan arc: a triple-planed deep seismic zone. J Geophys Res 106:2177–2191

63. Ivandic M, Grevemeyer I, Berhorst A, Flueh ER, McIntosh K (2008) Impact of bending related faulting on the seismic properties of the incoming oceanic plate offshore of Nicaragua. J Geophys Res 113:B05410. doi:10.1029/2007JB005291
64. Iyer K, Rüpke LH, Morgan JP, Grevemeyer I (2012) Controls of faulting and reaction kinetics on serpentinization and double Benioff zones. Geochem Geophys Geosyst 13:Q09010. doi:10.1029/2012GC004304
65. Jiao W, Silver PG, Fei Y, Prewitt CT (2000) Do intermediate- and deep-focus earthquakes occur on preexisting weak zones? An examination of the Tonga subduction zone. J Geophys Res 105:28125–28138
66. Jung H, Green HW (2004) Experimental faulting of serpentinite during dehydration: implications for earthquakes, seismic low-velocity zones, and anomalous hypocenter distributions in subduction zones. Int Geol Rev 46:1089–1102
67. Jung H, Green HW, Dobrzhinetskaya F (2004) Intermediate-depth earthquake faulting by dehydration embrittlement with negative volume change. Nature 428:545–549
68. Jung H, Fei YW, Silver PG, Green HW (2006) System for detecting acoustic emissions in multianvil experiments: application to deep seismicity in the Earth. Rev Sci Instrum 77:7
69. Jung H, Fei Y, Silver P, Green HW (2009) Frictional sliding in serpentine at very high pressure. Earth Planet Sci Lett 277:273–279. doi:10.1016/jepsl200810019
70. Kao H, Chen WP (1994) The double seismic zone in Kuril-Kamchatka: the tale of two overlapping single zones. J Geophys Res 99:6913–6930. doi:10.1029/92JB03409
71. Kao H, Rau RJ (1999) Detailed structures of the subducted Philippine Sea plate beneath northeast Taiwan: a new type of double seismic zone. J Geophys Res 104:1015–1033. doi:10.1029/1998JB900010
72. Karato S, Ridel MR, Yuen DA (2001) Rheological structure and deformation of subducted slabs in the mantle transition zone: implications for mantle circulation and deep earthquakes. Phys Earth Planet Inter 127:83–108
73. Kawakatsu H (1985) Double zone in Tonga. Nature 316:53–55. doi:10.1038/316053a0
74. Kawakatsu H (1986) Downdip tensional earthquakes beneath the Tonga arc: a double seismic zone? J Geophys Res 91:6432–6440
75. Kawakatsu H (1991) Insignificant isotropic component in the moment tensor of deep earthquakes. Nature 351:50–53. doi:10.1038/35100a0
76. Kawakatsu H (1996) Observability of the isotropic component of a moment tensor. Geophys J Int 126:525–544
77. Kelemen PB, Hirth G (2007) A periodic shear-heating mechanism for intermediate-depth earthquakes in the mantle. Nature 446:787–790. doi:10.1038/nature05717
78. Kimura G (1986) Oblique subduction and collision: forearc tectonics of the Kuril arc. Geology 14:404–407
79. Kirby SH (1987) Localize polymorphic phase transformations in high-pressure faults and applications to the physical mechanism of deep earthquakes. J Geophys Res 92:13789–13800
80. Kirby SH (1995) Intraslab earthquakes and phase changes in subducting lithosphere. US Ntl Rep Int Union Geod Geophys, 1991–1994. Rev Geophys 33:287–297
81. Kirby SH, Engdahl ER, Denlinger R (1996) Intermediate-depth intraslab earthquakes and arc volcanism as physical expressions of crustal and uppermost mantle metamorphism in subducting slabs. In: Bebout GE, Scholl D, Kirby SH, Platt JP (eds) Subduction, top to bottom. Geophys Mono Ser, 96. AGU, Washington, DC, pp 347–355
82. Kita S, Okada T, Nakajima J, Matsuzawa T, Hasegawa A (2006) Existence of a seismic belt in the upper plane of the double seismic zone extending in the along-arc direction at depths of 70–100 km beneath NE Japan. Geophys Res Lett 33:L24310. doi:10.1029/2006GL028239
83. Kita S, Okada T, Hasegawa A, Nakajima J, Matsuzawa T (2010a) Anomalous deepening of a seismic belt in the upper-plane of the double seismic zone in the Pacific slab beneath the Hokkaido corner: possible evidence for thermal shielding caused by subducted forearc crust materials. Earth Planet Sci Lett 290:415–426
84. Kita S, Okada T, Hasegawa A, Nakajima J, Matsuzawa T (2010b) Existence of interplane earthquakes and neutral stress boundary between the upper and lower planes of the

double seismic zone beneath Tohoku and Hokkaido, northeastern Japan. Tectonophysics 496(1):68–82. doi:10.1016/jtecto201010010
85. Ko SC, Olgaard DL, Wong TF (1997) Generation and maintenance of pore pressure excess in a dehydrating system 1: experimental and microstructural observations. J Geophys Res 102:825–839
86. Kodaira S, Takahashi N, Park J, Mochizuki K, Shinohara M, Kimura S (2000) Western Nankai Trough seismogenic zone: results from a wide-angle ocean bottom seismic survey. J Geophys Res 105(B3):5887–5905
87. Kodaira S, Iidake T, Kato A, Park J, Iwasaki T, Kaneda Y (2004) High pore fluid pressure may cause silent slip in the Nankai Trough. Science 304:1295–1298
88. Korenaga J (2017) On the extent of mantle hydration caused by plate bending. Earth Planet Sci Lett 457:1–9
89. Kuge K, Kawakatsu H (1993) Significance of non-double couple components of deep and intermediate-depth earthquakes: implications from moment tensor inversions of long-period seismic waves. Phys Earth Planet Inter 75:243–266
90. Leith A, Sharpe JA (1936) Deep-focus earthquakes and their geological significance. J Geol 44:877–917
91. Matsuzawa T, Umino N, Hasegawa A, Takagi A (1986) Upper mantle velocity structure estimated from PS-converted wave beneath the northeastern Japan arc. Geophys J Int 86:767–787
92. Matsuzawa T, Kono T, Hasegawa A, Takagi A (1990) Subducting plate boundary beneath the northeastern Japan arc estimated from SP converted waves. Tectonophysics 181:123–133
93. Meade C, Jeanloz R (1991) Deep-focus earthquakes and recycling of water into the earth's mantle. Science 252:68–72
94. Mishra OP, Zhao D (2004) Seismic evidence for dehydration embrittlement of the subducting Pacific slab. Geophys Res Lett 31:L09610. doi:10.1029/2004GL019489
95. Nakajima J, Hasegawa A (2007) Tomographic evidence for the mantle upwelling beneath southwestern Japan and its implications for arc magmatism. Earth Planet Sci Lett 254:90–105
96. Nakajima J, Hasegawa A (2010) Cause of M7 earthquakes beneath the Tokyo metropolitan area, Japan: possible evidence for a vertical tear at the easternmost portion of the Philippine Sea slab. J Geophys Res. 115. doi:10.1029/2009JB006863
97. Nakajima J, Matsuzawa T, Hasegawa A, Zhao D (2001) Three-dimensional structure of Vp, Vs, and Vp/Vs beneath northeastern Japan: implications for arc magmatism and fluids. J Geophys Res 106(B10):21,843–21,858. doi:10.1029/2000JB000008
98. Nakajima J, Tsuji Y, Hasegawa A (2009a) Seismic evidence for thermally-controlled dehydration reaction in subducting oceanic crust. Geophys Res Lett 36:L03303. doi:10.1029/2008GL036865
99. Nakajima J, Hirose F, Hasegawa A (2009b) Seismotectonics beneath the Tokyo metropolitan area: effect of slab–slab contact and overlap on seismicity. J Geophys Res 114:B08309. doi:10.1029/2008JB006101
100. Nakajima J, Tsuji Y, Hasegawa A, Kita S, Okada T, Matsuzawa T (2009c) Tomographic imaging of hydrated crust and mantle in the subducting Pacific slab beneath Hokkaido, Japan: evidence for dehydration embrittlement as a cause of intraslab earthquakes. Gondwana Res 16:470–481
101. Nakajima J, Hasegawa A, Kita S (2011) Seismic evidence for reactivation of a buried hydrated fault in the Pacific slab by the 2011 M90 Tohoku Earthquake. Geophys Res Lett 38:L16318. doi:10.1029/2011GL048754
102. Nakajima J, Uchida N, Shiina T, Hasegawa A, Hacker BR, Kirby SH (2013) Intermediate-depth earthquakes facilitated by eclogitization-related stresses. Geology 41:659–662
103. Nishiyama T (1992) Mantle hydrology in a subduction zone: a key to episodic geologic events, double Wadati–Benioff zones and magma genesis. Math Seismol VII, R Inst Stat Math Tokyo 34:31–67
104. Ogawa M (1987) Shear instability in a viscoelastic material as the cause of deep focus earthquakes. J Geophys Res 92:13801–13810

105. Okal EA (1996) Radial modes from the great 1994 Bolivian earthquake: no evidence for anisotropic component to the source. Geophys Res Lett 23:431–434. doi:10.1029/96GL00375
106. Okazaki K, Hirth G (2016) Dehydration of lawsonite could directly trigger earthquakes in subducting oceanic crust. Nature 530. doi:10.1038/nature16501.
107. Omori S, Kita S, Maruyama S, Santosh M (2009) Pressure–temperature conditions of ongoing regional metamorphism beneath the Japanese Islands. Gondwana Res 16:458–469
108. Peacock SM (2001) Are the lower planes of double seismic zones caused by serpentine dehydration in subducting oceanic mantle? Geology 29:299–302
109. Peacock SM, Wang K (1999) Seismic consequences of warm v. cool subduction metamorphism: examples from southwest and northeast Japan. Science 286:937–939
110. Proctor B, Hirth G (2015) Role of pore fluid pressure on transient strength changes and fabric development during serpentine dehydration at mantle conditions: implications for subduction-zone seismicity. Earth Planet Sci Lett 421:1–12
111. Qin Y, Singh SC (2015) Seismic evidence of a two-layer lithospheric deformation in the Indian Ocean. Nat Commun 6:8298. doi:10.1038/ncomms9298
112. Raleigh CB (1967) Tectonic implications of serpentinite weakening. Geophys J Int 14:113–118
113. Raleigh CB, Paterson MS (1965) Experimental deformation of serpentinite and its tectonic implications. J Geophys Res 70:3965–3985
114. Ranero CR, Sallares V (2004) Geophysical evidence for alteration of the crust and mantle of the Nazca Plate during bending at the north Chile trench. Geology 32:549–552
115. Ranero CR, Morgan JP, McIntosh K, Reicher C (2003) Bending-related faulting, and mantle serpentinization at the Middle America trench. Nature 425:367–373
116. Ranero CR, Villaseñor A, Morgan JP, Weinrebe W (2005) Relationship between bend-faulting at trenches and intermediate-depth seismicity. Geochem Geophys Geosyst 6:Q12002. doi:10.1029/2005GC000997
117. Reynard B, Nakajima J, Kawakatsu H (2010) Earthquakes and plastic deformation of anhydrous slab mantle in double Wadati–Benioff zones. Geophys Res Lett 37:L24309. doi:10.1029/2010GL045494
118. Reyners M, Robinson R, McGinity P (1997) Plate coupling in the northern South Island and southernmost North Island, New Zealand, as illuminated by earthquake focal mechanisms. J Geophys Res 102:15197–15210. doi:10.1029/97JB00973
119. Reyners M, Eberhart-Phillips D, Stuart G (1999) A three-dimensional image of shallow subduction: crustal structure of the Raukumara Peninsula, New Zealand. Geophys J Int 137:873–890
120. Rondenay S, Abers GA, van Keken PE (2008) Seismic imaging of subduction zone metamorphism. Geology 36:275–278
121. Rondenay S, Montési L, Abers GA (2010) New geophysical insight into the origin of the Denali volcanic gap. Geophys J Int 182:613–630
122. Rüpke LH, Morgan JP, Hort M, Connolly JAD (2004) Serpentine and the subduction zone water cycle. Earth Planet Sci Lett 223:17–34
123. Russakoff D, Ekstrom G, Tromp J (1997) A new analysis of the great 1970 Colombia earthquake and its isotropic component. J Geophys Res 102:20423–20434. doi:10.1029/97JB01645
124. Rutter EH, Brodie KH (1988) Experimental syntectonic dehydration of serpentinite under conditions of controlled pore water-pressure. J Geophys Res 93:4907–4932
125. Sawai M, Katayama I, Hamada A, Maeda M, Nakashima S (2013) Dehydration kinetics of antigorite using in situ high-temperature infrared microspectroscopy. Phys Chem Minerals 40:319–330. doi:10.1007/s00269-013-0573-9
126. Schmidt MW, Poli S (1998) Experimentally based water budgets for dehydrating slabs and consequences for arc magma generation. Earth Planet Sci Lett 163:361–379. doi:10.1016/S0012-821X(98)00142-3
127. Seno T, Maruyama S (1984) Paleogeographic reconstruction and origin of the Philippine Sea. Tectonophysics 102:53–84

128. Seno T, Yamanaka Y (1996) Double seismic zones, compressional deep trench–outer rise events and superplumes. In: Bebout GE, Scholl D, Kirby S, Platt JP (eds) Subduction: top to bottom, Geophys Mono Ser, 96. AGU, Washington, DC, pp 347–355
129. Shelly DR, Beroza GC, Zhang H, Thurber CH, Ide S (2006) High-resolution subduction zone seismicity and velocity structure beneath Ibaraki Prefecture, Japan. J Geophys Res 111:B06311. doi:10.1029/2005JB004081
130. Shiina T, Nakajima J, Matsuzawa T (2013) Seismic evidence for high pore pressure in the oceanic crust: implications for fluid-related embrittlement. Geophys Res Lett 40:2006–2010. doi:10.1002/grl50468
131. Smith SW, Knapp JS, McPherson RC (1993) Seismicity of the Gorda Plate, structure of the continental margin, and an eastward jump of Mendocino triple junction. J Geophys Res 98:8153–8171
132. Staudigel H, Davies GR, Hart SR, Marchant KM, Smith BM (1995) Large scale isotopic Sr, Nd and O isotopic anatomy of altered oceanic crust: DSDP/ODP sites 417/418. Earth Planet Sci Lett 130:105–120
133. Suzuki S, Sasatani T, Motoya Y (1983) Double seismic zone beneath the middle of Hokkaido, Japan, in the southwestern side of the Kurile Arc. Tectonophysics 96:59–76
134. Syracuse EM, Abers GA, Fischer K, MacKenzie L, Rychert C, Protti M, Gonza'lez V, Strauch W (2008) Seismic tomography and earthquake locations in the Nicaraguan and Costa Rican upper mantle. Geochem Geophys Geosyst 9:Q07S08. doi:10.1029/2008GC001963
135. Syracuse EM, van Keken PE, Abers GA (2010) The global range of subduction zone thermal models. Phys Earth Planet Inter 183:73–90. doi:10.1016/jpepi201010021004
136. Takei Y (2002) Effect of pore geometry on Vp/Vs: from equilibrium geometry to crack. J Geophys Res 107(B2):2043. doi:10.1029/2001JB000522
137. Tsuji Y, Nakajima J, Hasegawa A (2008) Tomographic evidence for hydrated oceanic crust of the Pacific slab beneath northeastern Japan: implications for water transportation in subduction zones. Geophys Res Lett 35:L14308. doi:10.1029/2008GL034461
138. Tsumura K (1973) Microearthquake activity in the Kanto District. In: publications for the 50th anniversary of the Great Kanto Earthquake, 1923. Earthq Res Inst, Tokyo Univ. pp 67–87. (in Japanese)
139. Uchida N, Nakajima J, Hasegawa A, Matsuzawa T (2009) What controls interplate coupling?: Evidence for abrupt change in coupling across a border between two overlying plates in the NE Japan subduction zone. Earth Planet Sci Lett 283:111–121
140. Uchida N, Matsuzawa T, Nakajima J, Hasegawa A (2010) Subduction of a wedge-shaped Philippine Sea plate beneath Kanto, central Japan, estimated from converted waves and small repeating earthquakes. J Geophys Res 115. doi:10.1029/2009JB006962.
141. Ulmer P, Trommsdorff V (1995) Serpentine stability to mantle depths and subduction-related magmatism. Science 268:858–861
142. Umino N, Hasegawa A (1975) On the two-layered structure of deep seismic plane in northeastern Japan arc. J Seismol Soc Jpn 28:125–139 (in Japanese with English abstract)
143. van Keken PE, Kiefer B, Peacock S (2002) High-resolution models of subduction zones: implications for mineral dehydration reactions and the transport of water into the deep mantle. Geochem Geophys Geosyst 3(10):1056. doi:10.1029/2001GC000256
144. Wada I, Wang K (2009) Common depth of slab-mantle decoupling: reconciling diversity and uniformity of subduction zones. Geochem Geophys Geosyst 10:Q10009
145. Wang K (2002) Unbending combined with dehydration embrittlement as a cause for double and triple seismic zones. Geophys Res Lett 29:1889. doi:10.1029/2002GL015441
146. Warren LM, Hughes AN, Silver PG (2007) Earthquake mechanics and deformation in the Tonga-Kermadec subduction zone from fault plane orientations of intermediate- and deep-focus earthquakes. J Geophys Res 112:B05314. doi:10.1029/2006JB004677.1
147. Warren LM, Langstaff MA, Silver PG (2008) Fault plane orientations of intermediate-depth earthquakes in the Middle America Trench. J Geophys Res 113:B01304. doi:10.1029/2007JB005028
148. Wunder B, Schreyer W (1997) Antigorite: high-pressure stability in the system MgO–SiO_2–H_2O (MSH). Lithos 41:213–227

149. Yamasaki T, Seno T (2003) Double seismic zone and dehydration embrittlement of the subducting slab. J Geophys Res 108. doi:10.1029/2002JB001918.
150. Zhang H, Thurber CH (2003) Double-difference tomography: the method and its application to the Hayward Fault, California. Bull Seismol Soc Amer 93:1875–1889
151. Zhang H, Thurber CH, Shelly D, Ide S, Beroza GC, Hasegawa A (2004) High-resolution subducting-slab structure beneath northern Honshu, Japan, revealed by double-difference tomography. Geology 32:361–364
152. Zhao D, Hasegawa A (1993) P-wave tomographic imaging of the crust and upper mantle beneath the Japan Islands. J Geophys Res 98:4333–4353
153. Zhao D, Hasegawa A, Horiuchi S (1992) Tomographic imaging of P and S wave velocity structure beneath northeastern Japan. J Geophys Res 97:19909–19928
154. Zhao D, Matsuzawa T, Hasegawa A (1997) Morphology of the subducting slab boundary in the northeastern Japan arc. Phys Earth Planet Inter 102:89–104. doi:10.1016/S0031-9201(96)03258-X

CHAPTER 7

A regional geodetic network to monitor ground surface response to resource extraction in the northern Surat Basin, Queensland

M. C. Garthwaite*, M. Hazelwood, S. Nancarrow, A. Hislop & J. H. Dawson

ABSTRACT

This paper describes the purpose, design and implementation of a new regional-scale geodetic network that has been installed in the Surat Basin, Queensland, and completed in November 2014. The network incorporates 65 survey marks and 40 radar corner reflectors to enable the combination of Global Navigation Satellite System measurements with remotely sensed surface deformation maps derived using the Interferometric Synthetic Aperture Radar technique. The combination of these geodetic techniques in this region will bring an enhanced understanding of how resource extraction affects the ground surface.

Keywords: geodetic network, GNSS, survey mark, InSAR, corner reflector, Surat Basin, resource extraction, environmental monitoring, ground deformation.

1. INTRODUCTION

Internationally, we rely on the Earth's crust to provide mineral, energy and water resources to support an evolving and growing society. Increasingly, we are also looking to use the crust as a repository for hazardous waste products that we generate such as radioactive waste and carbon dioxide. As extraction continues, it becomes increasingly important to have an understanding of the remaining resources in addition to the evolving storage capacity of the crust in order to secure sustainable, ethical and low-cost resourcing for our society. To this end, the Australian Government invested $23 million in 2010 to develop an Australian Geophysical Observing System (AGOS) infrastructure through funding from the Education Investment Fund (EIF) Round 3. Administered by AuScope Ltd, the purpose of AGOS is to enable collection of new baseline data including surface geospatial and subsurface imaging and monitoring data, to provide an understanding of the physical state of the accessible crust of the Australian continent, particularly focussed on emerging energy issues. Geoscience Australia (GA) has been responsible for implementing the AGOS geospatial observatory, which features:

A geodetic survey mark network, including co-located radar corner reflectors, to enable the precise measurement of surface deformation at a regional scale using Interferometric Synthetic Aperture Radar (InSAR) and Global Navigation Satellite System (GNSS) techniques.

Four high-precision GNSS continuously operating reference station (CORS) monuments installed at Mitchell (Queensland), King Island (Tasmania), Blinman (South Australia) and at the Murchison Radio Observatory, Boolardy (Western Australia). These supplement 101 more GNSS CORS funded by a National Collaborative Research Infrastructure Strategy (NCRIS) award to AuScope.

A robotic GNSS antenna calibration facility; the only one of its kind in the southern hemisphere.

A deployable pool of GNSS instruments for episodic campaign surveys in Australia. This includes 80 GNSS instruments, 10 ionospheric receivers and three Real Time Kinematic (RTK) kits.

An open-access repository of Synthetic Aperture Radar (SAR) data acquired by the 'European Remote-sensing Satellite' (ERS) satellites over the Australian territory, suitable for InSAR analysis to detect ground surface deformation.

Geodetic networks comprising discrete surface observation sites are essential for determining a coordinate reference system at any scale, be it at the jurisdictional level, nationally or globally. Collectively, these coordinate systems provide the fundamental datum with which all other geospatial datasets can be referenced. The Geocentric Datum of Australia 1994 (GDA94; ICSM 1998) is the official Australian coordinate system that underpins all geospatial activity in Australia including surveying, mapping and remote sensing. Implemented in 1995 by the Intergovernmental Committee on Surveying and Mapping (ICSM), and adopted as the national datum in 2000, GDA94 is a static geospatial referencing system defined initially on a network of eight GNSS CORS that are considered to be geologically stable across Australia. The static nature of the datum means that GDA94 coordinates remain unchanged with time despite ongoing tectonic motion of the Australian Plate (Altamimi et al. 2011), intra-continental deformation (Watson et al. 2010; Tregoning et al. 2013) and natural and anthropogenic deformation processes evident at the local and regional scales (e.g. Dawson et al. 2008; Featherstone et al. 2012). With an increasing societal demand for improved positioning for geospatial applications, the ICSM are currently undertaking research towards the implementation of the next Australian geodetic datum, which is anticipated to incorporate time-variable coordinates. The inclusion of regional-scale crustal deformation models in addition to the well-defined plate movement model for Australia is key to ensuring this new datum meets the requirements of all users (Dawson &Woods 2010).

Following recent densification, the national GNSS CORS network maintained by GA now contains 121 stations with a mean minimum station spacing of »209 km (and a standard deviation of 181 km). Where there is local- or regional-scale crustal deformation occurring at spatial scales of less than 200 km (e.g. owing to resource extraction in a sedimentary basin) it is likely that this deformation will not be sampled by a GNSS CORS, and therefore the coordinate system will be inaccurate in regions affected by such deformation. Complementary geodetic networks covering localised deforming regions are therefore crucial for enabling the widespread accuracy of any new national datum. In the modern era of multi-constellation satellite geodesy and remote sensing, geodetic networks are also used exclusively as an observatory for monitoring dynamic deformation of the Earth's surface. For instance, the Plate Boundary Observatory, a network of over 1000 GNSS CORS sites in the western USA, is designed to measure the complex tectonic strain field on the margin of the North American and Pacific plates. Surface deformation can also occur when resources are extracted from the Earth's crust, although no geodetic networks have been established in Australia that span an entire sedimentary basin and all the extraction activities that exist within it. Although a CORS network provides the most accurate observations of surface movement, they are expensive to construct and require ongoing maintenance. Therefore, a regional high-density geodetic network will usually make use of permanent survey marks that can be re-observed periodically during campaign GNSS surveys.

In this paper, we describe the design of the new regional AGOS geodetic network that has been established in the northern Surat Basin, Queensland, Australia. This geodetic network spans the full east-to-west extent of the Surat Basin in Queensland and contains 65 permanent GNSS survey marks with the innovative inclusion of 40 radar corner reflectors co-located at a subset of the survey mark sites. The purpose of the new geodetic network is to enable the combination of traditional ground network-based geodetic techniques (such as GNSS) with modern remotely sensed surface deformation maps derived using the InSAR technique to yield temporal estimates of regional scale surface deformation with millimetre-to-centimetre level accuracy and spatial resolution on the order of 100 m.

2. RESOURCE EXTRACTION IN THE SURAT BASIN

The Surat Basin is a large (»300 000 km2) Jurassic_Cretaceous age basin that occupies southern Queensland and northern New South Wales (Exon 1976). The Middle Jurassic Walloon Subgroup contains shallow coal beds that have been found to have sufficient unconventional gas content (i.e. gas trapped throughout a formation rather than concentrated in a structural trap) to be economically viable (Scott et al. 2007). Production of coal seam gas from the Walloon Subgroup coal beds began in January 2006 from the Kogan area of the Darling Downs of Queensland (»50 km west of Dalby), typically from between 300 and 600 m depth (Queensland State Government 2015). By the time of the 2013_2014 Australian fiscal year, production from the Surat Basin had grown to 166 Petajoules (Queensland State Government 2015). With proved and probable (2P) reserves of over 32 000 Petajoules (Queensland State Government 2015) the Surat Basin is the largest onshore reserve of gas in Australia (Geoscience Australia and BREE 2012). Therefore, the Surat Basin represents a significant energy resource for Australia and production could increase in the future.

In order to extract gas from the coal beds, a well is drilled into the coal seam, and a large volume of groundwater (on the order of megalitres per annum) is pumped out in order to reduce pressure in the formation and allow the gas to desorb from the coal. It is generally understood that extracting groundwater from the subsurface can result in surface subsidence (i.e. lowering of the ground surface) owing to the compaction of susceptible layers within the aquifer after the pressure is lowered (Galloway & Burbey 2011). In the case of an impermeable coal bed, well stimulation techniques such as hydraulic fracturing may be used to increase gas production. Hydraulic fracturing involves injection of a fluid into the formation that could potentially result in ground surface uplift, similar to the situation observed in carbon capture and storage projects when CO2 is injected (e.g. Onuma & Ohkawa 2009). However, because of the permeability of the Walloon Subgroup, to date well stimulation has generally not been used in the Surat Basin.

It is possible to map the spatial patterns of subsidence induced by groundwater and other fluid extraction on a regional scale using space-based InSAR techniques (e.g. Fielding et al. 1998; Galloway et al. 1998; Schmidt & B€urgmann 2003; Bell et al. 2008) and even to predict aquifer-system behaviour using these measurements (Chaussard et al. 2014; Reeves et al. 2014). The magnitude of subsidence will depend on the specific conditions in a basin, and so for this reason, the impacts of groundwater extraction in the Surat Basin (and other gas-producing basins and aquifers in Australia) are poorly known. Understanding the temporal and spatial evolution of surface movement is an important constraint on what is happening at depth, but can also help develop our understanding of the environmental impacts of extraction. For instance, differential subsidence could change the local surface slope morphology and thus impact on established patterns of surface water runoff. Furthermore, patterns of vertical movement can indicate the pathways taken by subsurface fluids and therefore lead to bettertargeted extraction well locations. It is therefore expected that the AGOS geodetic network could contribute to enhanced resource production and better environmental management in the Surat Basin

3. SITE SELECTION

Initial geodetic site locations were identified by a desk-top study based on the access, site security and land-use, as well as providing a good spatial sampling of the expected pattern of ground deformation in the region. In the Darling Downs, between Dalby and Miles, existing gas production wells are concentrated in a NW–SE oriented band approximately 30 km wide (Figure 1). Predictions of surface deformation based on homogenous poroelastic modelling and projected hydraulic draw-down indicate a long-term vertical deformation signal on the order of a decimeter (Brown et al. 2014). The time period of this signal is unknown and will vary spatially depending on local gas production rates. The spatial pattern of deformation has a wavelength of around 100 km that correlates with the location of the production wells. We therefore sought to locate geodetic sites at around 20 km spacing to fully sample this signal and also to locate sites where deformation is not expected.

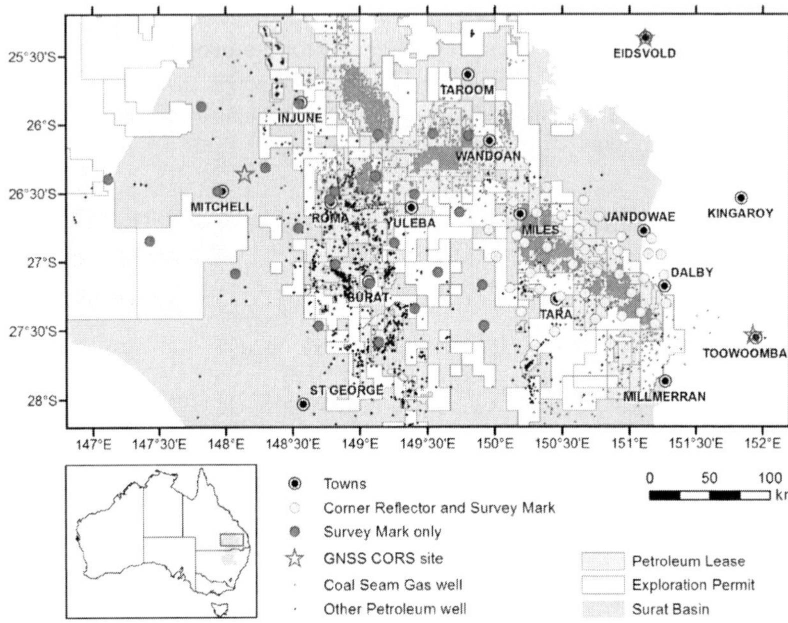

Figure 1 Distribution of 65 new geodetic sites in southern Queensland and locations of three nearby GNSS CORS sites at Mitchell, Eidsvold and Toowoomba. Also plotted are land parcels designated as petroleum leases for production and exploration permits for petroleum, coal seam gas (CSG) wells and other petroleum wells (these data obtained on 11 December 2014 from the Queensland State Department of Natural Resources and Mines' 'Queensland Globe' service). Background grey shading indicates the spatial extent of the Surat Basin.

In order to establish an understanding of the reconnaissance and installation costs, the first phase of the project was confined to the Darling Downs region. Contractors were engaged in February 2014 to undertake field evaluation, land access and cultural heritage clearances for the sites identified in the initial desktop study. Almost immediately, the contractor, responsible for site evaluations and land access, encountered resistance from local landowners, citing 'stakeholder fatigue' as the primary reason for their resistance. This was largely due to previous sustained engagement with the resource extraction industry, which was seeking access to property to establish coal seam gas infrastructure. Based on this feedback, the focus of the site location activity changed to utilise the vast road reserve network, including proposed road corridors.

Once the initial site selection and evaluation were completed, the spatial locations of the proposed sites on private land and road reserves were sent to the Queensland State Department of Natural Resources and Mines for querying against the Queensland Heritage Register. This was in order to determine the proximity of each site to already-known cultural heritage features and to identify the Aboriginal groups on whose native-title claim the sites were located. Additionally, all Aboriginal groups were sent a Notice (pursuant to section 24KA of the Commonwealth of Australia Native Title Act, 1993) that provided details of the project and gave the registered native title claimants the opportunity to make a submission or comment. For the Western Downs area, the principal Aboriginal group were the Barunggam People, represented by the Warner and Daylight family groups. Representatives from GA and the cultural heritage consultants met with the representatives of the Barunggam People to provide an overview of the project and to request assistance in discharging GA's responsibilities under the Queensland State Government Aboriginal Cultural Heritage Act 2003 Duty of Care Guidelines ('the Guidelines'; ACHA 2003). The first outcome of these discussions was the recognition that a Cultural Heritage Management Plan (CHMP) between GA and the Barunggam People would be required for the AGOS project. A CHMP is an agreement that establishes the mechanisms by which the proponent (in this case GA) will use to avoid harm to cultural heritage identified by the Barunggam People during the project. The CHMP was successfully negotiated by the parties and came into effect on 24 March 2014.

The second outcome of the discussions was the classification of the installation of radar corner reflectors as either a Category 4 or 5 activity under the Guidelines (ACHA 2003; pp. 8_10). Therefore, further investigation of the potential cultural heritage impact of the project was required. The spatial locations of the 40 proposed sites for co-located corner reflectors and survey marks were made available to the technical advisor appointed by the Barunggam People. The desktop assessment performed by the technical advisor identified that 15 of the 40 proposed sites had a moderate to high probability of impacting cultural heritage. This meant that an initial cultural heritage assessment (ICHA) of these sites would be required. This ICHA was undertaken in May 2014 by a representative from the Warner and Daylight families, the technical advisor and a representative from GA. The ICHA was able to clear all 15 sites for installation of AGOS infrastructure.

Once the cultural heritage clearances for all corner reflector (and survey mark) sites in the Western Downs region were completed, the land access agreements with each of the participating landowners were finalised. The agreement determined the specific location of the corner reflector and survey mark that would be installed, and the terms and conditions of access for the installation activities and subsequent observations.

At the completion of this first phase of the project, funding was available for a further 25 survey mark-only sites located to the west in the South West region of Queensland. Gas-production wells in the South West region are currently concentrated to the northeast of Roma in three distinct areas (Figure 1). The site spacing in the second phase was increased to around 40–50 km in order to achieve the widest coverage across the Surat Basin including regions where gas production is not currently occurring or has recently begun. Based on the lessons learned from the land access for the first 40 sites, the new sites were primarily located on road reserves and public land adjacent to several regional airports (e.g. Roma, Mitchell and Morven), and on private land where a willing landowner was identified through property visits or cold calling. With only survey marks being installed on these 25 sites, the potential impact of installation on cultural heritage was re-assessed, based on legal advice provided to GA, as Category 2 (Activities causing No Additional Surface Disturbance) under the Guidelines (ACHA 2003, p. 7). This meant that GA could proceed with the site selection and installation without the need for a CHMP with the Aboriginal groups, providing that these sites were located on areas where ground disturbance had already occurred (e.g. grazed paddocks and existing road corridors).

Construction of the geodetic network was completed on 21 November 2014. The positions of the 65 geodetic sites in the network are shown in Figure 1, and the GDA94 positions (with heights referenced to the GRS80 ellipsoid) of each survey mark and corner reflector are listed in the Supplementary Paper. Additionally, we describe the survey methods used to determine the positions in the Supplementary Paper. In the constructed geodetic network, 38 sites were established on private property with the remaining 27 sites located on road reserves administered by the State of Queensland. Forty co-located survey mark and corner reflector sites are concentrated in a 150 km × 150 km area in the Darling Downs region of the Surat Basin with a site spacing of 10–30 km. An additional network of 25 survey mark sites extends further west in the South West region to the western margin of the Surat Basin with an increased spacing of 30–70 km. One of the four previously mentioned AGOS GNSS CORS was established within the geodetic network near to Mitchell, Queensland with data logging commencing on 9 April 2014. The next closest GNSS CORS are located approximately 80 km east and 150 km northeast of the geodetic network at Toowoomba and Eidsvold, respectively (Figure 1).

4. SURVEY MARK DESIGN

Permanent survey marks for the AGOS geodetic network were designed to adhere to the recommendations laid out by the ICSM (2014a, b). Sites with a 360° unobstructed sky view 10° above the horizon were chosen. At a small number of sites where obstructions were unavoidable, the survey mark was placed to the north of that obstruction, since no GNSS satellites are in the sky to the south in an arc governed by the orbital inclinations.

Ideally, each survey mark should be directly coupled to stable *in situ* rock outcrops that are just above or below the ground surface. However, in order for the geodetic network to effectively sample spatial patterns of deformation at the basin scale, survey marks must also be located within the soil profile of the sedimentary basin. Vertosols are the dominant soil type in the northern Surat Basin. This soil type (commonly termed 'black soils') exhibits shrink-swell properties that could impart a signal correlated with changing moisture patterns on to surface-mounted survey marks. This shallow-soil signal component will imprint on any signal originating from the crust and will ensure that the overall geodetic signal measured on these survey marks will not uniquely tell us about deformation processes in the crust. The survey mark for installation in the soil profile was therefore designed to minimise the effect of shallow-soil signals and is specified as follows (Figure 2).

A hole of 450 mm diameter was drilled to a depth of 6.0 m below the ground surface unless bedrock or indurated material was reached at a shallower level. Survey mark hole depths are given in the Supplementary Paper. This hole was then sleeved to within 300 mm of the ground surface with a continuous section of PN6 PVC pressure pipe with external diameter of 400 mm. The gap between the pipe and hole excavation was in-filled with washed sand. A 25 mm diameter stainless steel rod (assembled from 2.0 m, 1.0 m, 0.5 m and 0.2 m lengths) was driven into the centre of the excavated hole to a refusal depth of at least 1.0 m for soil or at least 200 mm into indurated material depending on the resistance. Where bedrock was encountered rather than indurated material, the rod was secured into a hole drilled in the bedrock to a depth of up to 250 mm using epoxy mortar. At all sites, the verticality of the stainless steel rod was measured using both a staff level and screw-on bubble level to ensure consistency of the verticality tolerance across the network. A reinforcement cage was inserted into the hole to provide horizontal stability to the stainless steel rod and to help maintain the structural strength and integrity of the concrete. The hole was then filled with pre-mixed concrete with a compressive strength of 32 MPa after 28 days. A 300 mm length of smaller 265 mm-diameter PVC pressure pipe was secured into the upper 150 mm of the hole. This short length of pipe was kept free of concrete in order to allow unimpeded access to the length of stainless steel rod housed within (see Figure 2b). The top 300 mm of the hole was further excavated to accommodate a valve box cover with its top surface flush with the ground surface.

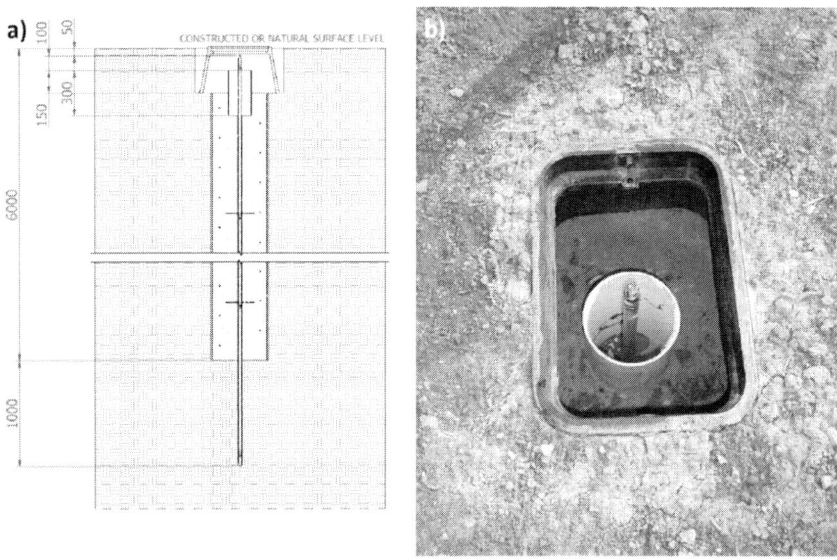

Figure 2 (a) Engineering drawing of the soil-profile survey mark design. Dimensions in millimetres. (b) Installed soil profile survey mark at Site 8 with valve box lid removed.

At two sites (30 and 33), bedrock exists at the surface, and a different survey mark design was installed (Figure 3). In the bedrock design, a 12 mm threaded rod was secured with epoxy mortar into a drilled hole in the bedrock to a depth of at least 250 mm. Three 6 mm levelling screws were also secured into the bedrock, and used to level the exposed instrument disc with reference to a bubble level temporarily screwed on to the centralised thread of the instrument disc. Following levelling, the three levelling screws are locked with nuts and cut off to the level of the disc.

A removable instrument holder was also designed that will be used to attach a GNSS antenna to the *in situ*survey mark during campaign measurements (see Figure 3b). The instrument holder consists of a 500 mm rod that elevates the antenna above the ground surface and an adaptor ring that can be rotated to ensure the antenna is appropriately oriented (i.e. in the north direction). By elevating the GNSS antenna above the ground the impact of multi-path signal interference owing to the reflection of radio waves from the ground surface is reduced. The antenna connects to the instrument holder using a 5/8" British Standard Whitworth (BSW) thread, which is the standard in all surveying forced centrings, including GNSS instrumentation.

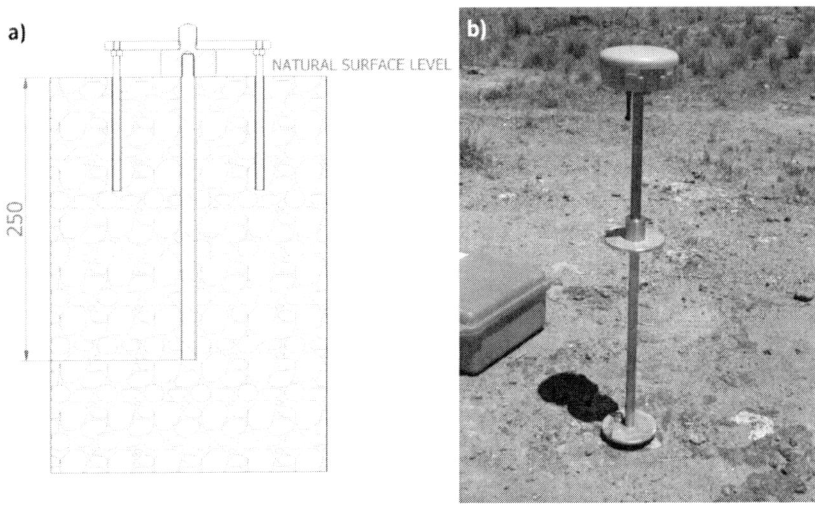

Figure 3 (a) Engineering drawing of bedrock survey mark design. Dimensions in millimetres. (b) Installed bedrock survey mark at Site 30 with attached instrument holder and GNSS antenna during a survey (the pink length of rod is an additional component of the particular GNSS antenna model used here).

Each AGOS survey mark installed within the soil profile cost approximately $8000 (excluding GST). Each AGOS survey mark installed onto bedrock cost approximately $5000 (excluding GST).

5. CORNER REFLECTOR DESIGN

Corner reflectors were chosen as the most suitable radar target for deployment in the AGOS geodetic network. A corner reflector is a passive device that reflects incoming electromagnetic energy directly back to the source of that energy, similar to an optical reflecting prism. In the case of a trihedral corner reflector, this is facilitated by the bouncing of the incident wave from three mutually orthogonal plates. The orthogonality of all three plates ensures that the corner reflector resembles the corner of a cube (and hence the name 'corner reflector'). Corner reflectors have traditionally been used as a target suitable for calibration of SAR images (Freeman 1992) but they are also popular as targets with a temporally stable phase response suitable for accurately measuring ground deformation using InSAR techniques (e.g. Hanssen 2001; Ketelaar 2009; Strozzi et al. 2013). Previous validation experiments have found that artificial reflectors can yield displacement estimates from Persistent Scatterer InSAR analysis with sub-millimetre accuracy in both vertical and east–west directions (Ferretti et al. 2007).

Through a prototyping exercise a single corner reflector design and size was sought that would provide a bright response in X-, C- and L-band SAR imagery and that would exhibit stable phase characteristics over long (nominally decadal) time periods for permanent deployment in the landscape (Garthwaite et al. 2015). Following this exercise, a triangular trihedral design with an inner-leg length (i.e. the non-hypotenuse edge of the right-angled isosceles triangle) of 1.5 m was chosen because of the relative simplicity of manufacture, long term structural rigidity, relative stability for large radar cross-section and a three decibel beamwidth of approximately 40_ (Curlander & McDonough 1991). This last property ensures that the permanently deployed corner reflectors can be aligned in an average orientation for all orbiting SAR sensors (on either an ascending or descending orbital pass) yet the signal-to-noise ratio of the corner reflector response in SAR imagery for each sensor will still be high.

Corner reflector panels were manufactured from 6 mm aluminium sheet, with cross-braces of 6 mm-thick, 50 mm-width aluminium L-section to improve structural rigidity (Figure 4). Particular care was taken through the manufacture and installation process to maintain flatness of each panel and minimise surface irregularities that might adversely impact the reflective performance of the corner reflector. As such, the L-section cross-braces were attached to each panel using insert studs that sit flush on the reflective side of each panel. Each assembled panel was powder-coated matt grey in order to enhance the longevity of the panel and to reduce the impact of thermal expansion. Three almost-identical corner reflector panels are mounted to a stand manufactured from steel hollow section that has been welded together and then powder-coated. The design of the stand includes azimuth and elevation adjustability that enables 360° azimuth adjustment and base-plate elevations between 0 and 40° (with respect to local horizontality). This adjustment range ensures that once permanently deployed, the corner reflector can be oriented for most imaging modes of present and future SAR sensors regardless of its location within radar imaging swathes. Each AGOS corner reflector assembly (three almost-identical panels plus mounting stand) cost approximately $5400 (excluding GST) to manufacture.

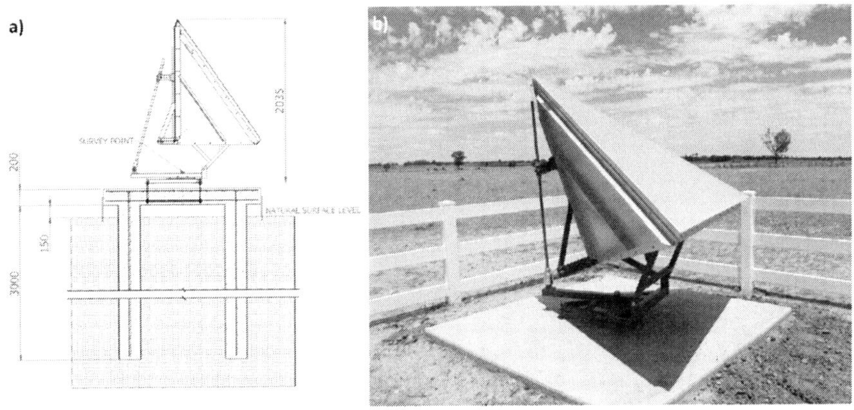

Figure 4 (a) Engineering drawing of the 1.5 m triangular trihedral corner reflector (side-on view) and concrete slab design for construction in the soil profile. Dimensions in millimetres. (b) 1.5 m corner reflector installed at Site 26. Concrete slab dimension is 2 m square

Corner reflectors were generally placed to the north of co-located survey marks with baseline lengths between 20 and 40 m to ensure the corner reflector does not encroach on the sky view of the survey mark. For sites installed into the soil profile a 200 mm thick, 2 m × 2 m suspended concrete slab supported on four 250 mm diameter PVC-cased concrete pillars was constructed. A centrally located reinforcing bar in the concrete pillars was tied to the reinforcing mesh within the slab, and concrete with a compressive strength of 32 MPa after 28 days was poured in one pour. All four pillars were required to be equivalent lengths up to 3.0 m and sitting on the same ground type. If bedrock was encountered at a depth shallower than 3.0 m, then the reinforcing bar was secured into the bedrock to a depth of 250 mm. Average depths for the four pillars at each site are given in the Supplementary Paper. The suspended slab is poured over collapsible formwork that will decompose over time to leave an air void between the slab and ground surface of 150 mm. The aim here is to isolate the corner reflector from the influence of shrinking and swelling in the shallow soil layer. After curing of the concrete, the sub-slab air void was blocked off using 300 mm-tall Compressed Fibre Cement sheeting to stop vermin gaining access to the void following formwork decomposition. The edges of the slab were then backfilled to a gentle slope using material from the hole excavation. At site 30, bedrock existed at the surface so the slab was poured directly on to the rock surface. In place of the four sub-surface concrete pillars, four 20 mm-diameter, 250 mm-deep holes were drilled into the rock into which a 16 mm reinforcing bar was secured with epoxy-mortar. These reinforcing bars were tied to the mesh layers within the slab. All slabs were poured so that one edge was oriented in the true north direction. The corner reflector assembly was secured to the slab by way of three 16 mm threaded rods that protruded above the slab surface and were connected to the bolt cage and reinforcing mesh set within the concrete.

At 33 sites, the corner reflectors were installed within a 5 m × 5 m fenced enclosure in order to protect them from interference by grazing livestock (and avoid injury to livestock caused by the sharp corners). The fences are manufactured from PVC in order to minimise their impact (if any) on the radar response of the corner reflector. Given that most corner reflectors are installed on a slab whose top surface is 350 mm above ground level and on a stand that supports the bottom panel a minimum of 535 mm above the slab, the 1.2 m high fence does not obstruct the sky view of the corner reflector whose boresight is elevated ~35° above the bottom panel. The ground infrastructure for each corner reflector cost approximately $8200, excluding GST.

6. FUTURE USE OF THE GEODETIC NETWORK

The deployed network of corner reflectors and surrounding region will be imaged periodically by orbiting SAR satellites. These SAR images will be analysed using differential InSAR techniques to produce interferograms and subsequently yield deformation maps of the height change through time. Typical InSAR interferograms contain noise from a multitude of sources but with a dominant contribution from the atmosphere. Garthwaite *et al.* (2013) found that the magnitude of noise is 20 mm or less in 48 out of a stack of 51 interferograms covering the Tibetan Plateau. When producing the deformation map the impact of noise is mitigated by 'stacking' multiple interferograms to reduce the variance of the noise by a factor proportional to the number of independent interferograms (Emardson *et al.*2003) Given 20 mm of noise in the interferograms, 10 mm/yr of deformation could be resolved after 1.2 years of monthly SAR acquisitions, which increases to 1.6 years for 30 mm of noise (Emardson *et al.*2003) Since InSAR is a differential technique, deformation maps are referenced relative to a particular acquisition epoch but not an absolute coordinate reference system. The network of co-located corner reflectors and survey marks will be used as the connection between the two to ensure that the absolute height of InSAR-derived products is accurate.

Annual campaign GNSS surveys will be conducted that observe all 65 AGOS survey marks (including the 40 co-located with corner reflectors). The first of these campaign surveys will be undertaken in 2015. Each survey mark will be occupied for a period of at least a week, which is a sufficient occupation length to reduce the height uncertainty to a few centimetres.

Brockmann (1997) derived relationships that can be used to estimate the precision of estimated velocities and coordinates from geodetic observations given defined measurement uncertainties. Importantly, these precisions are independent of the deformation rate and are proportional to the measurement uncertainty. Assuming linear deformation, a vertical measurement uncertainty of 0.1 m and our campaign strategy, the precision of the velocity estimation would drop below 1 mm/yr after six campaign occupations (i.e. 5 years). However, given that the magnitude of this height uncertainty is a worst-case scenario and not representative of all sites, this period can be considered an upper-bound on the time needed to achieve this precision from the new geodetic network. It should be noted that our defined campaign strategy gives a precision that is around 15% of that achievable from a CORS network (Brockmann 1997)

Observed data from the AGOS survey marks will be included in network adjustments with other survey marks and GNSS CORS to determine coordinates consistent with the new Australian national datum and heights with respect to a stable vertical reference frame (Bevis *et al.* 2013). The co-located corner reflectors will enable the accuracy of the determined three-dimensional coordinates to be propagated into the InSAR analysis. The geodetic surface observations combined in this way will then be used as a constraint to geophysical inversions that make inferences about the temporal evolution of the poroelastic properties of the Walloon Subgroup of the Surat Basin (e.g. Brown *et al.* 2014).

ACKNOWLEDGEMENTS

This work was funded by the AuScope Australian Geophysical Observing System (AGOS) through funding from the Education Investment Fund (EIF) Round 3. The implementation of the AGOS geodetic network was a complicated project that could not have been achieved without the assistance of the following: AuScope Ltd; the Queensland State Department of Natural Resources and Mines, and Department of Transport and Main Roads; the Western Downs, Maranoa and Murweh regional councils; CS Energy, Linc Energy, Origin Energy, QGC, Santos and Yancoal; team members of the Geodesy Section, Geoscience Australia for many contributions to this work. We are grateful to the private landowners who agreed to host a geodetic site, and the traditional owners of the lands, in particular: the representatives of the Warner and Daylight families; the Barunggam People (all sites cleared for installation under a Cultural Heritage Management Plan established on 24 March 2014); the Iman People (all sites inspected and cleared for installation by an elder, Uncle Patrick, in September 2014); and the Mandandanji, Gunggari and Bigambul peoples (all sites installed under Duty of Care Category 2). McCall Consulting performed the site reconnaissance and secured land access, Mipec installed the survey marks and corner reflector slabs, J&H Williams manufactured the corner reflectors and Twin City Engineering manufactured the survey mark components. The paper was improved by the comments of Nicholas Brown, Anna Riddell and two anonymous reviewers. This paper is published with the permission of the CEO, Geoscience Australia.

REFERENCES

1. ACHA. 2003. Aboriginal cultural heritage act 2003 duty of care guidelines. Queensland State Government Department of Aboriginal and Torres Strait Islander and Multicultural Affairs, Brisbane Qld
2. Altamimi Z., Collilieux X. & Métivier L. 2011. ITRF2008: an improved solution of the international terrestrial reference frame. *Journal of Geodesy* **85**, 457–473.
3. Bell J. W., Amelung F., Ferretti A., Bianchi M. & Novali F. 2008. Permanent scatterer InSAR reveals seasonal and long-term aquifer-system response to groundwater pumping and artificial recharge. *Water Resources Research* **44**, W02407.
4. Bevis M., Brown A. & Kendrick E. 2013. Devising stable geometrical reference frames for use in geodetic studies of vertical crustal motion. *Journal of Geodesy* **87**, 311–321.
5. Brockmann E. 1997. Combination of solutions for geodetic and geodynamic applications of the Global Positioning System (GPS). *Geodesy-Geophysics Arb. Schweiz* **55**, 211, Zurich Switzerland
6. Brown N. J., Woods A. W., Neufeld J. A. & Richardson C. 2014. Constraining surface deformation predictions resulting from coal seam gas extraction. *Geoscience Australia* **2014/44**, Canberra ACT.
7. Chaussard E., Bürgmann R., Shirzaei M., Fielding E. J. & Baker B. 2014. Predictability of hydraulic head changes and characterization of aquifer-system and fault properties from InSAR-derived ground deformation. *Journal of Geophysical Research: Solid Earth* **119**, 6572–6590.
8. Curlander J. C. & McDonough R. N. 1991. *Synthetic aperture radar systems and signal processing*, John Wiley & Sons, Inc, New York NY.
9. Dawson J., Cummins P., Tregoning P. & Leonard M. 2008. Shallow intraplate earthquakes in Western Australia observed by Interferometric Synthetic Aperture Radar. *Journal of Geophysical Research: Solid Earth* **113**, B11408.
10. Dawson J. & Woods A. 2010. ITRF to GDA94 coordinate transformations. *Journal of Applied Geodesy* **4**, 189.
11. Emardson T. R., Simons M. & Webb F. H. 2003. Neutral atmospheric delay in interferometric synthetic aperture radar applications: Statistical description and mitigation. *Journal of Geophysical Research: Solid Earth* **108**, 2231.
12. Exon N. F. 1976. Geology of the Surat Basin in Queensland. *Bureau of Mineral Resources, Geology and Geophysics*Bulletin 166, 0642019479, Canberra ACT.
13. Featherstone W., Filmer M., Penna N. T., Morgan L. M. & Schenk A. 2012. Anthropogenic land subsidence in the Perth Basin: challenges for its retrospective geodetic detection. *Journal of the Royal Society of Western Australia* **95**, 53–62.
14. Ferretti A., Savio G., Barzaghi R., Borghi A., Musazzi S., Novali F., Prati C. & Rocca F. 2007. Submillimeter Accuracy of InSAR Time Series: Experimental Validation. *IEEE Transactions on Geoscience and Remote Sensing* **45**, 1142–1153.
15. Fielding E. J., Blom R. G. & Goldstein R. M. 1998. Rapid subsidence over oil fields measured by SAR interferometry. *Geophysical Research Letters* **25**, 3215–3218.

16. Freeman A. 1992. SAR calibration: an overview. *IEEE Transactions on Geoscience and Remote Sensing* **30**, 1107–1121.
17. Galloway D. L. & Burbey T. J. 2011. Review: regional land subsidence accompanying groundwater extraction. *Hydrogeology Journal* **19**, 1459–1486.
18. Galloway D. L., Hudnut K. W., Ingebritsen S. E., Phillips S. P., Peltzer G., Rogez F. & Rosen P. A. 1998. Detection of aquifer system compaction and land subsidence using interferometric synthetic aperture radar, Antelope Valley, Mojave Desert, California. *Water Resources Research* **34**, 2573–2585.
19. Garthwaite M. C., Nancarrow S., Hislop A., Thankappan M., Dawson J. H. & Lawrie S. 2015. Design of radar corner reflectors for the Australian Geophysical Observing System. *Geoscience Australia*, 2015/03, Canberra ACT.
20. Garthwaite M. C., Wang H. & Wright T. J. 2013. Broadscale interseismic deformation and fault slip rates in the central Tibetan Plateau observed using InSAR. *Journal of Geophysical Research: Solid Earth* **118**, 5071–5083.
21. Geoscience Australia and BREE. 2012. Australian Gas Resource Assessment 2012. *Geoscience Australia and BREE.* <http://www.bree.gov.au/documents/publications/_other/gasresourceassessment.pdf>, Canberra ACT.
22. Hanssen R. F. 2001. *Radar interferometry. data interpretation and error analysis*, Kluwer Academic Publishers, Dordrecht, the Netherlands.
23. ICSM. 1998. Geocentric datum of Australia, technical manual. *Intergovernmental Committee on Surveying and Mapping (ICSM)*, Canberra ACT.
24. ICSM. 2014a. Guidelines for the Installation and Documentation of Survey Control Marks. *Intergovernmental Committee on Surveying and Mapping, Permanent Committee on Geodesy (PCG)* **Special Publication 1**, Canberra ACT.
25. ICSM. 2014b. Standard for the Australian Survey Control Network. *Intergovernmental Committee on Surveying and Mapping (ICSM), Permanent Committee on Geodesy (PCG)* **Special Publication 1**, Canberra ACT.
26. Ketelaar V. B. H. 2009. *Satellite radar interferometry. subsidence monitoring techniques*, Springer, Berlin Germany.
27. Onuma T. & Ohkawa S. 2009. Detection of surface deformation related with CO2 injection by DInSAR at In Salah, Algeria. *Energy Procedia* **1**, 2177–2184.
28. Queensland State Government 2015. Queensland's petroleum and coal seam gas 2013–14. *Department of Natural Resources and Mines.* https://www.dnrm.qld.gov.au/_data/assets/pdf_file/.../petroleum.pdf, Brisbane Qld.
29. Reeves J. A., Knight R., Zebker H. A., Kitanidis P. K. & Schreüder W. A. 2014. Estimating temporal changes in hydraulic head using InSAR data in the San Luis Valley, Colorado. *Water Resources Research* **50**, 4459–4473.
30. Schmidt D. A. & Bürgmann R. 2003. Time-dependent land uplift and subsidence in the Santa Clara valley, California, from a large interferometric synthetic aperture radar data set. *Journal of Geophysical Research: Solid Earth* **108**, 2416.
31. Scott S., Anderson B., Crosdale P., Dingwall J. & Leblang G. 2007. Coal petrology and coal seam gas contents of the Walloon Subgroup — Surat Basin, Queensland, Australia. *International Journal of Coal Geology* **70**, 209–222.
32. Strozzi T., Teatini P., Tosi L., Wegmüller U. & Werner C. 2013. Land subsidence of natural transitional environments by satellite radar interferometry on artificial reflectors. *Journal of Geophysical Research: Earth Surface* **118**, 1177–1191.
33. Tregoning P., Burgette R., McClusky S. C., Lejeune S., Watson C. S. & McQueen H. 2013. A decade of horizontal deformation from great earthquakes. *Journal of Geophysical Research: Solid Earth* **118**, 2371–2381.
34. Watson C., Burgette R., Tregoning P., White N., Hunter J., Coleman R., Handsworth R. & Brolsma H. 2010. Twentieth century constraints on sea level change and earthquake deformation at Macquarie Island. *Geophysical Journal International* 182, 781–796.

CHAPTER 8

Geomagnetic storms of cycle 24 and their solar sources

Shinichi Watari

National Institute of Information and Communications Technology, 4-2-1, Nukuikita, Koganei, Tokyo 184-8795, Japan

ABSTRACT

Solar activity of cycle 24 following the deep minimum between cycle 23 and cycle 24 is the weakest one since cycle 14 (1902–1913). Geomagnetic activity is also low in cycle 24. We show that this low geomagnetic activity is caused by the weak dawn-to-dusk solar wind electric field ($E_{d\text{-}d}$) and that the occurrence rate of $E_{d\text{-}d} > 5$ mV/m decreased in the interval from 2013 to 2014. We picked up seventeen geomagnetic storms with the minimum Dst index of less than −100 nT and identified their solar sources in cycle 24 (2009–2015). It is shown that the relatively slow coronal mass ejections contributed to the geomagnetic storms in cycle 24.

Keywords: Geomagnetic storm Rising and maximum phases, Two peaks, Solar cycle, 24 Coronal mass ejection, Coronal hole

1. INTRODUCTION

The solar minimum period between cycle 23 and cycle 24 was the lowest and the longest one since the minimum between cycle 14 and cycle 15, and is called as the 'deep minimum' (Russell et al. 2010; Richardson and Cane

2012a; Richardson 2013; McComas et al. 2013). The solar activity of the current solar cycle (No. 24), following the extraordinary low minimum, is also low (Kamide and Kusano 2013; Gopalswamy et al. 2014; Watari et al. 2015). The maximum of cycle 24, determined by the 13-month smoothed monthly sunspot number (SSN), occurred in April 2014. The maximum SSN was 116.4 according to the World Data Center for Sunspot Index and Long-term Solar Observation (WDC-SILSO), Royal Observatory of Belgium, Brussels. This is the smallest one ever observed since the maximum of cycle 14 (SSN of 107.1 in February 1906). The SSNs of cycle 24 show two peaks: 98.3 in March 2012 and 116.4 in April 2014 (Svalgaard and Kamide 2013; Gopalswamy et al. 2015). On a two-peak variation of geomagnetic activities seen in past solar cycles, Gonzalez et al. (1990) and Echer et al. (2011) noted that the first peak, appearing in the maximum phase, is caused by coronal mass ejections (CMEs) and the second peak, appearing in the declining phase, is caused by high-speed streams from coronal holes. Gopalswamy (2008) pointed out latitudinal distribution of CMEs had a close connection to the two-peak characteristics of geomagnetic activities.

To study the nature of the weak activity of cycle 24, we examined long-term variations of geomagnetic activities, expressed by the Dst index, comparing with the SSNs and the solar wind data. We also investigated the solar sources of the geomagnetic storms in the rising and the maximum phases of cycle 24 using the solar and the solar wind data.

There are numerous studies on geomagnetic storms and their solar sources in past cycles (Zhang et al. 2007; Echer et al. 2008, 2011; Richardson et al. 2006; Richardson and Cane 2012a, b; references therein). According to the previous studies, the principal solar sources of intense geomagnetic storms (minimum Dst < −100 nT) were identified to be CMEs and approximately 11–14% of the storms were associated with high-speed streams from coronal holes (see Fig. 4 in Zhang et al. (2007) and Fig. 10 in Echer et al. (2008)). We will compare these results of the previous cycles with our study of cycle 24.

2. DATA SOURCES

We used the Dst index data, provided by the WDC for Geomagnetism, Kyoto, as the parameter of the level of geomagnetic activity. There are three categories: the real-time, the provisional, and the final indices. The real-time Dst is derived from the unverified raw data. The provisional Dst is calculated using the data visually screened for artificial noises. The WDC provides the final Dst between 1957 and 2011 and the provisional Dst between 2012 and 2014 and for March 2015 at the time of this analysis. Long-term variation of geomagnetic activity, represented by the Dst data, will be compared with the SSN data, provided by the WDC-SILSO, and the NASA/OMNI solar wind data (http://omniweb.gsfc.nasa.gov).

3. DATA ANALYSIS AND DISCUSSION

3.1 Long-term variation of geomagnetic activity since 1957

Because the comprehensive dataset of the Dst index is available since 1957, we can see the long-term correlation characteristics between solar and geomagnetic activities in the interval including six solar maxima. Figure 1 shows the yearly and the 13-month smoothed monthly SSNs and occurrence rates of the daily minimum Dst of less than −100, −200, and −300 nT, respectively for each year. The smoothed curve of the monthly SSNs of cycle 24 shows a peak in 2012 and 2014, respectively. Also seen in this figure is a decrease of the geomagnetic activity (Dst < −100 nT) in 2013–2014. This decrease reflects a decreasing tendency of the number of halo CMEs in 2013 and 2014 as noted by Gopalswamy et al. (2015). Cycle 22 also shows two peaks of SSNs taking place in 1989 and in 1991, respectively, and there is a decrease of the geomagnetic activity (Dst < −100 nT) between these peaks, namely in 1990.

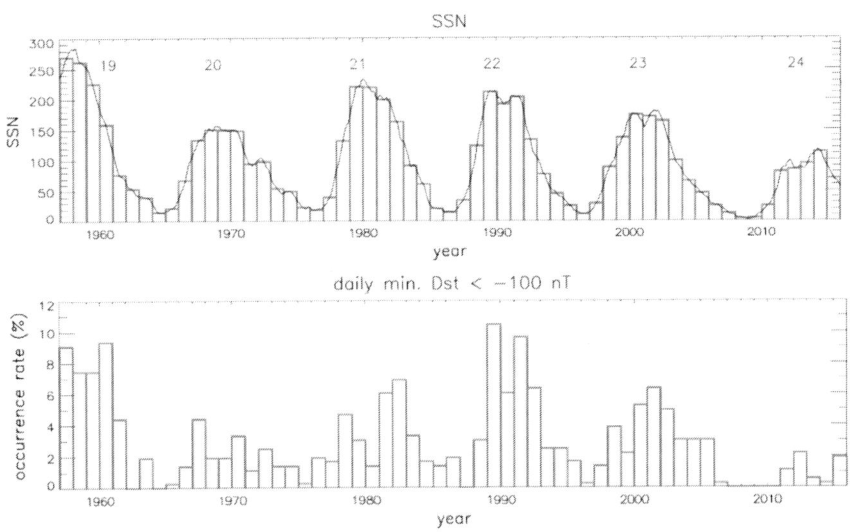

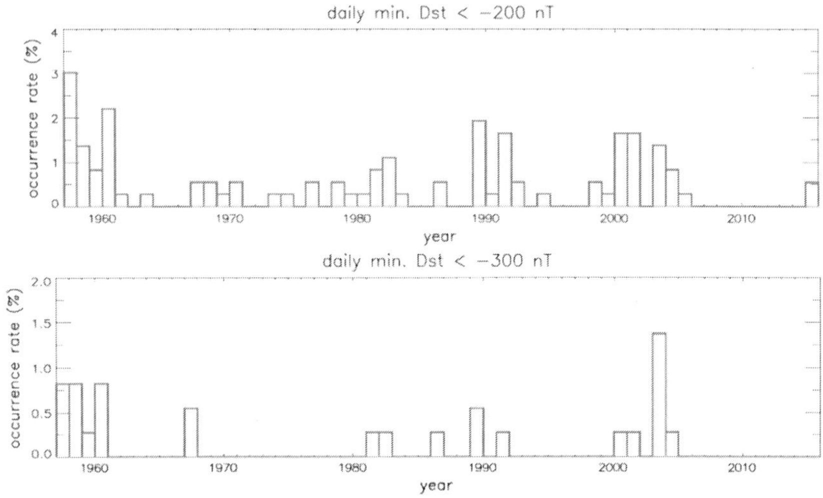

Figure. 1 *Top panel* is the yearly SSNs (*red line*) with the 13-month smoothed monthly SSNs (*black line*). The *second, third, fourth panels* show yearly occurrence rates of the daily minimum Dst of less than −100, −200, and −300 nT, respectively

3.2 Correlation between geomagnetic activity and SSN

According to Fig. 1, the level of the geomagnetic activity tends to be proportional to that of the solar activity. Figure 2 shows scatter plots of occurrence rates of the daily minimum Dst of less than −50, −100, −200, and −300 nT, respectively for each year, against the yearly averaged SSNs. For the data points in each panel for the Dst of <−50, <−100, and <−200 nT, respectively, in Fig. 2, we can see the presence of an upper border line having the positive inclination, namely the occurrence rate tends to be increased with respect to the yearly SSN. Similar tendency can be seen also in the case of strong geomagnetic activities (Dst < −300 nT) because the 'no-event' points are nested in the low SSN part of the diagram.

3.3 Correlation between geomagnetic activity and solar wind electric field

It is known that the rate of energy injection into the magnetospheric ring current, affecting change of Dst, is a function of the dawn-to-dusk solar wind electric field, $E_{d\text{-}d}$ (Burton et al. 1975; O'Brien and McPherron 2000). Figure 3 shows scatter plots of the daily minimum of Dst versus the daily maximum $E_{d\text{-}d}$ in the rising–maximum phases of cycle 23 (1996–2002) and cycle 24 (2009–2015). In this analysis, we use the 1-h averaged solar wind data since 1996 because the time resolution of Dst is 1 h and coverage of the solar wind data is approximately 100% since this year.

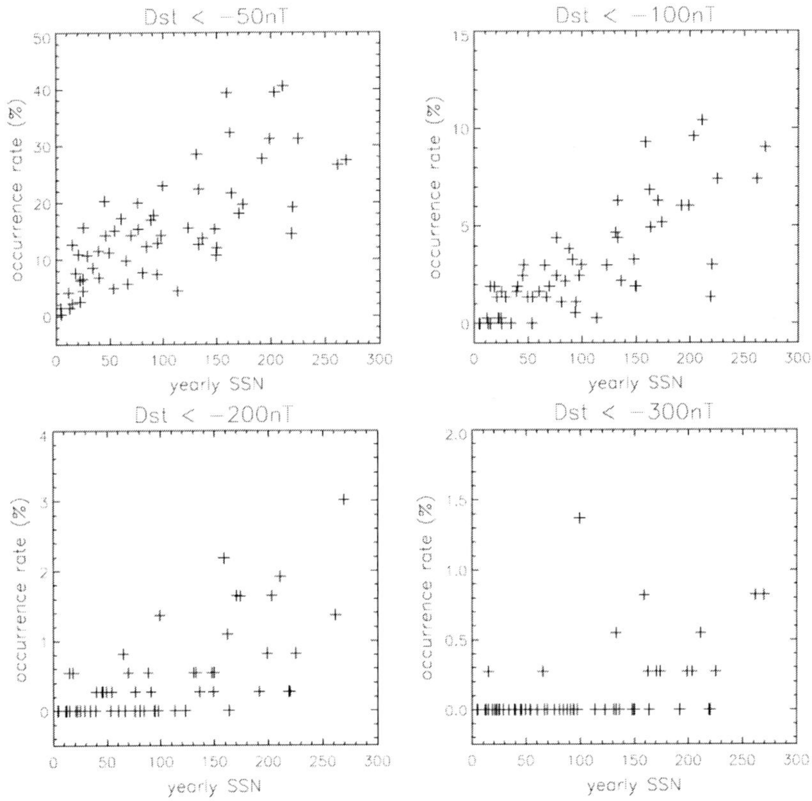

Figure. 2 Yearly SSNs and the yearly occurrence rates of the daily minimum Dst of less than −50, −100, −200, and −300 nT, respectively

The number of the data points of Dst < −100 nT and $E_{d\text{-}d}$ > 5 mV/m is 60 points in cycle 23 and 18 points in cycle 24. $E_{d\text{-}d}$ of larger than 20 mV/m or Dst of less than −250 nT were not observed in the rising–maximum phases of cycle 24. It is also seen in Fig. 3 that there exists a remarkable tendency in the cases of strong electric fields or strong geomagnetic storms. This scattering tendency is produced by the data points obtained in the growth or the recovery phase of individual geomagnetic storms. For example, the data point corresponding to $E_{d\text{-}d}$ = 14.3 mV/m and Dst = −118 nT in the right panel of Fig. 3 is associated with the shock and the sheath of a CME, which was identified to be the source of the 26 September 2011 storm. On the other hand, the data point of $E_{d\text{-}d}$ = 17.8 mV/m and Dst = −121 nT in the right panel of Fig. 3 is associated with the growth phase of the 22 June 2015 storm.

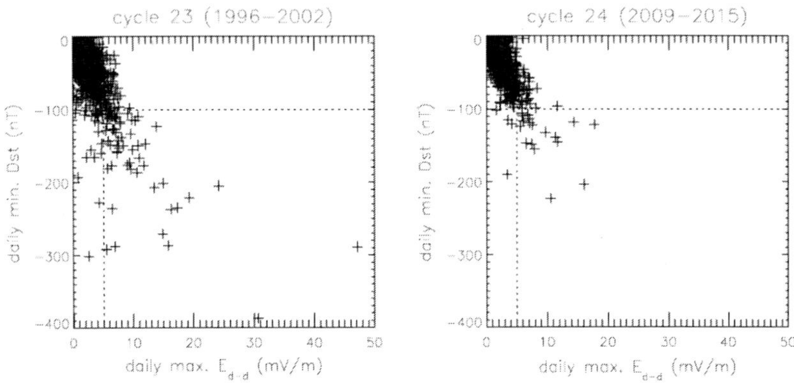

Figure. 3 Daily maximum dawn-dusk solar wind electric fields (E_{d-d}) and the daily minimum Dst for the rising–maximum phases of cycle 23 (1996–2002, the *left panel*) and cycle 24 (2009–2015, the *right panel*). The *horizontal dotted line* shows the daily minimum Dst of −100 nT, and the *vertical dotted line* shows the daily maximum E_{d-d} of 5 mV/m

Figure 4, respectively, shows the yearly and the 13-month smoothed monthly SSNs (the top panel) and the occurrence rates of $|E_{d-d}| > 5$ mV/m (the second panel), of $E_{d-d} > 5$ mV/m (the third panel), and of $E_{d-d} < -5$ mV/m (the forth panel) in the 1-h averaged solar wind data of the year. The bottom panel shows yearly occurrence ratios of $E_{d-d} < -5$ mV/m to $|E_{d-d}| > 5$ mV/m. The occurrence rates of $|E_{d-d}| > 5$ mV/m in the rising–maximum phases of cycle 24 are generally smaller than those in the same phase of cycle 23 and the occurrence rate of $|E_{d-d}| > 5$ mV/m decreased in 2013–2014. The averaged ratio of $E_{d-d} < -5$ mV/m to $|E_{d-d}| > 5$ mV/m enhanced remarkably in 2014, namely in the maximum phase of cycle 24, probably due to an increase of the number of magnetic clouds with fields rotating north–south (NS-type) in even cycles. Bothmer and Rust (1997) noted that the NS-type magnetic clouds are expected to be observed more frequently in even solar cycles than in odd cycles. Northward magnetic field of front-side of the NS-type magnetic clouds tends to be strengthened by compression during their propagation from Sun to Earth. This will increase the occurrence rate of the strong positive electric filed. On the other hand, Kilpua et al. (2012) pointed out that there is no obvious difference in the geoeffectiveness between SN-type and NS-type magnetic clouds. To examine the difference of solar wind characteristics controlling E_{d-d}, we analysed solar wind parameters related to E_{d-d} in the following subsection.

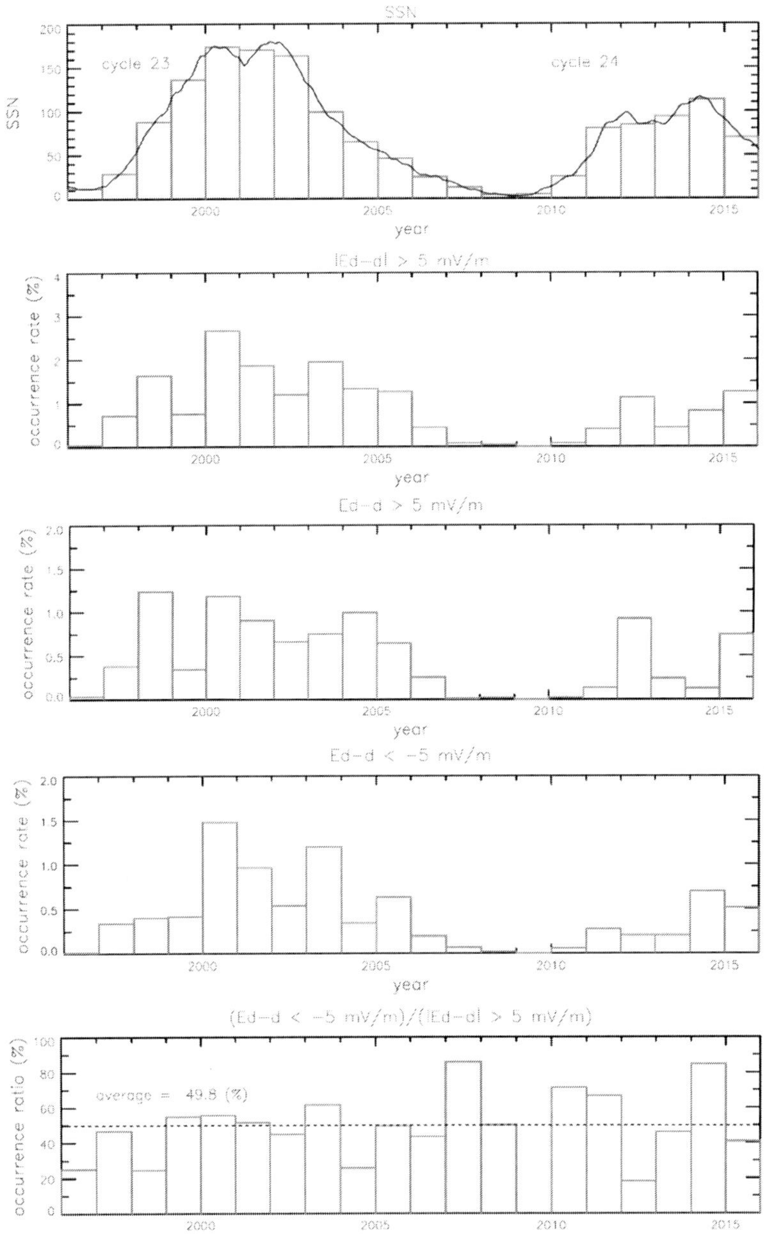

Figure. **4** *Top panel* is the yearly SSNs (*red line*) with the 13-month smoothed monthly SSNs (*black line*). The *second*, *third*, and *forth panels* are the yearly occurrence rates of $|E_{d\text{-}d}| > 5$ mV/m, of $E_{d\text{-}d} > 5$ mV/m, and of $E_{d\text{-}d} < -5$ mV/m in the 1-year data. The *bottom panel* is yearly occurrence ratios of $E_{d\text{-}d} < -5$ mV/m to $|E_{d\text{-}d}| > 5$ mV/m

3.4 Correlation between $E_{d\text{-}d}$ and solar wind parameters

The $E_{d\text{-}d}$ is given by a product of the solar wind speed (*V*) and the north–south component of solar wind magnetic field (Bz). Figure 5 shows scatter plots of Bz versus *V* in the rising–maximum phases of cycle 23 (1996–2002) and cycle 24 (2009–2015). Doted curves in the plots show $E_{d\text{-}d}$ of ±5, ±10, ±15, and ±20 mV/m, respectively.

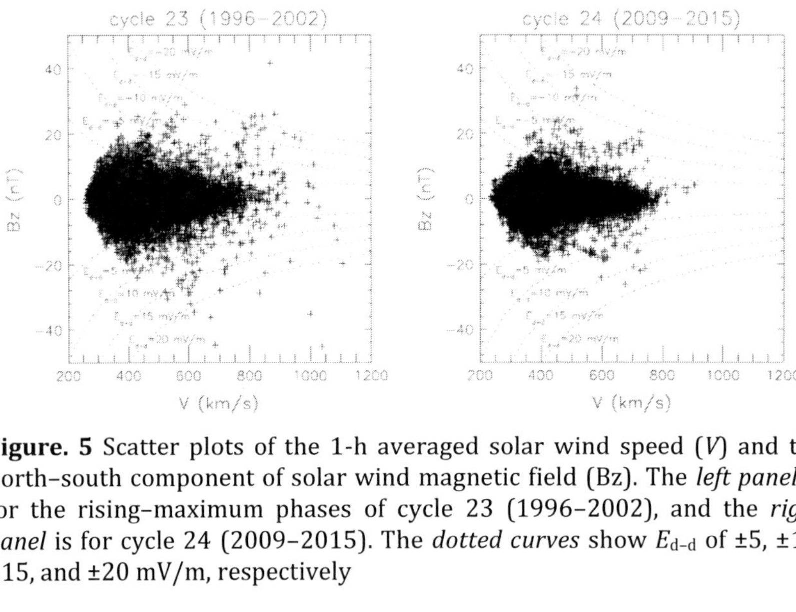

Figure. 5 Scatter plots of the 1-h averaged solar wind speed (*V*) and the north–south component of solar wind magnetic field (Bz). The *left panel* is for the rising–maximum phases of cycle 23 (1996–2002), and the *right panel* is for cycle 24 (2009–2015). The *dotted curves* show $E_{d\text{-}d}$ of ±5, ±10, ±15, and ±20 mV/m, respectively

The number of data points (193) corresponding to the cases of $E_{d\text{-}d} > 5$ mV/m in cycle 24 is smaller than that (415) in cycle 23, and no data point can be seen in the area of both $V > 750$ km/s and Dst < −10 nT in cycle 24. This suggests that the solar wind disturbances associated with geomagnetic storms in cycle 24 are weaker than those in cycle 23. Averaged *V* and Bz for the data points corresponding to $E_{d\text{-}d} > 5$ mV/m are 532 km/s and −14 nT for cycle 23, and 503 km/s and −14 nT for cycle 24, respectively.

3.5 Geomagnetic storms in the rising–maximum phases of cycle 24

We selected seventeen geomagnetic storms with the minimum Dst of less than −100 nT in the list of geomagnetic storms (2009–2015) provided from the Kakioka Magnetic Observatory, the Japan Meteorological Agency (http://www.kakioka-jma.go.jp/obsdata/metadata/en/products/list/event/ kak). Their solar sources are identified using the NASA/OMNI solar wind data and the solar data obtained by the Solar and Heliosphere Observatory (SOHO)/the Large Angle and Spectrometric Coronagraph (LASCO) (ESA/NASA) and by the Solar Dynamics Observatory (SDO)/Atmospheric Imaging Assembly (AIA) (NASA, http://sdo.gsfc.nasa.gov/).

Table 1 shows characteristics of the selected geomagnetic storms with information on their solar sources. The most intense geomagnetic storm in the rising–maximum phases of cycle 24 is the 17 March 2015 storm, currently called as 'the St. Patrick's Day storm', with the minimum Dst index of −223 nT (Kamide and Kusano 2015). Kataoka et al. (2015) noted that this storm was intensified by interaction of a CME and following high-speed stream shortly before its arrival at the Earth. According to Table 1, occurrence of the geomagnetic storms in cycle 24 showed two-peak characteristics. There are six storms in 2012 and five storms in 2015, but only three storms occurred between 2013 and 2014.

Table 1 Geomagnetic storms (minimum Dst < −100 nT) in the rising–maximum phases of cycle 24 with their characteristics

No.	Date	Min. Dst (nT)	Type	Solar sources	Speed at 1AU (km/s)
1	2011/08/05 17:50–2011/08/06 15:00	−115*1	SC	Full halo CME (sheath)	611
2	2011/09/26 12:35–2011/09/28 17:00	−118*1	SC	Full halo CME (sheath)	704
3	2011/10/24 18:31–2011/10/25 21:00	−147*1	SC	Full halo CME (sheath, MC)	534
4	2012/03/08 11:03–2012/03/10 19:00	−145*2	SC	Full halo CME (sheath, MC) multiple	737
5	2012/04/23 03:20–2012/04/26 16:00	−120*2	SC	Partial halo CME (sheath, MC)	394/720
6	2012/07/14 18:10–2012/07/17 12:00	−139*2	SC	Full halo CME (sheath, MC)	667
7	2012/09/30 11:32/23:05–2012/10/01 16:00	−122*2	SC	Full halo CMEs (sheath, MC) multiple	410
8	2012/10/08 05:16–	−109*2	SC	Partial halo CME	447/526

	2012/10/09 24:00			(sheath, MC) multiple	
9	2012/11/12 23:12–2012/11/14 19:00	−108*2	SC	Partial halo CME (sheath, MC) multiple	467
10	2013/03/17 06:00–2013/03/18 12:00	−132*2	SC	Full halo CME (sheath, MC) multiple	725
11	2013/05/31 16:17–2013/06/02 21:00	−119*2	SC	(MC?) coronal hole	402/774
12	2014/02/18 13:54–2014/02/19 23:00	−116*2	GC	Partial halo CME (MC) multiple	530
13	2015/03/17 04:45–2015/03/21 15:00	−223*2	SC	Partial halo CME (sheath, MC) coronal hole?	683
14	2015/06/22 18:33–2015/06/24 12:00	−204*3	SC	Full halo CME (sheath, MC) multiple	742
15	2015/10/07 04:24–2015/10/10 02:00	−124*3	GC	(MC) coronal hole	460/775
16	2015/12/19 16:17–2015/12/22 02:00	−155*3	SC	Full halo CME (MC)	497
17	2015/12/31 00:49–2016/01/01 16:00	−110*3	SC	Partial halo CME (MC)	485

SC sudden commencement, *GC* gradual commencement, *MC* magnetic cloud

*1Final Dst index

*2Provisional Dst index

*3Real-time Dst index

Also seen in Table 1 is that the geomagnetic storms were mainly caused by CMEs in the studied period of cycle 24 and that relatively slow CMEs contributed to the geomagnetic storms. Two geomagnetic storms in this table were mainly produced by high-speed solar wind from coronal holes.

SUMMARY

We made a comprehensive data analysis of long-term variations of geomagnetic activities with the SSNs and the solar wind data to investigate characteristics of the low geomagnetic activity of cycle 24. We identified the solar sources of the geomagnetic storms in the rising–maximum phases of cycle 24 using the solar and the solar wind data. Geomagnetic activity of the studied period of cycle 24 (2009–2015) was in the lowest level in the recent six solar cycles since 1957. There were only seventeen geomagnetic storms (minimum Dst < −100 nT). The results of our analysis are summarized below.

1) There is an upper limit of the yearly occurrence rate of the geomagnetic activities for a given SSN. The level of the limit is approximately proportional to the yearly SSN.
2) The storm time daily minimum Dst is approximately proportional to the daily maximum value of $E_{\mathrm{d\text{-}d}}$.
3) The occurrence rate of $|E_{\mathrm{d\text{-}d}}|$ of larger than 5 mV/m in cycle 24 is lower than the occurrence rate in cycle 23 and the occurrence rate of $|E_{\mathrm{d\text{-}d}}| >$ 5 mV/m decreased in 2013–2014. The averaged ratio of $E_{\mathrm{d\text{-}d}} < -5$ mV/m to $|E_{\mathrm{d\text{-}d}}| > 5$ mV/m enhanced remarkably in 2014.
4) Concerning the 1-h averaged solar wind data in the rising–maximum phases of cycle 24, no data point of Bz < −10 nT is recorded in the cases of $V > 750$ km/s. For the cases of $E_{\mathrm{d\text{-}d}} > 5$ mV/m, averaged V and Bz of the data points ($E_{\mathrm{d\text{-}d}} > 5$ mV/m) are 532 km/s and −14 nT for cycle 23, and 503 km/s and −14 nT for cycle 24, respectively.
5) Geomagnetic activities in cycle 24 showed two-peak characteristics.
6) The geomagnetic storms in the studied period of cycle 24 were mainly caused by CMEs. Only two storms have been identified to be associated with high-speed solar wind from coronal holes. The rate of solar sources is similar to those of other cycles estimated by Zhang et al. (2007) and Echer et al. (2008).

We need to continuously watch geomagnetic activity in declining-minimum phase of cycle 24 for further understanding of this weak cycle.

ACKNOWLEDGEMENTS

We thank the Kakioka Magnetic Observatory for the list of geomagnetic storms, the WDC for Geomagnetism, Kyoto, for the Dst index data, and the

WDC-SILSO for the SSN data, the SDO/AIA data and the OMNI solar wind data for NASA, the SOHO/LASCO data for ESA and NASA. This work was supported by JSPS KAKENHI Grant Number 15H05815.

REFERENCES

1. Bothmer V, Rust DM (1997) The field configuration of magnetic clouds and the solar cycle. In: Crooker N, Joselyn JA, Feynman J (eds) Coronal mass ejections. Geophysical monograph. AGU, Washington
2. Burton RK, McPherron RL, Russell CT (1975) An empirical relationship between interplanetary conditions Dst. J Geophys Res 80(31):4204–4214
3. Echer E, Gonzalez WD, Tsurutani BT, Gonzalez LC (2008) Interplanetary conditions causing intense geomagnetic storms (Dst ≤ −100 nT) during solar cycle 23 (1996–2006). J Geophys Res 113:A052221. doi:10.1029/2007JA012744
4. Echer E, Gonzalez WD, Tsurutani BT (2011) Statistical studies of geomagnetic storms with peak Dst ≤ −50 nT from 1957 to 2008. J Atmos Sol Terr Phys73(11–12):1454–1459. doi:10.1016/j.jastp.2011.04.21
5. Gonzalez WD, Gonzalez ALC, Tsurutani BT (1990) Dual-peak solar cycle distribution of intense geomagnetic storms. Planet Space Sci 38(2):181–187
6. Gopalswamy N (2008) Solar connections of geoeffective magnetic structures. J Atmos Sol Terr Phys 70(17):2078–2100. doi:10.1016/j.jastp.2008.06.010
7. Gopalswamy N, Xie H, Akiyama S, Makela PA, Yashiro S (2014) Major solar eruptions and high-energy particle events during solar cycle 24. Earth Planets Space 66:104. doi:10.1186/1880-5981-66-104
8. Gopalswamy N, Makela P, Akiyama S, Yashiro S, Thakur N (2015) CMEs during the two peaks in cycle 24 and their space weather consequences. Sun Geosph 10(2):111–118
9. Kamide Y, Kusano K (2013) Is something wrong with the present solar maximum? Space Weather 11:140–1411. doi:10.1002/ewe.20045
10. Kamide Y, Kusano K (2015) No major solar flares but the largest geomagnetic storm in the present solar cycle. Space Weather 13:365–367. doi:10.1002/2015SW001213
11. Kataoka R, Shiota D, Kilpua E, Keika K (2015) Pileup accident hypothesis of magnetic storm on 17 March 2015. Gephys Res Lett 42:5155–5156.doi:10.1002/215GL064816
12. Kilpua EKJ, Li Y, Luhmann JG, Jian LK, Russell CT (2012) On the relationship between magnetic cloud field polarity and geoeffectiveness. Ann Geophys 30:1037–1050. doi:10.5194/angeo-30-1037-2012
13. McComas DJ, Angold N, Elliott HA, Livadiotis G, Schwadron NA, Skoug RM, Smith CW (2013) Weakest solar wind of the space age and the current “Mini” solar maximum. Astrophys J. doi:10.1088/0004-637X/779/1/2
14. O’Brien TP, McPherron L (2000) Forecasting the ring index Dst in real time. JASTP 62:1295–1299
15. Richardson IG (2013) Geomagnetic activity during the rising phase of solar cycle 24. J. Space Weather Space Clim. 3:A08. doi:10.1051/swsc/2013031
16. Richardson IG, Cane HV (2012a) Solar wind driver of geomagnetic storms during more than four solar cycles. J. Space Weather Space Clim 2:A01. doi:10.1051/swsc/2012001
17. Richardson IG, Cane HV (2012b) Near-Earth solar wind flows and related geomagnetic activity during more than dour solar cycles (1963–2011). J Space Weather Space Clim 2:A02. doi:10.1051/swsc/2013003
18. Richardson IG, Webb DF, Zhang J, Berdichevsky DB, Biesecker DA, Kasper JC, Kataoka R, Steinberg JT, Thompson BJ, Wu C-C, Zhukov AN (2006) Solar and interplanetary sources of major geomagnetic storms (Dst ≤ −100nT) generated by corotating interaction regions. J Geophys Res 111:A07S09. doi:10.1029/2005JA011476

19. Russell CT, Luhmann JG, Jian LK (2010) How unprecedented a solar minimum? Rev Geophys 48:RG2004. doi:10.1029/2009RG000316
20. Svalgaard L, Kamide Y (2013) Asymmetric solar polar field reversal. Astrophys J. doi:10.1088/0004-637X/763/1/23
21. Watari S, Kato H, Yamamoto K (2015) Hit rate of space weather forecasts of the Japanese forecast center and analysis of problematic events on the forecasts between June 2014 and March 2015. Sun Geosph 10(2):163–171
22. Zhang J, Richardson IG, Webb DF, Gopalswamy N, Huttunen E, Kasper JC, Nitta NV, Poomvises W, Thompson BJ, Wu C-C, Yashiro S, Zhukov AN (2007) Solar and interplanetary sources of major geomagnetic storms (Dst ≤ −100 nT) during 1996–2005. J Geophys Res 112:A10102. doi:10.1029/2007JA012321

CHAPTER 9

The spatial distribution of earthquake stress rotations following large subduction zone earthquakes

Jeanne L. Hardebeck

U.S. Geological Survey, Menlo Park, CA, USA

ABSTRACT

Rotations of the principal stress axes due to great subduction zone earthquakes have been used to infer low differential stress and near-complete stress drop. The spatial distribution of coseismic and postseismic stress rotation as a function of depth and along-strike distance is explored for three recent $M \geq 8.8$ subduction megathrust earthquakes. In the down-dip direction, the largest coseismic stress rotations are found just above the Moho depth of the overriding plate. This zone has been identified as hosting large patches of large slip in great earthquakes, based on the lack of high-frequency radiated energy. The large continuous slip patches may facilitate near-complete stress drop. There is seismological evidence for high fluid pressures in the subducted slab around the Moho depth of the overriding plate, suggesting low differential stress levels in this zone due to high fluid pressure, also facilitating stress rotations. The coseismic stress rotations have similar along-strike extent as the mainshock rupture. Postseismic stress rotations tend to occur in the same locations as the coseismic stress rotations, probably due to the very low remaining differential stress following the near-complete coseismic stress drop. The spatial complexity of the observed stress changes suggests that an analytical solution for finding the differential stress from the coseismic stress rotation may be overly simplistic, and that modeling of the full spatial distribution of the mainshock static stress changes is necessary.

Keywords: Stress, Subduction zone, Fault, strength

1. INTRODUCTION

The stresses in the Earth's lithosphere are the driving forces of crustal deformation. The stresses acting on the subduction zone plate interface influence the dynamics of great megathrust earthquakes, as well as seismic and aseismic deformation throughout the seismic cycle. Direct measurement of stress in boreholes is costly and limited to relatively shallow depths, so indirect measurements must also be used. Earthquake moment tensors can be inverted for the orientation and relative amplitude of the principal stress axes, but cannot determine the absolute amplitude of the differential stress. Coseismic stress rotations due to major earthquakes are one way to indirectly measure the differential stress level: If the differential stress is on the order of earthquake stress drop we would expect to see a coseismic stress change expressed as a stress rotation, while if the differential stress is substantially larger we would not expect any observable stress changes.

Rotations of the principal stress axes have been observed following recent great subduction zone earthquakes (e.g., Hasegawa et al. 2011, 2012; Hardebeck 2012; Chiba et al. 2012; Yang et al. 2013). These rotations usually take the form of a change in the plunge of the maximum compressive stress axis, $\sigma 1$, and can also involve a horizontal rotation of the trend of the sub-horizontal $\sigma 1$ axis. These stress rotations imply that the absolute stress level in subduction zones is low, on the order of earthquake stress drops of ~10 MPa, and that stress drop in great earthquakes is nearly complete. Stress rotations would not be observed if the differential stress was on the order of 100 MPa, as predicted from laboratory observations of fault frictional strength assuming hydrostatic pore pressure. Low differential stress may be related to high pore fluid pressure or other mechanisms reducing the frictional strength of faults. Relating stress rotations to other properties of subduction zones may therefore lead to further understanding of the role of elevated fluid pressure and near-complete stress drop in subduction zone dynamics.

Lay et al. (2012) propose a depth-dependent model of the slip behavior of subduction megathrust faults. The topmost zone, called zone A, extends from the trench to 15 km depth and produces long-period tsunami earthquakes. Zone B, from 15 to 35 km depth, hosts large patches of large slip in great earthquakes, with little high-frequency radiated energy. Zone C, from 35 to 55 km depth, exhibits moderate coseismic slip and radiates more high-frequency energy from smaller slip patches. Zone D, which hosts tremor and slow slip events, exists in only some subduction zones and overlaps zone B and zone C at 30–45 km depths. The boundary between zones B and C roughly corresponds to the Moho depth in the overriding plate. There is seismological evidence that high fluid pressure exists in the slab at and above the Moho depth of the overriding plate (e.g., Shelly et al. 2006; Audet et al. 2009). Hardebeck (2012) observed that $M \geq 8$ subduction zone earthquakes in zone B generally exhibit stress rotations, while those in zone C generally did not, potentially linking stress rotations to both radiated energy and fluid pressure. However, that study was limited by assigning a single depth to each great earthquake, while in reality the rupture of a $M \geq 8$ subduction zone earthquake has a large down-dip width.

Stress rotations have been analyzed by some researchers (e.g., Hasegawa et al. 2011; Hardebeck 2012) using the theoretical framework of Hardebeck and Hauksson (2001). A simple analytic solution relates the average stress rotation over a given rupture area to the ratio between the earthquake stress drop and the differential stress. However, this may oversimplify the spatial patterns of the stress rotations. In particular, the coseismic stress changes are modeled as the tensor representation of the mainshock stress drop, implicitly assuming that the observed stress changes are mainly sampling the stress shadow (e.g., Harris and Simpson 1996) of relieved stress around the mainshock rupture surface. If much of the observed stress rotation occurs outside of the stress shadow, i.e., beyond the ends of the earthquake rupture, the Hardebeck and Hauksson (2001) analytical solution may not be a good approximation.

Therefore, better imaging of the spatial areas where stress changes occur following great megathrust earthquakes is needed. Constraints on the down-dip extent of the stress changes would relate near-complete stress drop to other observables, such as radiated energy (e.g., Lay et al. 2012) and seismological evidence for fluids (e.g., Audet et al. 2009), and inform physical models. Testing whether the stress rotations are contained within the rupture area or extend past the ends of the rupture would illuminate whether simple models relating stress rotations to differential stress are generally applicable.

Some observed coseismic stress rotations are followed by a back-rotation to the pre-mainshock stress orientation within the first few months to years following the mainshock (e.g., Hardebeck 2012). The back-rotation reflects stress reloading of the mainshock rupture area, and therefore may provide insight into postseismic processes. The observed reloading is too rapid to be due to tectonic loading alone and must therefore be due to transient processes such as afterslip and viscoelastic relaxation. Additionally, afterslip in the mainshock rupture zone may continue unloading the fault and may cause a postseismic rotation in the same direction as the coseismic rotation. Comparing the spatial distribution of the postseismic stress rotation to the locations of observed afterslip may shed light on the relative importance of afterslip in the reloading and/or continued unloading of the stress on the mainshock rupture zone. Additionally, the coseismic rotations may be partially obscured by the postseismic rotations, if the postmainshock time period contains a rapid rotation. Understanding the mechanism for the postseismic rotation may help correct for this problem and better capture the coseismic rotation.

In this study, I explore the spatial distribution of coseismic and postseismic stress rotations as a function of depth and along-strike distance for three recent $M \geq 8.8$ subduction zone earthquakes.

2. METHODS

I consider three megathrust earthquakes: the 2011 M9.0 Tohoku-Oki earthquake (e.g., Hirose et al. 2011), the 2010 M8.8 Maule earthquake (e.g., Madariaga et al. 2010), and the 2004 M9.3 Sumatra–Andaman earthquake (e.g., Lay et al. 2005). Each of these earthquakes has been shown to have significantly altered the stress field (e.g., Hasegawa et al. 2011, 2012; Hardebeck 2012). In this study, I determine the spatial distribution of the coseismic and postseismic stress rotations from inversion of earthquake moment tensors. I compare the spatial extent of the coseismic rotation to published finite rupture models of the mainshock, both along-strike and down-dip, and with the depth-dependent megathrust zones defined by Lay et al. (2012). I also compare the postseismic rotations to published models of the spatial distribution of afterslip.

I use moment tensors from the NIED catalog (Kubo et al. 2002) for the Tohoku-Oki earthquake sequence, the Global CMT (GCMT) catalog (Ekström et al. 2012) for the Sumatra–Andaman earthquake sequence, and a combined catalog of GCMT solutions and regional moment tensors from Hayes et al. (2013) for the Maule earthquake sequence. The Tohoku-Oki earthquake has the densest data coverage, while the Sumatra–Andaman earthquake has the sparsest coverage. A spatial area around each mainshock rupture is selected, and all earthquakes within that region and within 20 km above or below the SLAB1.0 plate interface model (Hayes et al. 2012) are used. This is intended to capture the stress field in which the rupture surface is embedded, and may include events from the upper and lower plates, as well as the plate interface. Inverting these events together assumes that the plate interface is not a stress boundary. If there is a substantial stress orientation change across the plate interface, it will appear in the inversion results as stress heterogeneity.

To determine the pre-mainshock stress, all events from the beginning of the catalog (January 1997 for NIED, January 1977 for GCMT) until the time of the mainshock are used. The postmainshock time extends through September 2016 for the Tohoku-Oki earthquake, July 2016 for the Sumatra–Andaman earthquake, and for the Maule earthquake, until the time of the nearby 2015 M8.3 Ilapel earthquake. The postmainshock periods for the Tohoku-Oki and Maule sequences are divided into three time periods with roughly the same number of aftershocks (the Tohoku-Oki aftershocks are divided into the first 3 months, the next 1 year, and the next ~4 years. The Maule aftershocks are divided into the first 2 months, the next 4 months, and the next ~5 years). For the Sumatra–Andaman earthquake, the postmainshock period is divided into four time periods separated by the nearby 2005 M8.6 Nias–Simeulue earthquake (~3 months later), 2007 M8.4 Mentawi earthquake (another ~2.5 years later), and 2012 M8.6 Indian Ocean earthquake (another ~4.5 years later), to isolate any possible stress changes due to these events.

The moment tensors are inverted for stress orientation using the Spatial and Temporal Stress Inversion (SATSI) method of Hardebeck and Michael (2006). The earthquakes are binned with 0.5° spacing in latitude and longitude, and into the pre-mainshock and postmainshock time periods described above. Each spatial–temporal bin is inverted for stress orientation, with a flattening constraint to minimize the difference in stress orientation between adjacent spatial bins and between consecutive time periods. The spatial and temporal damping parameters are found from trade-off curves between the data misfit and model length. All spatial–temporal bins are included in the inversion, and results are reported for those with at least 6 earthquakes. The uncertainty in the stress tensor is found by inverting 3000 bootstrap resamplings of the data. One of the two nodal planes of each moment tensor is selected randomly for each bootstrap resampling to account for the nodal plane ambiguity.

The mean misfit of the moment tensors to the best fitting stress orientation is 10°–15° for the Maule and Sumatra–Andaman catalogs, and 35° for the Tohoku-Oki catalog. This suggests some unmodeled stress variations in Tohoku, although the misfit for all three catalogs is low enough to indicate that the inversion successfully retrieved the homogeneous part of the stress field. Michael (1991) used synthetic tests to show that the homogeneous part of the stress tensor can be retrieved if it has larger amplitude than the heterogeneous part. For moment tensors with errors of ~15°, like the NIED and GCMT catalogs (Kubo et al. 2002; Frohlich and Davis 1999), Michael (1991) found that the homogeneous part of the stress tensor is larger than the heterogeneous part and can be retrieved if the mean misfit is <45°.

I define the coseismic rotation as the difference between the pre-mainshock and the first postmainshock time periods. I use the rotation between each pair of consecutive postmainshock time periods to characterize the postseismic rotation. I determine both the vertical rotation, i.e., the change in plunge of the maximum compressive stress axis, $\sigma 1$, and the horizontal rotation, i.e., the change in the trend of the maximum horizontal stress direction, SHmax. Uncertainty in the vertical and horizontal rotations is found from the 3000 realizations of the stress inversion with bootstrap resampling.

3. RESULTS

For all three mainshocks, the coseismic and postseismic stress rotations occur in generally the same area as the largest mainshock slip (Figures. 1, 2, 3). For the Tohoku earthquake (Figure. 1), coseismic stress rotations are observed offshore in an area extending from 35°N to 39.5°N latitude, with the largest rotations corresponding to and south of the region where the largest coseismic slip was observed. This is similar to results of Hasegawa et al. (2012) which show significant stress rotations in the hanging wall between about 36.5°N and 39°N latitude. The largest vertical rotations occur within and down-dip of the largest mainshock rupture area, while the largest horizontal rotations occur at a range of depths. The largest postseismic rotations cover a similar spatial area, with both the vertical and horizontal rotations generally up-dip of the area of largest afterslip.

For the Maule earthquake (Figure. 2), the coseismic stress rotations span 38°S–34°S, the full length of the observed mainshock rupture, with the postseismic stress rotations spanning a similar region. The rotations occur within the depth range of the largest rupture area. For the Sumatra earthquake (Figure. 3), large coseismic rotations are observed between 4°N and 7°N, where the largest mainshock slip occurred, while some rotations extend to the northern end of the rupture at about 13°N. Postseismic rotations occur over this zone as well as south to 0°N, although the southernmost stress perturbations are likely associated with the 2005 M8.6 Nias–Simeulue earthquake.

In the down-dip direction, the largest coseismic stress rotations are generally found in zone B of Lay et al. (2012). For the Tohoku-Oki earthquake (Figure. 4a), the largest coseismic stress rotation is the lower part of zone B and the rotation extends into the upper part of zone C.

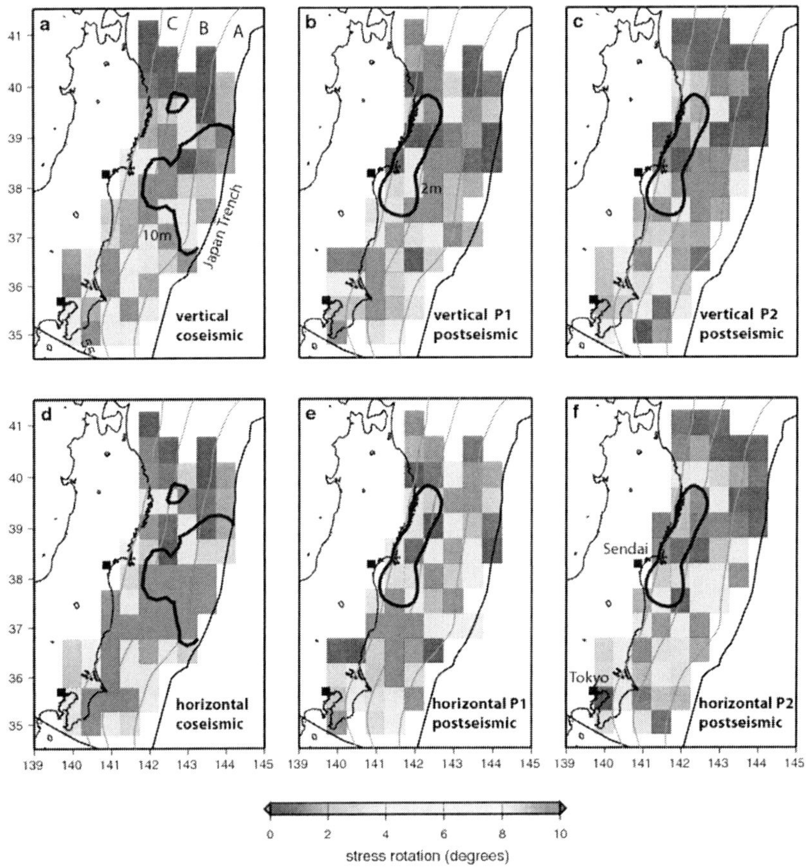

Figure. 1 Stress rotations following the 2011 M9.0 Tohoku-Oki earthquake. Shown are both the vertical rotation, i.e., the change in plunge of the $\sigma 1$ axis, and the horizontal rotation, i.e., the change in trend of the SHmax direction, for 0.5° × 0.5° spatial bins. **A** Vertical coseismic rotation, the change between the pre-mainshock (1997-2-16–2011-3-11) and first postmainshock (2011-3-11–2011-6-11) time periods. *Thick black lines* indicate the 10-m contour of the mainshock slip model of Minson et al. (2014), **B** vertical postseismic rotation P1, between the first and second (2011-6-11–2012-6-11) postmainshock time periods. *Thick black lines* indicate the 2-m contour of the postseismic slip model of Ozawa et al. (2011), **C** vertical postseismic rotation P2, between the second and third (2012-6-11–2016-9-17) postmainshock time periods, **D** horizontal coseismic rotation, **E** horizontal postseismic rotation P1, **F** horizontal postseismic rotation P2

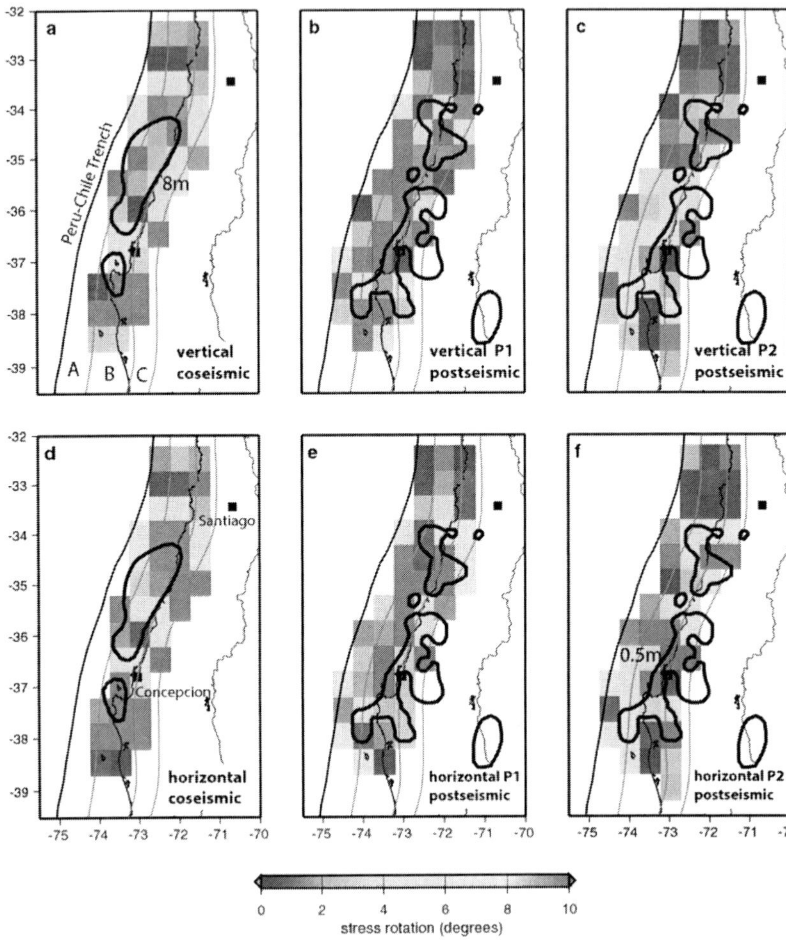

Figure. 2 Stress rotations following the 2010 M8.8 Maule earthquake. **A** Vertical coseismic rotation, the change between the pre-mainshock (1977-5-20–2010-2-27) and first postmainshock (2010-2-27–2010-4-30) time periods. *Thick black lines* indicate the 8-m contour of the mainshock slip model of Moreno et al. (2012), **B** vertical postseismic rotation P1, between the first and second (2010-4-30–2010-8-31) postmainshock time periods. *Thick black lines* indicate the 0.5-m contour of the postseismic slip model of Lin et al. (2013), **C** vertical postseismic rotation P2, between the second and third (2010-8-31–2015-9-16) postmainshock time periods, **D** horizontal coseismic rotation, **E** horizontal postseismic rotation P1, **F** horizontal postseismic rotation P2

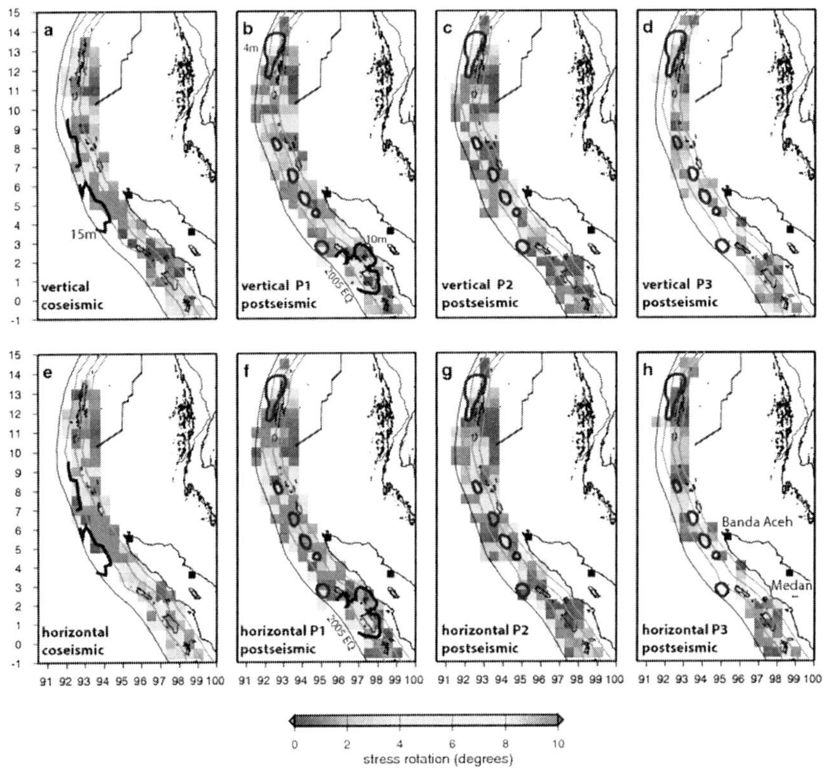

Figure. 3 Stress rotations following the 2004 M9.3 Sumatra–Andaman earthquake. **A** Vertical coseismic rotation, the change between the pre-mainshock (1977-5-20–2004-12-26) and first postmainshock (2004-12-26–2005-3-28) time periods. *Thick black lines* indicate the 15-m contour of the 2004 M9.3 mainshock slip model of Rhie et al. (2007), **B** vertical postseismic rotation P1, between the first and second (2005-3-28–2007-9-12) postmainshock time periods. *Thick black lines* indicate the 4-m contour of the postseismic slip model of Chlieh et al. (2007), and the southernmost contour is the 10-m contour of the 2005 M8.6 Nias–Simeulue earthquake slip model of Ji (2005), **C** vertical postseismic rotation P2, between the second and third (2007-9-12–2012-4-11) postmainshock time periods, **D** vertical postseismic rotation P3, between the third and fourth (2012-4-11–2016-9-17) postmainshock time periods, **E** horizontal coseismic rotation, **F** horizontal postseismic rotation P1, **G** horizontal postseismic rotation P2, **H** horizontal postseismic rotation P3

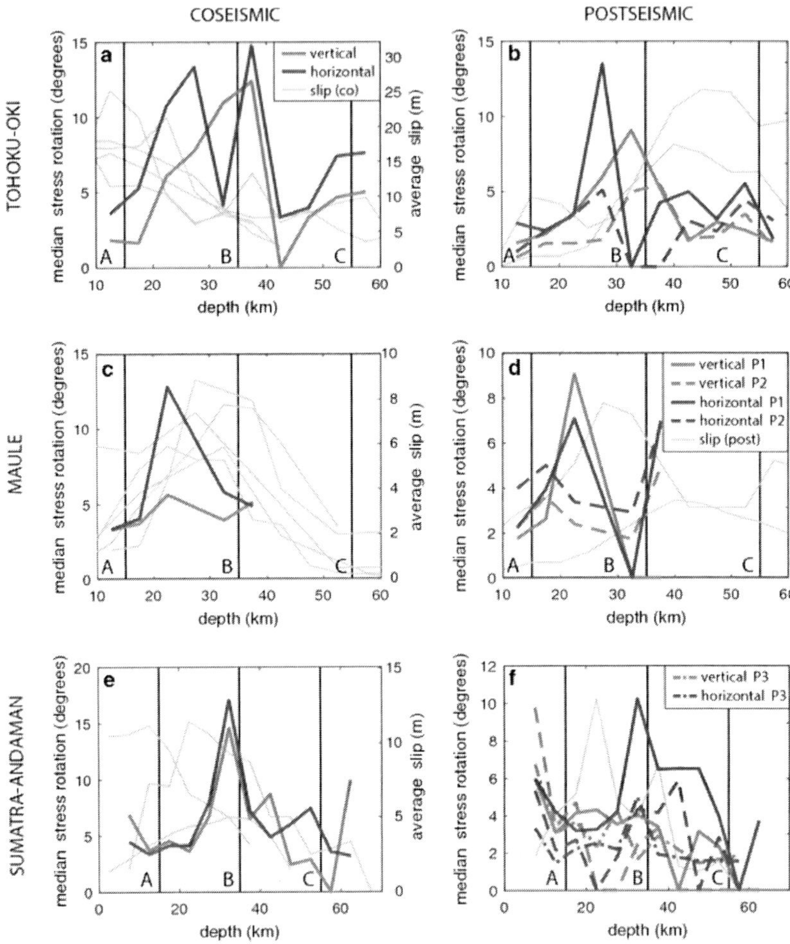

Figure. 4 Stress rotation as a function of subduction interface depth. The median vertical (*red*) and horizontal (*blue*) stress rotations are found for all inversion bins in 5-km depth intervals. Multiple realizations are performed by sampling a rotation for each bin from its bootstrap uncertainty, and the median is set to zero if it is not significantly different from zero at 1-sigma uncertainty. *Vertical black lines* and *letters* separate zones *A*, *B*, and *C*, as defined by Lay et al. (2012). **A** Tohoku-Oki earthquake coseismic stress rotation. *Gray curves* indicate average mainshock slip as a function of depth from slip models of Shao et al. (2011), Ide et al. (2011), Yagi and Fukahata (2011), Yamazaki et al. (2011) and Minson et al. (2014). **B** Tohoku-Oki earthquake postseismic stress rotation, P1 shown as *solid lines* and P2 as *dashed lines*. *Gray curves* indicate average afterslip as a function of depth from models of Ozawa et al. (2011) and Yamagiwa et al. (2015). **C** Maule earthquake coseismic stress rotation. *Gray curves* indicate mainshock slip

models of Delouis et al. (2010), Lorito et al. (2011), Moreno et al. (2012), Pollitz et al. (2011) and Yue et al. (2014). **D** Maule earthquake postseismic stress rotation, P1 shown as *solid lines* and P2 as *dashed lines*. *Gray curves* indicate afterslip models of Bedford et al. (2013) and Lin et al. (2013). **E** Sumatra–Andaman earthquake coseismic stress rotation. *Gray curves* indicate mainshock slip models of Ammon et al. (2005), Rhie et al. (2007) and Chlieh et al. (2007), **F** Sumatra–Andaman earthquake postseismic stress rotation, P1 shown as *solid lines*, P2 as *dashed lines*, and P3 as *dash-dot lines*. *Gray curve* indicates afterslip model of Chlieh et al. (2007), black curve the 2005 M8.6 Nias–Simeulue earthquake slip model of Ji (2005)

The Moho depth of the overriding plate is 30–35 km, shallowing toward the trench (Zhao et al. 1994), meaning that much of the rotation is below the Moho depth. In contrast, the largest slip in most finite rupture models is in the upper part of zone B and in zone A (e.g., Shao et al. 2011; Yagi and Fukahata 2011; Yamazaki et al. 2011; Minson et al. 2014), while the model of Ide et al. (2011) places a second peak of slip at the base of zone C. For the Tohoku-Oki earthquake, the largest coseismic stress rotations therefore appear to occur at the down-dip edge of the shallow area of large slip. For the Sumatra earthquake (Figure. 4e), the largest coseismic rotation is also observed at the base of zone B, with rotation extending into the upper part of zone C. There is no good agreement on the depth of earthquake rupture between different models, with the depth range of the largest slip ranging from zone A to the top of zone C (e.g., Ammon et al. 2005; Rhie et al. 2007; Chlieh et al. 2007). The Maule earthquake (Figure. 4c) exhibits a more shallow stress rotation than the other two events, with the largest coseismic stress rotation in the middle of zone B. The largest slip of the Maule rupture models is generally in the middle to lower part of zone B (e.g., Delouis et al. 2010; Lorito et al. 2011; Pollitz et al. 2011; Moreno et al. 2012; Yue et al. 2014). For the Maule earthquake, the largest coseismic stress rotations therefore appear to occur in the middle or near the up-dip edge of the largest rupture area.

The coseismic stress rotations tend to have similar along-strike extent as the mainshock rupture. For the Tohoku earthquake (Figure. 5a), the largest slip is concentrated between 36.4°N and 38.5°N. For the Maule earthquake (Figure. 5c), the slip models generally exhibit a bimodal slip distribution with a low at about 36°S. The stress rotations exhibit a similar bimodal distribution with a low at 36°S, with peaks at 34°S–34.5°S and 37.5°S–38°S similar to the peaks in the rupture. For the Sumatra earthquake (Figure. 5e), the stress rotations and two of the three slip distributions (Ammon et al. 2005; Rhie et al. 2007) peak at around 5°N and drop off rapidly to the south while tapering off more slowly to the north. An additional area of stress rotation is observed just south of the rupture zone at 1.5°N.

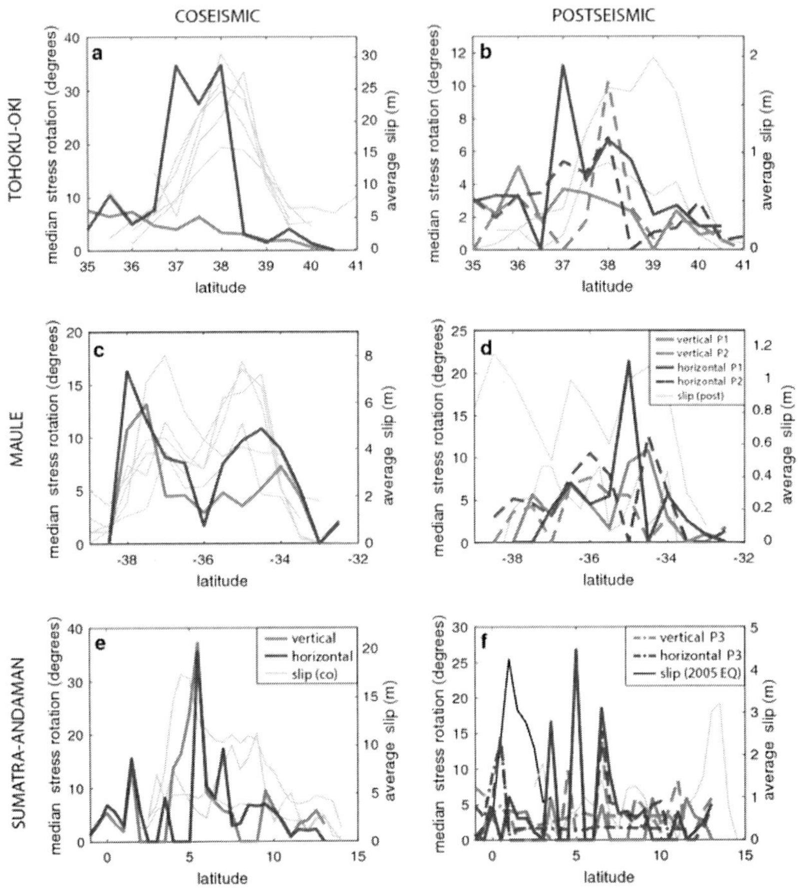

Figure. 5 Stress rotation as a function of latitude, a proxy for along-strike distance. The median vertical (*red*) and horizontal (*blue*) stress rotation is found for all inversion bins in 0.5° latitude intervals. Multiple realizations are performed by sampling a rotation for each bin from its bootstrap uncertainty, and the median is set to zero if it is not significantly different from zero at 1-sigma uncertainty. *Gray curves* indicate average mainshock slip or average afterslip as a function of latitude, from the same models as in Figure. 4. **A** Tohoku-Oki earthquake coseismic stress rotation, **B** Tohoku-Oki earthquake postseismic stress rotation, P1 shown as *solid lines* and P2 as *dashed lines*, **C** Maule earthquake coseismic stress rotation, **D** Maule earthquake postseismic stress rotation, P1 shown as *solid lines* and P2 as *dashed lines*, **E** Sumatra–Andaman earthquake coseismic stress rotation. **F** Sumatra–Andaman earthquake postseismic stress rotation, P1 shown as *solid lines*, P2 as *dashed lines*, and P3 as *dash-dot lines*

Postseismic stress rotations tend to occur in the same along-strike and down-dip locations as the coseismic stress rotations, rather than being clearly related to the spatial extent of observed afterslip. For the Tohoku earthquake, the postseismic stress rotations are largest at the base of zone B and the top of zone C (Figure. 4b), while the largest afterslip occurs in the middle of zone C (Ozawa et al. 2011; Yamagiwa et al. 2015). The postseismic stress rotation extends south of the coseismic rotation, while the afterslip is centered somewhat north of the mainshock slip (Figure. 5b). For the Maule earthquake, the postseismic rotations mainly occur in the middle of zone B (Figure. 4d), while the afterslip models range from the lower part of zone B to zone C (Bedford et al. 2013; Lin et al. 2013). Along-strike, the postseismic rotations are largest in the north of the two mainshock slip patches, while the afterslip is distributed more evenly across the along-strike extent of the mainshock (Figure. 5d). For the Sumatra earthquake, the largest postseismic rotations are in the lower part of zone B and the upper part of zone C, with another area of rotation in zone A (Figure. 4f), while the largest afterslip occurred in the middle of zone B (Chlieh et al. 2007). Along-strike, the largest postseismic rotations occur in the same locations as the coseismic rotation (Figure. 5f), with an additional rotation at 0°N that is probably related to the occurrence of the 2005 M8.6 Nias–Simeulue earthquake in that area. The largest afterslip, on the other hand, is near the northern end of the rupture zone.

For the Tohoku-Oki earthquake, ~2/3 of bins with a significant coseismic rotation ≥3° experience a postseismic rotation of the opposite direction as the coseismic rotation, indicating reloading of the mainshock rupture area. For the Maule and Sumatra–Andaman earthquakes, only ~1/3 of such bins experience a postseismic rotation indicative of reloading. This implies that while reloading of the mainshock rupture occurs in some places, in many other places the stress on the rupture zone continues to be unloaded during the postseismic period.

4. DISCUSSION

For all three $M \geq 8.8$ earthquakes, the largest coseismic stress rotations are found in the center to the lower end of zone B (15–35 km depth) of Lay et al. (2012), with rotation often extending into the upper part of zone C (35–55 km depth). The coseismic stress rotation is clearly deeper than the area of largest rupture for the Tohoku-Oki earthquake and may be shallower for the Maule earthquake. This suggests that the mechanisms causing the stress rotation are closely connected with the properties of the lower end of zone B.

There are at least two possible physical mechanisms that could explain the stress rotations within the lower end of zone B. The first mechanism is that the differential stress level may be substantially lower there because of high fluid pressure. Audet et al. (2009) interpret receiver function images in Cascadia showing that the plate interface is a low-permeability barrier in the lower part of zone B, confining fluids in the slab at high pressure, while this barrier is not present in zone C. Seismic tomography has also imaged high fluid pressures near the B–C boundary (e.g., Shelly et al. 2006). Furthermore, seismic tomography in South America (Husen and Kissling 2001) shows what appear to be fluids rising from the slab into the crust of the overriding plate after a large earthquake fractures the low-permeability seal at the plate interface. The buildup and release of high-pressure fluids would repeat each seismic cycle (Sibson 2013). The periodic high fluid pressures would reduce fault strength in the slab and overriding plate, keeping the differential stress low. Stress rotations roughly scale with the ratio of the stress change to the differential stress, so a reduction in the differential stress would facilitate stress rotations.

The second mechanism is that near-complete stress drop may be easier to achieve in zone B because of the large coherent slip patches that Lay et al. (2012) propose as the explanation for the lack of high-frequency radiation. The large patches of slip may represent areas of smooth fault surface, which may allow greater dynamic weakening during a large earthquake (e.g., Fang and Dunham 2013). Flat, smooth fault regions also appear to produce the largest earthquakes (Bletery et al. 2016). Near-complete stress drop would increase the ratio of the stress change to the differential stress, also facilitating stress rotation. While the stress drop may be near-complete in smooth fault regions, they are not necessarily the locations of the largest stress drop. If the background differential stress is relatively low, perhaps kept low by the greater dynamic weakening, the stress drop need not be unusually high to be near-complete. The stress drop cannot be fully complete, however, as the stress axes of the postmainshock stress field are resolvable and aftershocks occur. It is unclear why the large stress rotations would tend to occur near the base of zone B in this model, instead of throughout zone B.

The depth range of the stress rotations, from the lower part of zone B to the top of zone C, is similar to the depth distribution of tremor and slow slip (e.g., Obara 2002; Rogers and Dragert 2003). These phenomena have not been observed in the same fault sections as the $M \geq 8.8$ earthquakes that are considered in this study, despite dense enough instrumentation to have detected tremor and slow slip elsewhere in Japan (Obara 2002) and Chile (Gallego et al. 2013). This suggests a shared physical mechanism that promotes tremor and slow slip in some cases and earthquakes with near-complete stress drop in others. High fluid pressure is a candidate mechanism, although this leaves unanswered the question of why some subduction zones produce tremor and slow slip while some do not.

For all three mainshocks, the coseismic stress rotations generally cover a spatial area similar to the area of largest mainshock slip. However, for the Tohoku-Oki earthquake, the depth distribution of the stress rotation clearly differs from the depth distribution of the earthquake rupture. This suggests that the Hardebeck and Hauksson (2001) analytical solution, which assumes that the stress rotation occurs in the mainshock stress shadow, may not be a good approximation. Additionally, the observed horizontal rotations and the spatial complexity of the vertical rotations are not consistent with the simple Hardebeck and Hauksson (2001) model, and suggest that heterogeneous earthquake slip contributes substantially to the observed stress rotations.

An alternative approach is to model the full spatial distribution of static stress changes due to the mainshock, and to map the observed stress changes with adequate resolution to compare to the modeled stress patterns of the direction and amplitude of the rotation (e.g., Yoshida et al. 2012; Hasegawa et al. 2012). Numerical inversion (Wesson and Boyd 2007; Yoshida et al. 2012; Yang et al. 2013) or forward modeling (e.g., Yoshida et al. 2014, 2015) can then be used to determine the background differential stress. This approach requires spatially mapping the stress rotation with a resolution that may not be possible for smaller or more poorly recorded events. Modeling difficulties may also be encountered close to the rupture where model discretization impacts the computed stress changes.

Postseismic stress rotations tend to occur in the same along-strike and down-dip locations as the coseismic stress rotations, rather than being clearly related to the spatial extent of observed afterslip. This may be explained by the near-complete coseismic stress drop, which leaves a region of very low differential stress just after the earthquake. Stress rotations scale with the ratio between the stress change and the differential stress. Therefore, the very low differential stress can facilitate large stress rotation in these areas, even if they are not the locations of greatest postseismic loading or unloading.

CONCLUSIONS

I explore the distribution of coseismic and postseismic stress rotations as a function of depth and along-strike distance for the 2011 M9.0 Tohoku-Oki, 2010 M8.8 Maule, and 2004 M9.3 Sumatra–Andaman earthquakes. The largest coseismic stress rotations are found in the mid- to lower part of zone B, and rotations extend into the upper part of zone C, as defined by Lay et al. (2012). Lay et al. (2012) found decreased high-frequency radiated energy in zone B, suggesting large patches of continuous slip. These large slip patches may facilitate the near-complete stress drop implied by the coseismic stress rotations. Seismological studies (e.g., Shelly et al. 2006; Audet et al. 2009) imply that high-pressure fluids are present at the base of zone B. High fluid pressure reduces the differential stress, facilitating coseismic stress rotations. The coseismic stress rotations are therefore potentially related to multiple physical properties of the subduction interface. The postseismic stress rotations tend to occur in the same locations as the coseismic rotations, which may be the result of the very low differential stress following the coseismic rotation.

The stress rotations are generally contained within the mainshock rupture area in the along-strike direction, although they can differ in their depth distribution. This suggests that the Hardebeck and Hauksson (2001) assumption that the stress rotation occurs in the mainshock stress shadow may not be generally applicable. Additionally, the observed horizontal rotations, as well as the spatial irregularity of the vertical rotations, suggest that heterogeneous earthquake slip contributes substantially to the observed stress rotations. This implies that the Hardebeck and Hauksson (2001) method for finding the differential stress from the coseismic stress rotation may be overly simplistic. More accurate results may be obtained by modeling the full spatial distribution of the mainshock static stress changes and finding the differential stress through inversion or forward modeling.

ACKNOWLEDGEMENTS

I am grateful to two anonymous reviewers, and David Shelly and Patricia McCrory for constructive reviews of an earlier version of this manuscript.

REFERENCES

1. Ammon CJ, Ji C, Thio HK, Robinson D, Ni S, Hjorleifsdottir V, Kanamori H, Lay T, Das S, Helmberger D, Ichinose G (2005) Rupture process of the 2004 Sumatra–Andaman earthquake. Science 308(5725):1133–1139
2. Audet P, Bostock MG, Christensen NI, Peacock S (2009) Seismic evidence for overpressured subducted oceanic crust and megathrust fault sealing. Nature 457(7225):76–78
3. Bedford J, Moreno M, Baez JC, Lange D, Tilmann F, Rosenau M, Heidbach O, Oncken O, Bartsch M, Rietbrock A, Tassara A (2013) A high-resolution, time-variable afterslip model for the 2010 Maule Mw = 8.8, Chile megathrust earthquake. Earth Planet Sci Lett 383:26–36
4. Bletery Q, Thomas AM, Rempel AW, Karlstrom L, Sladen A, De Barros L (2016) Mega-earthquakes rupture flat megathrusts. Science 354(6315):1027–1031

5. Chiba K, Iio Y, Fukahata Y (2012) Detailed stress fields in the focal region of the 2011 off the Pacific coast of Tohoku Earthquake—implication for the distribution of moment release. Earth Planets Space 64(12):1157–1165. doi:10.5047/eps.2012.07.008
6. Chlieh M, Avouac JP, Hjorleifsdottir V, Song TRA, Ji C, Sieh K, Sladen A, Hebert H, Prawirodirdjo L, Bock Y, Galetzka J (2007) Coseismic slip and afterslip of the great Mw 9.15 Sumatra–Andaman earthquake of 2004. Bull Seismol Soc Am 97(1A):S152–S173
7. Delouis B, Nocquet JM, Vallée M (2010) Slip distribution of the February 27, 2010 Mw = 8.8 Maule earthquake, central Chile, from static and high-rate GPS, InSAR, and broadband teleseismic data. Geophys Res Lett. doi:10.1029/2010GL043899
8. Ekström G, Nettles M, Dziewonski AM (2012) The global CMT project 2004–2010: centroid-moment tensors for 13,017 earthquakes. Phys Earth Planet Inter 200–201:1–9
9. Fang Z, Dunham EM (2013) Additional shear resistance from fault roughness and stress levels on geometrically complex faults. J Geophys Res 118(7):3642–3654
10. Frohlich C, Davis SD (1999) How well constrained are well-constrained T, B, and P axes in moment tensor catalogs? J Geophys Res 104:4901–4910
11. Gallego A, Russo RM, Comte D, Mocanu V, Murdie RE, VanDecar JC (2013) Tidal modulation of continuous nonvolcanic seismic tremor in the Chile triple junction region. Geochem Geophys Geosyst 14(4):851–863
12. Hardebeck JL (2012) Coseismic and postseismic stress rotations due to great subduction zone earthquakes. Geophys Res Lett. doi:10.1029/2012GL053438
13. Hardebeck JL, Hauksson E (2001) Crustal stress field in southern California and its implications for fault mechanics. J Geophys Res 106(B10):21859–21882
14. Hardebeck JL, Michael AJ (2006) Damped regional-scale stress inversions: methodology and examples for southern California and the Coalinga aftershock sequence. J Geophys Res. doi:10.1029/2005JB004144
15. Harris RA, Simpson RW (1996) In the shadow of 1857-the effect of the Great Ft. Tejon Earthquake on subsequent earthquakes in southern California. Geophys Res Lett 23(3):229–232
16. Hasegawa A, Yoshida K, Okada T (2011) Nearly complete stress drop in the 2011 Mw 9.0 off the Pacific coast of Tohoku Earthquake. Earth Planets Space 63(7):703–707. doi:10.5047/eps.2011.06.007
17. Hasegawa A, Yoshida K, Asano Y, Okada T, Iinuma T, Ito Y (2012) Change in stress field after the 2011 great Tohoku-Oki earthquake. Earth Planet Sci Lett 355–356:231–243. doi:10.1016/j.epsl.2012.08.042
18. Hayes GP, Wald DJ, Johnson RL (2012) Slab1.0: a three-dimensional model of global subduction zone geometries. J Geophys Res. doi:10.1029/2011JB008524
19. Hayes GP, Bergman E, Johnson KL, Benz HM, Brown L, Meltzer AS (2013) Seismotectonic framework of the 2010 February 27 Mw 8.8 Maule, Chile earthquake sequence. Geophys J Int 195(2):1034–1051
20. Hirose F, Miyaoka K, Hayashimoto N, Yamazaki T, Nakamura M (2011) Outline of the 2011 off the Pacific coast of Tohoku Earthquake (Mw 9.0)—Seismicity: foreshocks, mainshock, aftershocks, and induced activity. Earth Planets Space 63(7):513–518. doi:10.5047/eps.2011.05.019
21. Husen S, Kissling E (2001) Postseismic fluid flow after the large subduction earthquake of Antofagasta, Chile. Geology 29(9):847–850
22. Ide S, Baltay A, Beroza GC (2011) Shallow dynamic overshoot and energetic deep rupture in the 2011 Mw 9.0 Tohoku-Oki earthquake. Science 332(6036):1426–1429
23. Ji C (2005) Preliminary Result 05/03/28 (Mw 8.7), Sumatra Earthquake. www.tectonics.caltech.edu/slip_history/2005_sumatra/sumatra.html. Accessed 1 July 2013
24. Kubo A, Fukuyama E, Kawai H, Nonomura K (2002) NIED seismic moment tensor catalogue for regional earthquakes around Japan: quality test and application. Tectonophysics 356:23–48
25. Lay T, Kanamori H, Ammon CJ, Nettles M, Ward SN, Aster RC, Beck SL, Bilek SL, Brudzinski MR, Butler R, DeShon HR, Ekström G, Satake K, Sipkin S (2005) The great Sumatra–Andaman earthquake of 26 December 2004. Science 308(5725):1127–1133

26. Lay T, Kanamori H, Ammon CJ, Koper KD, Hutko AR, Ye L, Yue H, Rushing TM (2012) Depth-varying rupture properties of subduction zone megathrust faults. J Geophys Res. doi:10.1029/2011JB009133
27. Lin YNN, Sladen A, Ortega-Culaciati F, Simons M, Avouac JP, Fielding EJ, Brooks BA, Bevis M, Genrich J, Rietbrock A, Vigny C (2013) Coseismic and postseismic slip associated with the 2010 Maule Earthquake, Chile: characterizing the Arauco Peninsula barrier effect. J Geophys Res 118(6):3142–3159
28. Lorito S, Romano F, Atzori S, Tong X, Avallone A, McCloskey J, Cocco M, Boschi E, Piatanesi A (2011) Limited overlap between the seismic gap and coseismic slip of the great 2010 Chile earthquake. Nat Geosci 4(3):173–177
29. Madariaga R, Métois M, Vigny C, Campos J (2010) Central Chile finally breaks. Science 328(5975):181–182
30. Michael AJ (1991) Spatial variations in stress within the 1987 Whittier Narrows, California, aftershock sequence: new techniques and results. J Geophys Res 96:6303–6320
31. Minson SE, Simons M, Beck JL, Ortega F, Jiang J, Owen SE, Moore AW, Inbal A, Sladen A (2014) Bayesian inversion for finite fault earthquake source models—II: the 2011 great Tohoku-Oki, Japan earthquake. Geophys J Int 198(2):922–940
32. Moreno M, Melnick D, Rosenau M, Baez J, Klotz J, Oncken O, Tassara A, Chen J, Bataille K, Bevis M, Socquet A (2012) Toward understanding tectonic control on the Mw 8.8 2010 Maule Chile earthquake. Earth Planet Sci Lett 321:152–165
33. Obara K (2002) Nonvolcanic deep tremor associated with subduction in southwest Japan. Science 296(5573):1679–1681
34. Ozawa S, Nishimura T, Suito H, Kobayashi T, Tobita M, Imakiire T (2011) Coseismic and postseismic slip of the 2011 magnitude-9 Tohoku-Oki earthquake. Nature 475(7356):373–376
35. Pollitz FF, Brooks B, Tong X, Bevis MG, Foster JH, Bürgmann R, Smalley R, Vigny C, Socquet A, Ruegg JC, Campos J (2011) Coseismic slip distribution of the February 27, 2010 Mw 8.8 Maule, Chile earthquake. Geophys Res Lett. doi:10.1029/2011GL047065
36. Rhie J, Dreger D, Bürgmann R, Romanowicz B (2007) Slip of the 2004 Sumatra–Andaman earthquake from joint inversion of long-period global seismic waveforms and GPS static offsets. Bull Seismol Soc Am 97(1A):S115–S127
37. Rogers G, Dragert H (2003) Episodic tremor and slip on the Cascadia subduction zone: the chatter of silent slip. Science 300(5627):1942–1943
38. Shao G, Li X, Ji C, Maeda T (2011) Focal mechanism and slip history of the 2011 Mw 9/1 off the Pacific coast of Tohoku Earthquake, constrained with teleseismic body and surface waves. Earth Planets Space 63(7):559–564
39. Shelly DR, Beroza GC, Ide S, Nakamura S (2006) Low-frequency earthquakes in Shikoku, Japan, and their relationship to episodic tremor and slip. Nature 442:188–191. doi:10.1038/nature04931
40. Sibson RH (2013) Stress switching in subduction forearcs: implications for overpressure containment and strength cycling on megathrusts. Tectonophysics 600:142–152
41. Wesson RL, Boyd OS (2007) Stress before and after the 2002 Denali fault earthquake. Geophys Res Lett. doi:10.1029/2007GL029189
42. Yagi Y, Fukahata Y (2011) Rupture process of the 2011 Tohoku-Oki earthquake and absolute elastic strain release. Geophys Res Lett. doi:10.1029/2011GL048701
43. Yamagiwa S, Miyazaki SI, Hirahara K, Fukahata Y (2015) Afterslip and viscoelastic relaxation following the 2011 Tohoku-Oki earthquake (Mw9. 0) inferred from inland GPS and seafloor GPS/Acoustic data. Geophys Res Lett 42(1):66–73
44. Yamazaki Y, Lay T, Cheung KF, Yue H, Kanamori H (2011) Modeling near-field tsunami observations to improve finite-fault slip models for the 11 March 2011 Tohoku earthquake. Geophys Res Lett. doi:10.1029/2011GL049130
45. Yang YR, Johnson KM, Chuang RY (2013) Inversion for absolute deviatoric crustal stress using focal mechanisms and coseismic stress changes: the 2011 M9 Tohoku-oki, Japan, earthquake. J Geophys Res 118(10):5516–5529

46. Yoshida K, Hasegawa A, Okada T, Iinuma T, Ito Y, Asano Y (2012) Stress before and after the 2011 great Tohoku-Oki earthquake and induced earthquakes in inland areas of eastern Japan. Geophys Res Lett. doi:10.1029/2011GL049729
47. Yoshida K, Hasegawa A, Okada T, Iinuma T (2014) Changes in the stress field after the 2008 M7.2 Iwate-Miyagi Nairiku earthquake in northeastern Japan. J Geophys Res 119:9016–9030. doi:10.1002/2014JB011291
48. Yoshida K, Hasegawa A, Okada T (2015) Spatially heterogeneous stress field in the source area of the 2011 Mw 6.6 Fukushima-Hamadori earthquake, NE Japan, probably caused by static stress change. Geophys J Int 201(2):1062–1071
49. Yue H, Lay T, Rivera L, An C, Vigny C, Tong X, Báez Soto JC (2014) Localized fault slip to the trench in the 2010 Maule, Chile Mw = 8.8 earthquake from joint inversion of high-rate GPS, teleseismic body waves, InSAR, campaign GPS, and tsunami observations. J Geophys Res 119(10):7786–7804
50. Zhao D, Hasegawa A, Kanamori H (1994) Deep structure of Japan subduction zone as derived from local, regional, and teleseismic events. J Geophys Res Solid Earth 99(B11):22313–22329

CHAPTER 10

Outcomes and challenges of global high-resolution non-hydrostatic atmospheric simulations using the K computer

Masaki Satoh[1*], Hirofumi Tomita[2], Hisashi Yashiro[2], Yoshiyuki Kajikawa[2,3], Yoshiaki Miyamoto[4,2], Tsuyoshi Yamaura[2], Tomoki Miyakawa[1], Masuo Nakano[5], Chihiro Kodama[5], Akira T. Noda[5], Tomoe Nasuno[5], Yohei Yamada[5] and Yoshiki Fukutomi[6]

AUTHORS' CONTRIBUTIONS

MS led the studies using NICAM in SPIRE and FLAGSHIP2020 and coordinated this review. HY wrote the "Computational aspects" section. YK and YM wrote the "Sub-kilometer global simulation" section, and the experiments described in this section were carried out by the team at the Advanced Institute for Computational Science: YM, YK, HY, TY, and HT. YK and HY wrote the "Asian summer monsoon" section, and HY conducted the experiments described in this section. TM wrote the "Madden-Julian oscillation" section and conducted the experiments described in this section. MN wrote the "Boreal summer intra-seasonal oscillation and tropical cyclogenesis" section and conducted the experiments described in this section. CK, ATN, TN, YY, and YF wrote the "AMIP-type experiments" section, and YY and MS wrote the "Future projection of tropical cyclones" section. CK, ATN, and YY conducted the experiments described in these two sections. HT, MS, and CK wrote the "Future perspectives" section. MS is responsible for the whole manuscript and wrote the "Introduction" and "Conclusions" sections. All authors read and approved the final manuscript.

ABSTRACT

This article reviews the major outcomes of a 5-year (2011–2016) project using the K computer to perform global numerical atmospheric simulations based on the non-hydrostatic icosahedral atmospheric model (NICAM). The K computer was made available to the public in September 2012 and was used as a primary resource for Japan's Strategic Programs for Innovative Research (SPIRE), an initiative to investigate five strategic research areas; the NICAM project fell under the research area of climate and weather simulation sciences. Combining NICAM with high-performance computing has created new opportunities in three areas of research: (1) higher resolution global simulations that produce more realistic representations of convective systems, (2) multi-member ensemble simulations that are able to perform extended-range forecasts 10–30 days in advance, and (3) multi-decadal simulations for climatology and variability. Before the K computer era, NICAM was used to demonstrate realistic simulations of intra-seasonal oscillations including the Madden-Julian oscillation (MJO), merely as a case study approach. Thanks to the big leap in computational performance of the K computer, we could greatly increase the number of cases of MJO events for numerical simulations, in addition to integrating time and horizontal resolution. We conclude that the high-resolution global non-hydrostatic model, as used in this five-year project, improves the ability to forecast intra-seasonal oscillations and associated tropical cyclogenesis compared with that of the relatively coarser operational models currently in use. The impacts of the sub-kilometer resolution simulation and the multi-decadal simulations using NICAM are also reviewed.

Keywords: K computer, NICAM, Intra-seasonal oscillations, Madden-Julian oscillation, Tropical cyclone, Global non-hydrostatic model

1. INTRODUCTION

The K computer began operation in September 2012 at RIKEN Advanced Institute for Computational Science, Japan. The Strategic Programs for Innovative Research (SPIRE), an initiative launched in October 2010 in

Japan, aims to leverage the unparalleled power of the K computer to produce the world's most cutting-edge simulation studies in five key strategic fields. SPIRE also aims more generally to boost the creation of promotional frameworks for computer science and technology and to yield significant social breakthroughs. Field 3 of SPIRE, "Advanced prediction research for natural disaster prevention and reduction," consists of weather and climate simulations, including earthquake and tsunami simulations (Oishi et al. 2015; Tsuboi et al. 2016; Ando et al. 2016; Hyodo et al. 2016). Studies using meso-scale atmospheric modeling (Saito et al. 2013; Kunii 2014a, 2014b; Chen et al. 2015a, 2015b; Duc et al. 2015; Ito et al. 2015; Kobayashi et al. 2015a; Seko et al. 2015; Yokota et al. 2016) and global modeling are also included under weather simulations. Here, we review the outcome of the global high-resolution atmospheric modeling studies conducted from 2012 to 2016 under Field 3 of SPIRE and preview the challenges for the next K computer project, called the FLAGSHIP2020 project (Future LAtency core-based General-purpose Supercomputer with HIgh Productivity). Running from 2014 to 2019, FLAGSHIP2020 is in preparation for the Post-K supercomputer, which was originally planned to start operation in 2020. (Recently, it was announced that the operation of the Post-K supercomputer has been delayed by at least 2 years.)

For global atmospheric modeling using the K computer, new research using the non-hydrostatic icosahedral atmospheric model (NICAM; Tomita and Satoh 2004; Satoh et al. 2008, 2014) has been pursued by enhancing horizontal resolution, ensemble sizes, or duration of integration time. In particular, we established the following research targets at the beginning of the project: (1) enable extended-range forecasts for a timeframe of 10–30 days, for better simulations of intra-seasonal oscillations (ISO) and (2) enable projection of tropical cyclones. Before the K computer became available, NICAM was known to more accurately reproduce individual ISOs in case studies at a horizontal grid spacing between 3.5 and 14 km (Miura et al. 2007a; Oouchi et al. 2009a, 2009b; Yanase et al. 2010). At that time, simulations such as these were performed mainly on the Earth Simulator at the Japan Agency for Marine-Earth Science and Technology (JAMSTEC). Multi-year simulations were also conducted with the 7 km grid spacing NICAM using the Athena computer (operated by the University of Tennessee's National Institute for Computational Science and hosted by Oak Ridge National Laboratory in the USA) and showed better simulations of boreal summer ISO (Satoh et al. 2012b; Kinter et al. 2013). The development of the K computer offered the opportunity to obtain more robust performance from NICAM by drastically increasing ensemble size and integration time together with resolution.

1.1 Review

This article reviews the outcomes of the NICAM studies using the K computer under Field 3 of SPIRE and previews challenges for the forthcoming Post-K supercomputer under the FLAGSHIP2020 project. In the "Computational aspects" section, we describe how NICAM was adapted for use on the K computer. Following this, we list major outcomes using NICAM with the K computer in the sections "Sub-kilometer global simulation," "Asian summer monsoon," "Madden-Julian oscillation," "Boreal summer intra-seasonal oscillation and tropical cyclogenesis," "Atmospheric Model Intercomparison Project-type experiments," and "Projection of tropical cyclones." The challenges to overcome in the near future are reviewed in "Future perspectives," and we conclude our review in the final section, "Conclusions."

2. COMPUTATIONAL ASPECTS

2.1 Background

The K computer was the first 10 petaflop supercomputer (10^{15} floating-point operations per second; Yokokawa et al. 2011). This massively parallel scalar machine was twice ranked as the world's top supercomputer in the TOP500 List in 2011 (Top 500 2011a, 2011b). Furthermore, it showed the top performance in new benchmarks such as the high-performance conjugate gradient (Kumahata et al. 2016) and Graph500 (Graph 500 2014). Two recipients of the Gordon Bell Prize used the K computer to perform their award-winning work (Hasegawa et al. 2011; Ishiyama et al. 2012). In the research area of earth science, which Field 3 of SPIRE also covers, an urban-scale earthquake simulation study using the K computer was selected as a finalist for the Gordon Bell Prize (Ichimura et al. 2014, 2015). These results reflect the K computer's high performance not only in benchmarks, but also in applications.

The K computer system possesses characteristics advantageous for weather and climate research, which require data-intensive simulation models. The performance of these models is limited not only by the number of calculations processed in the computer's central processing unit but also by data transfer processes, such as the memory-cache transfer, inter-node communication, and file input/output (I/O). The K computer has relatively high memory throughput (0.5 byte per floating-point operation) compared to the today's massively parallel scalar supercomputers. The asynchronous, distributed file I/O system of the K computer minimizes waiting time for frequent output of the atmospheric variables during the simulation. The K computer is also highly resilient and has a lower machine failure rate (Yamamoto et al. 2014) compared to other peta-scale supercomputers, which is very important for long-term climate simulations with a large number of computational nodes.

At an early stage of the development of the K computer, NICAM was selected as one of the target applications and was used in many of the computer's performance optimizations. In experiments with a small number of nodes, we achieved about 10% of the performance efficiency (Terai et al. 2014; Yashiro et al. 2016b), which is almost competitive with that of the state-of-the-art weather and climate models. NICAM also shows good scalability, up to almost the full-node of the K computer in the weak scaling test.

2.2 Single-node performance

For general model applications, a small number of major calculation parts occupy most of the floating-point operation and elapsed time. These parts are called hot spots, and we usually focus on these hot spots for the optimization of computational performance. However, most weather and climate models do not have a clear hot spot. Every part of the source code contains a small computation that consumes a little time, a characteristic called a flat profile. Weather and climate models generally have a flat profile, where every part of the source code has the potential to deteriorate the total performance. The effort to optimize NICAM on the K computer aimed to overcome the flat profile (Yashiro et al. 2016b). Optimizations were applied mainly to the calculation parts (kernels) such as divergence, gradient, and Laplacian operations. These kernels are a type of stencil calculation, which updates the grid point value using the value of surrounding grids. Because the finite-volume method is used for numerical discretization in NICAM, these kernels are the most important part for solving the fluid dynamics equations of the atmosphere. After optimization, each kernel showed improved performance (Terai et al. 2014). However, we achieved only a 1% gain in sustained performance over the total simulation, because the ratio of elapsed time of the kernels to total time decreased. This result was related to the effect of Amdahl's law (Amdahl 1967). From this experience, we realized that we needed to find the time-consuming, less computationally intensive parts of the source code and remove them. As a result of these efforts, we achieved about 10% of the sustained performance, which was appropriate to the memory performance of the K computer. We also succeeded in reducing the model execution time by 30%.

2.3 Scalability

Terai et al. (2014) performed a weak scaling test using NICAM on the K computer and showed good scalability of elapsed time and sustained performance of the NICAM (Fig. 1). The test was conducted by changing the number of nodes and the horizontal grid spacing. Horizontal grid spacing was decreased from 56 km to 440 m, which corresponded to an increase in the number of nodes from 5 to 81,920. The total time was divided into three parts: (1) the dynamical step, which covered the time taken to solve the fluid dynamics; (2) the physical step, which covered the physics components of the calculations, such as cloud microphysics, atmospheric radiation, and boundary layer turbulence; and (3) the communication step, which comprised the inter-node data transfer using message passing interface (MPI) (referred to as comm_data_transfer in Fig. 1). The computations and communications were completely separated using barrier statements of MPI (MPI_barrier), in which all tasks must synchronize. The result shows good scalability of total elapsed time. The timing of the dynamical step does not change as the number of nodes is increased, because the number of operations remains almost the same. The increase in the communication time as the number of nodes increases is not significant because most of the communications are conducted node-to-node using halo exchanges. The increase in the total elapsed time is mainly caused by increases in the time of the physical step due to the load imbalance between the processes. Terai et al. (2014) found that the imbalances occurred mainly in the cloud microphysics scheme. They found that the global map of the number of floating-point operations in the cloud microphysics showed inhomogeneous distribution, similar to that of ice hydrometeors. Overall, we achieved a maximum performance of 0.87 petaflops with 81,920 nodes (655,360 cores), as shown by Terai et al. (2014).

The results of the strong scaling test are different from those of the weak scaling test (Fig. 2). In the strong scaling test, we used NICAM with horizontal grid spacing of 14 km and 38 vertical layers. The number of grid points in each process decreased from 33,800 to 100 as the number of nodes increased from 80 to 40,960. The model showed good scalability up to 2560 nodes. The number of simulation days per wall-clock day increased in proportion to the number of nodes. However, the performance saturated beyond 2560 nodes, due to the relative increase in communication time; at 40,960 nodes, the communication time occupied over 50% of total elapsed time. This result led to production runs being conducted with 640 nodes in the studies described in the following sections. For long-term simulations in future studies, strong scalability will need to be improved. Reducing the number of subroutine calls within the MPI communication is one possible solution.

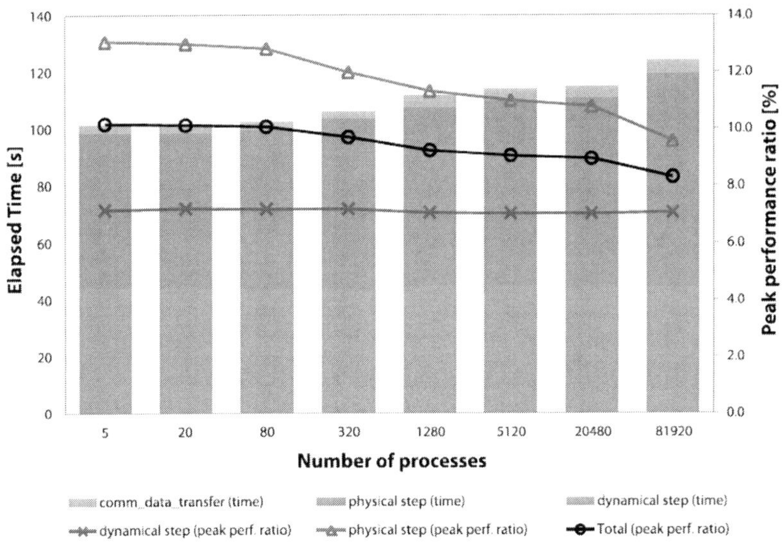

Figure 1 Results of the weak scaling test on the K computer. *Bars* show the elapsed time for the dynamical step, the physical step, and the communication (data transfer) step. *Lines* indicate the sustained performance. After Fig. 2 of Terai et al. (2014)

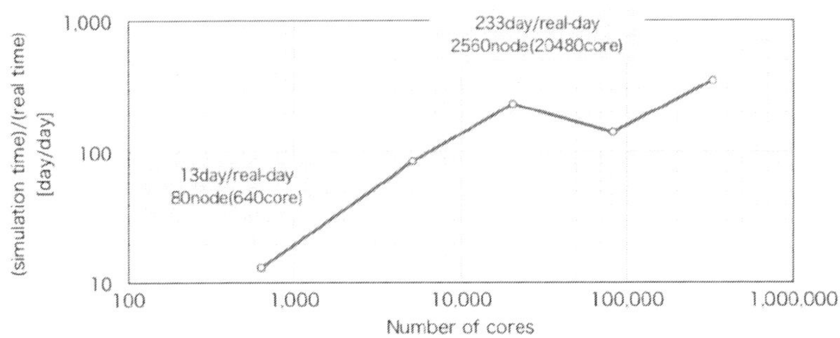

Figure 2 Results of the strong scaling test on the K computer for the NICAM simulations. *Vertical axis* is simulation time per real time [day day^{-1}]. *Horizontal axis* is the number of cores

2.4 Pre- and post-processing

Owing to the comprehensive optimizations of NICAM, we achieved high efficiency and good scalability. However, the elapsed pre- and post-processing times were relatively increased in the total simulation time for the following reasons. The remapping process has one of the heaviest workloads in post-processing. To use common analysis and visualization tools, we usually remap the meteorological variables from an icosahedral grid to a geodesic (latitude-longitude) grid. For example, the data size of the global 0.87 km horizontal grid spacing simulations was 6 TB for one snapshot of all 3-D and 2-D variables (Miyamoto et al. 2013, 2015; Kajikawa et al. 2016). The required time for remapping and analyzing all data was estimated to be more than 2 months using typical computer clusters, which is much longer than the time required for simulation on the K computer. To solve this problem, we developed programs for parallel analysis on the K computer without remapping, thereby reducing the post-processing time from several months to less than 1 h. This improvement was due to massively parallel I/O rather than parallel computation. In the future, simulation sizes are expected to increase even more drastically. The most important consideration is the total elapsed time of the workflow in simulation studies, which includes pre- and post-processing, analysis, and visualization. All computational work will need to be carried out on the supercomputer to keep this time to a minimum.

2.5 Future directions

NICAM has been selected as the target application for the development of Japan's next-generation flagship computer (the Post-K supercomputer), together with the local ensemble transform Kalman filter (LETKF). NICAM and LETKF play an important role in the process of application-architecture co-design. Generally, computations of weather and climate models are not expected to be carried out faster than improvement in microprocessors, because the rate of improvement in memory throughput is slower than the floating-point operation speed. Thus, the computational performance of NICAM will be improved by modifications of the algorithm and its optimization (e.g., changes of loop order, data layout, and call structure).

Active use of mixed-precision floating-point operation is one possible approach to further enhance computational performance. The bit length of floating-point value affects the data size of memory transfer and the compression ratio of the output data. The effect of less precise values on the simulation results should be investigated. In recent years, the development of components and schemes such as the atmosphere-ocean coupling system with coupling library Jcup (Arakawa et al. 2014) and the ensemble-based data assimilation system (Terasaki et al. 2015) have expanded the capabilities of the simulations using NICAM. While taking these developments into account, we must increase the speed of total workflow. Maximizing the power of future supercomputers will enable simulations with even more sophisticated physics and/or higher resolution and will ensure that it becomes a powerful tool for further understanding weather and climate phenomena.

3. SUB-KILOMETER GLOBAL SIMULATION

3.1 Background

One of the greatest challenges in climate modeling is achieving more accurate simulations of clouds (Stevens and Bony 2013; Palmer 2014). Among various types of naturally occurring clouds, deep moist convective clouds have important roles in determining the structure of atmospheric circulation through the release of latent heat. This deep convection is also part of the cloudy atmospheric disturbances that sometimes cause natural disasters. Since it has been impractical to simultaneously simulate all-scale phenomena in global atmospheric models, from deep convection on a scale of O(1) km to global circulation on a scale of $O(10^4)$ km, deep convection has been expressed as a parameterization in conventional general circulation models (GCMs) (e.g., Arakawa and Schubert 1974; Kain and Fritsch 1990). NICAM was developed to simulate the global atmospheric circulation without cumulus parameterization (Satoh et al. 2008, 2014). Previous studies using NICAM have shown advanced performance in simulating large-scale convective systems and disturbances mainly using the Earth Simulator (e.g., Miura et al. 2007a; Sato et al. 2009; Nasuno and Satoh 2011; and references in Satoh et al. 2014). However, their grid spacing (several kilometers) was coarser than or similar to that of the observed convection scale. The highest resolution simulations using the Earth Simulator were the 3.5 km horizontal grid spacing simulations by Tomita et al. (2005) and Miura et al. (2007a, 2007b). Meanwhile, studies using cloud-resolving models in a confined area, such as large eddy simulation (LES) models, have indicated a change in deep convection characteristics at grid spacing less than 1 km (Petch et al. 2002; Bryan et al. 2003). Under Field 3 of SPIRE, we successfully conducted the first-ever global atmospheric simulation with sub-kilometer grid spacing using NICAM, which opened the door for simulations resolving deep convective cores in global atmospheric models. This work also allowed us to highlight the resolution dependency of simulated convection in the global model as well as its statistical features.

3.2 Results

Miyamoto et al. (2013) conducted a set of grid-refinement experiments with grid spacing that varied from 14 to 0.87 km. Further analysis of these experiments is shown by Miyamoto et al. (2015), Kajikawa et al. (2016), and Yashiro et al. (2016a), and the details of the experimental settings are described in Miyamoto et al. (2013) and Kajikawa et al. (2016). Although the large-scale characteristics of global cloud distributions are almost unchanged between the simulations with 14, 3.5, and 0.87 km grid spacing, the simulation with sub-kilometer resolution provided a more detailed description of the smaller scale cloud structure (Fig. 3). The resolution dependency showed that the properties of simulated convection were qualitatively changed between 3.5 and 1.7 km grid spacing. We defined convective cores in the simulations and made a composite structure of convective cores averaged over the globe. In simulations at finer resolution than the 3.5 km grid spacing, the convective core is expressed by multiple grid points instead of a single grid point for the 3.5 km grid spacing; the dependency of the number of convective cores on resolution changes, with smaller dependency for finer resolutions (Fig. 4); and the minimum distance between convective cores is larger than four grid points for the finer resolutions than the 3.5 km grid spacing (Fig. 4). From these results, the horizontal grid spacing around 2 km can be viewed as the minimum required resolution for a deep convective resolving model over the entire Earth, although the required grid spacing might depend on numerical discretization methods.

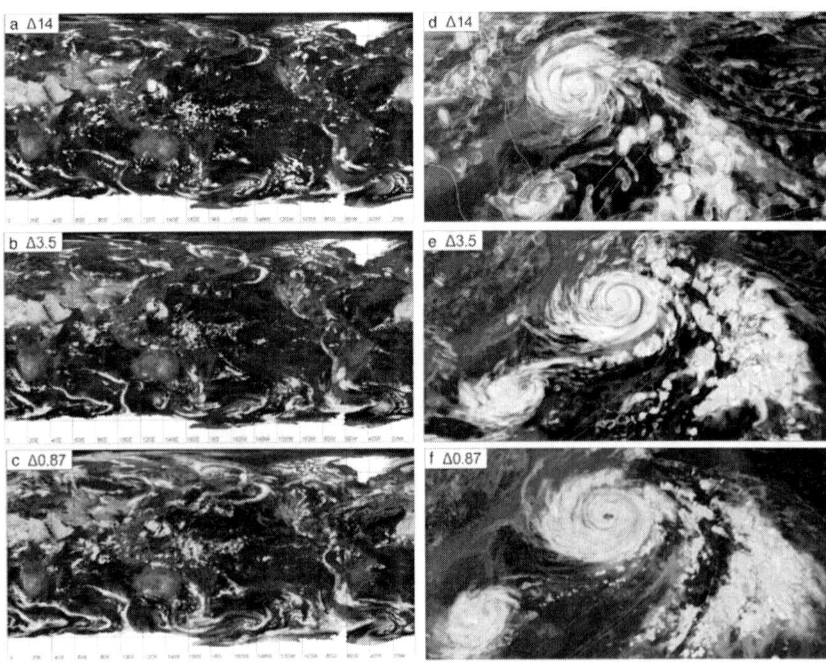

Figure 3 Resolution dependencies of simulated cloud distribution. Global view of simulations with **a** 14 km, **b** 3.5 km, and **c** 0.87 km grid spacing. **d–f** Focus over the western North Pacific region of (**a–c**), respectively. After Fig. 3 of Kajikawa et al. (2016)

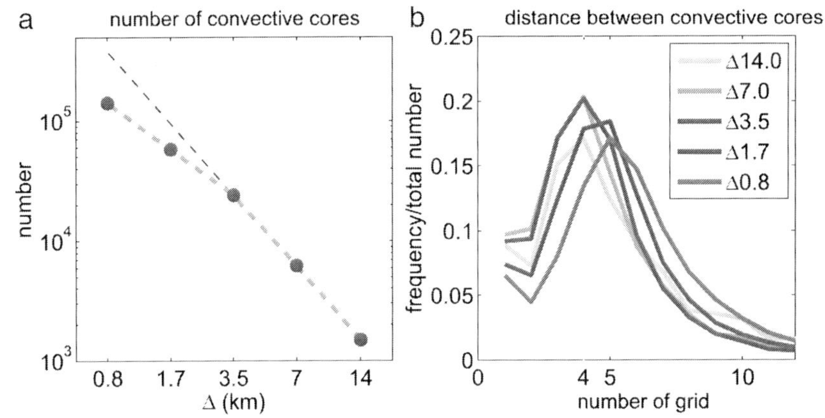

Figure 4 Resolution dependencies of convective features. **a** Number of convective cores, and **b** grid distance to convective cores. The *dashed line* in (**a**) indicates a log Δ^4 crossing at the point of simulation results with 14 km grid spacing as a reference. After Fig. 4 of Miyamoto et al. (2013)

Miyamoto et al. (2015) conducted detailed analysis of deep convection in the simulation with the finest 0.87 km grid spacing. They defined cloud disturbances such as the Madden-Julian oscillation (MJO), tropical cyclones (TCs), mid-latitude lows, and fronts and analyzed the statistical features of deep convection associated with these disturbances. The cloud-top height, strength of upward motion, and the environmental field of the convection were different in each type of disturbance. For example, the MJO convection was generated under high convective available potential energy (CAPE) and relatively weak lower convergence, whereas the TC convection was accompanied by low CAPE with strong convergence.

Kajikawa et al. (2016) analyzed the grid-refinement experiments to address the difference in the resolution dependence of the simulated convection by location and environment over the globe. This study extends the work of Miyamoto et al. (2013), which mainly discussed globally averaged convection properties. Kajikawa et al. (2016) showed that convective clouds over the tropics are more dependent on resolution than those in mid-latitudes; on the other hand, no significant difference is found in the resolution dependence of convection properties between land and ocean. Convection within cloud disturbances such as the MJO and TCs also shows large resolution dependency. Consequently, deep convection that is not categorized as cloud disturbances makes a large contribution to the global mean convection properties in Miyamoto et al. (2013). It should be stressed that most deep convection is expressed by multiple grid points in the 0.87 km grid spacing simulation even in tropical cloud disturbances.

Yashiro et al. (2016a) investigated the diurnal cycle of precipitation in this series of grid-refinement experiments and also found that the resolution dependency of its characteristics changed at a grid spacing around 2–3 km. The simulations using 1.7 and 0.87 km grid spacing better reproduced the observed peak time of the diurnal cycle compared to the experiments with coarser grid spacing. This could be a result of the better representation of deep convective cores, as shown by Miyamoto et al. (2013) and Kajikawa et al. (2016).

3.3 Future directions

We showed the resolution dependency of the simulated convection properties and their differences in various environments through a set of grid-refinement experiments. The simulation with sub-kilometer grid spacing revealed new results that differed from previous studies; for example, deep convective cores were expressed by multiple grid points. Higher spatial resolution and more realistic convection simulations are required to better understand cloud disturbances from the meso-scale to the synoptic scale (Prein et al. 2015). Moreover, longer integration at sub-kilometer grid spacing would enable us to investigate the interaction between convection and larger-scale cloud disturbances. Using the Post-K supercomputer, we estimate that we can perform experiments at further finer resolution, such as 220 m, to resolve more detailed structure of clouds.

4. ASIAN SUMMER MONSOON

4.1 Background

The Asian monsoon system is characterized by a seasonal cycle of convective activity and atmospheric circulations (e.g., Murakami and Matsumoto 1994). The variability of the monsoon system's effects on weather and climate over Asia is largely achieved through local convective activities. This variability is also affected by synoptic scale tropical disturbances. For example, the monsoon onset is often triggered by tropical disturbances or ISO. Recently, global climate modeling has made advances in many aspects of Asian summer monsoon simulations (Sperber et al. 2013; Ogata et al. 2014; Lee and Wang 2014). However, numerical representations of the seasonal transition of the monsoon, such as the monsoon onset, with scale interactions remain challenging. Meanwhile, NICAM has shown an advantage over conventional GCMs in simulating tropical disturbances, such as TCs (Fudeyasu et al. 2008; Taniguchi et al. 2010; Nakano et al. 2015) and MJO (Miura et al. 2007a; Miyakawa et al.

2014). Under Field 3 of SPIRE, we investigated the seasonal transition of the monsoon and anticipated a potential extension of its predictability using a large number of ensemble NICAM simulations, in contrast to previous studies that used simulations of only a single or a few members.

4.2 Experimental setup

Kajikawa et al. (2015) investigated the process of the Indian summer monsoon onset in 2012 based on ensemble simulations using NICAM with a grid spacing of about 14 km and 38 vertical levels. Moist processes were calculated using the NICAM single-moment water 6-cloud microphysics scheme (NSW6; Tomita, 2008) without any convective parameterization. Sea surface temperature (SST) was forecasted using a slab ocean model to allow diurnal variation and nudged to the persistent SST anomaly at the initial time with an e-folding time (that is, the time in which SST is nudged to specified values) of 7 days. The ensemble simulations were started at 00 UTC each day during the period from 10 May to 10 June 2012, and integrated for 30 days. Using the same set of NICAM ensemble simulations, Yamaura et al. (2013) clarified the role of TCs on the migration of the Baiu front.

4.3 Results

The onset of the 2012 Indian summer monsoon was announced on 5 June that year by the Indian Institute of Tropical Meteorology. The NICAM simulations using initial conditions 2 weeks before the onset (15 May) reproduced the onset very well (Fig. 5). Both the observations and simulations included the abrupt onset with northward and eastward migrating tropical disturbances over the Bay of Bengal and the Arabian Sea in late May. Based on a comparison with the Japan Meteorological Agency's operational Ensemble Prediction System, Kajikawa et al. (2015) pointed out that the better reproducibility of the tropical disturbances seen in the NICAM results extended the predictability of the Asian summer monsoon transition phase.

Yamaura et al. (2013) specifically used the NICAM simulations initialized at 29 and 30 May 2012 to clarify the role of a TC in the migration of the Baiu front. Observations revealed that the Baiu front shifted northward with the TC. A simulation integrated from 29 May reproduced both the TC and the northward migration of the Baiu front, whereas the Baiu front stayed stagnant in another simulation integrated from 30 May. Since the TC in the latter simulation developed farther eastward than the observed TC, Yamaura et al. (2013) suggested that the potential extension of the predictability of the Baiu front migration was due to the realistic reproducibility of TCs.

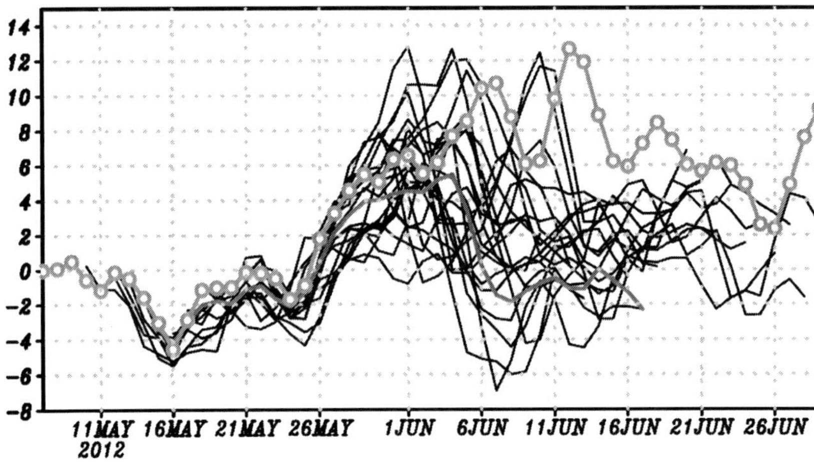

Figure 5 Time series of the Indian summer monsoon indices for the ensemble simulations. The *black curves* are for each run, the *blue curve* is for averaging five members with initial conditions covering May 15–19, and the *red curve* is for the actual observations. The Indian summer monsoon index is defined as the meridional shear of zonal winds at 850 hPa over the Indian continent (40°E–80°E, 5°N–15°N minus 70°E–90°E, 20°N–30°N). After Fig. 1 of Kajikawa et al. (2015)

Seasonal simulations of boreal summer using the global 7 km grid spacing NICAM have captured the Asian summer monsoon as a case study (Oouchi et al. 2009a, 2009b; Satoh et al. 2012b; Kinter et al. 2013). However, these simulations have large biases in precipitation and lower tropospheric wind fields, particularly over the western North Pacific. These biases were partly due to insufficient spin-up of the land surface model, which was upgraded in the experiments conducted during Field 3 of SPIRE.

Another achievement of this project was the first-ever 30-year simulations (see "Atmospheric Model Intercomparison Project-type experiments" section; Kodama et al. 2015). The 30-year simulation successfully simulated the climatology of the annual cycle of the western North Pacific monsoon and its inter-annual variability (see Figures 14 and 15 of Kodama et al. 2015). In Murakami and Matsumoto (1994), the climatological Asian monsoon was defined by differences between the annual maximum and minimum outgoing longwave radiation (OLR) (Fig. 6). The NICAM simulation adequately represents each monsoon sub-system in the observation (e.g., northern Australia-Indonesia monsoon, Southeast Asian monsoon, western North Pacific monsoon); although, the magnitude is stronger by approximately 20–40 W m^{-2}. The NICAM simulation also included a northward extension of the Southeast Asian monsoon and the western North Pacific monsoon which are related to the northward displacement of a low-level westerly and a subtropical high; this in turn led to earlier termination of the Baiu season and to a weaker monsoon trough (Kodama et al. 2015).

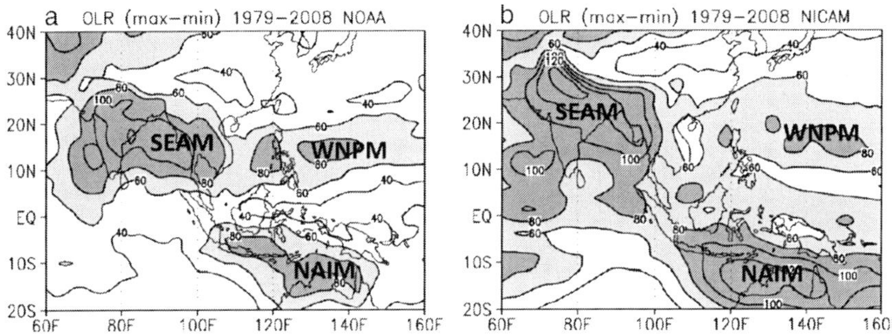

Figure 6 Horizontal distribution of OLR comparing Asian monsoon between simulations and observation. **a** National Oceanic and Atmospheric Administration (NOAA) OLR and **b** OLR of the NICAM AMIP-type simulation defined by the difference between the annual maximum and minimum OLR in the pentad mean climatology (1979–2008). The northern Australia-Indonesia monsoon (NAIM), Southeast Asian monsoon (SEAM), and western North Pacific monsoon (WNPM) are identified after Murakami and Matsumoto (1994)

4.4 Future directions

Both NICAM ensemble simulations and climate simulation showed realistic behavior of the Asian summer monsoon system including the relationship between TCs and the Baiu front. However, similar to other climate models, NICAM suffers from model biases. In particular, the Asian summer monsoon simulations are severely affected by the warm temperature bias over continents. In addition, the mid-latitude jets were stronger in the simulation than in observations partly because a gravity drag parameterization scheme was turned off in the 14 km grid spacing NICAM simulations. In future, simulations of Asian summer monsoons will need to reduce such biases in order to improve their performance; this is currently being tested through development of the land surface model and the introduction of the gravity wave drag scheme.

NICAM simulations with atmosphere and ocean coupling are also being investigated, because realistic responses of convection to basin-scale ocean surface conditions are strongly related to ocean-atmosphere coupling (Wang et al. 2005). In addition, process-oriented diagnosis in comparison with other climate models (Sperber et al. 2013; Ogata et al. 2014; Kusunoki and Arakawa 2015; Zou and Zhou 2015) is planned to understand the effects of basic-state biases on monsoon systems and to make the necessary improvements. The multi-scale processes of monsoons and the Baiu front under global warming conditions in cloud-system resolving simulations are another ongoing research topic (e.g,. Krishnamurthy et al. 2014).

5. MADDEN-JULIAN OSCILLATION

5.1 Background

The MJO, which is often described as an eastward propagating atmospheric pulse with a typical lifespan of 30–60 days, is one of the dominant disturbances in the equatorial atmosphere at the intra-seasonal timescale (Madden and Julian 1972). It is viewed as a key source of predictability for the extended-range forecasts that target the time range of 10–30 days. More than 40 years since its foundation, the majority of GCMs still struggle to produce a satisfactory MJO (Hung et al. 2013). NICAM had enjoyed success in producing a particular MJO event, maintaining a highly accurate eastward propagation speed along with realistic precipitation and zonal wind structures (Miura et al. 2007a). In Field 3 of SPIRE, we utilized the increased computational resource of the K computer to conduct ensemble simulations of MJOs using NICAM to statistically assess the model's ability to predict MJOs (Miyakawa et al. 2014). Applying a widely used evaluation method proposed by Gottschalck et al. (2010), NICAM was diagnosed to have a prediction limit of 26–28 days. The precipitation patterns associated with MJO phases also compared well with observations.

5.2 Experimental setup

The following procedure based on the real-time multivariate MJO (RMM) indices (Wheeler and Hendon 2004) was applied to the boreal winter (October–March) MJOs during an evaluation period (2003–2012) to assign initial dates for the experiment. We identified target MJO cases when a succession of observational RMM plots move counterclockwise through phase 2 to phase 5 (corresponding to the MJO convective envelope traveling from the western Indian Ocean to the eastern Indonesian Archipelago) without retreating more than one phase and had an average amplitude greater than 1 over the span between phases 2 and 5. For each of the MJO cases extracted, the first dates on which the RMM plot falls in phases 8, 1, and 2 are assigned as initial dates.For the 19 MJO cases identified, 54 initial dates were assigned.We used the European Centre for Medium-Range Weather Forecasts interim reanalysis (ERA-Interim) dataset (Dee et al. 2011) for initial conditions of the atmosphere and ocean. The resolution and the physical schemes of NICAM were the same as those described in the "Asian summer monsoon" section except for modified cloud microphysics parameters that control the formation and fall speeds of precipitation particles and minor updates of land-process/boundary layer scheme parameters. A mixed-layer ocean model was applied, and its SST was nudged to externally provided SST at an e-folding time of 7 days. The external SST was defined as the sum of the time-varying mean annual cycle and constant anomalous component. The difference from the mean annual cycle was averaged over the week before the initial date to derive the anomalous component. Thus, no information observed after the initial date was used in the simulations. Further details are found in Miyakawa et al. (2014).

5.3 Results

NICAM maintained a valid (>0.6) MJO skill score for 26–28 days depending on the initial MJO phase, and for 27 days when all 54 cases were included. Composites of observed and simulated precipitation anomalies for MJO phases 3, 5, and 7 are shown in Fig. 7. The horizontal structure of the simulated precipitation anomalies closely resembles the observations, even for phase 7, for which the average lead time was 28 days. However, precipitation anomalies were overestimated over the central Pacific in phase 5. Such a bias is occasionally found in 14 km grid spacing NICAM, perhaps due to under-resolved local vertical moisture transport, which could be a common problem for models that explicitly calculate cloud systems without cumulus parameterization. The simulation series includes an MJO case that occurred in November–December 2011 during the Cooperative Indian Ocean Experiment on Intraseasonal Variability in Year 2011 and Dynamics of the MJO field campaign (CINDY2011/DYNAMO; Yoneyama et al. 2013). Figure 8 shows the observed and simulated convective signals that clearly propagate at a very similar eastward speed over the tropical Indian to the western Pacific Ocean (10°S–10°N, 40°E–160°W). Time-height sections of observed and simulated zonal wind over Gan Island (not shown) further reveal that the model captures the depth and timing of the intrusion of the westerly winds as well as the deepening of moisture with time, both of which are key aspects of MJO propagation. Despite a lead time of nearly 4 weeks, the model also appears to capture the occurrence of the next MJO-like signal in mid-December. The successful prediction of the event provides the opportunity to compare a global non-hydrostatic model simulation with a major in situ observation for the same MJO event for the first time.

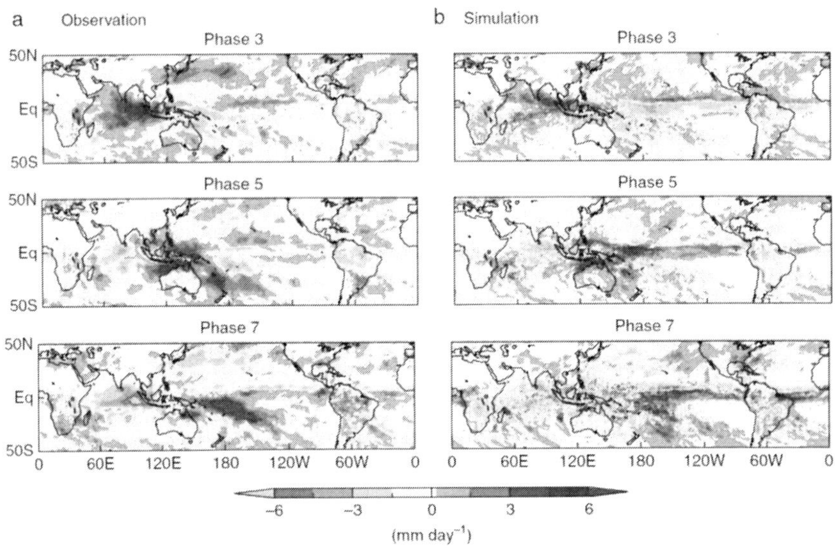

Figure 7 Precipitation anomaly composites for different MJO phases. **a** Observed 1° daily product version 1.2 by the Global Precipitation Climatology Project (Huffman et al. 2001), and **b** simulation using NICAM. The average number of days from the initial dates of the simulations are 16 days (phase 3), 25 days (phase 5), and 28 days (phase 7). Resolution of the simulation output is lowered to 1° mesh to equal the resolution of the observational data set. Anomalies are calculated as deviations from 40-day mean values in each case. After Fig. 2 of Miyakawa et al. (2014)

5.4 Future directions

Although NICAM appears to be one of the unique models that can produce MJO events with high accuracy, we believe that the model still does not live up to its potential in terms of MJO prediction. The limit in available resources has prevented us from applying ensemble simulations of significant size for each event. However, a next-generation supercomputer designed to be roughly 20–100 times more powerful than the K computer should be able to accommodate such simulations. The initialization process of this experiment was a simple interpolation from the ERA-Interim dataset, which allows significant initial shock, but a new data assimilation system specifically designed for NICAM is being developed. The use of a simple mixed-layer ocean model limits the model skill in situations in which dynamical features of the ocean drastically affect the SST conditions. A fully coupled version of NICAM with a three-dimensional dynamical ocean is now available and currently being tested (NICAM-COCO coupled model; Satoh et al. 2014).

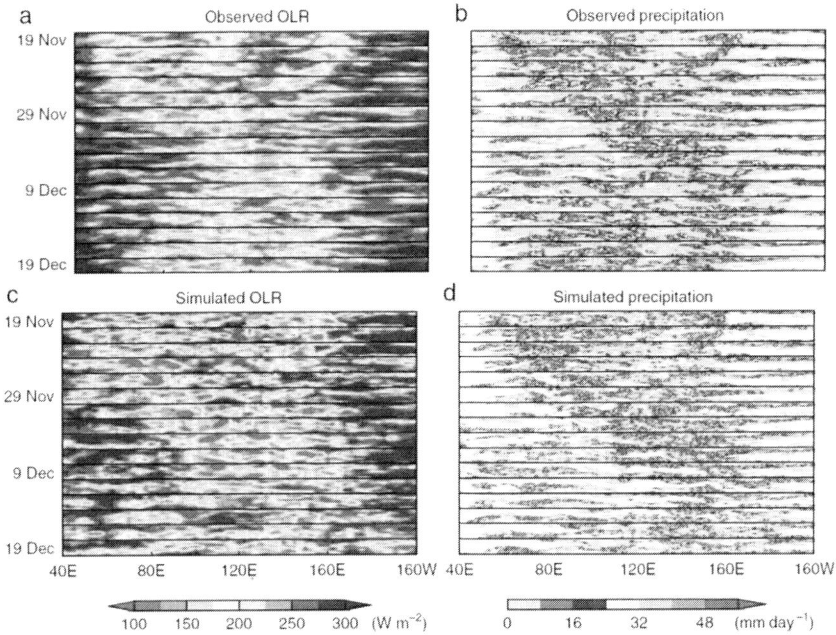

Figure 8 Precipitation anomaly composites for different Madden-Julian oscillation (MJO) phases. **a** Interpolated OLR by NOAA polar-orbiting series of satellites (NOAA-OLR; Liebmann and Smith, 1996) and **c** OLR by a single simulation initialized at 00 UTC 17 November 2011. **b** Precipitation by the 3B42 product of the Tropical Rainfall Measuring Mission (TRMM) satellite (Huffman et al. 2007) and **d** precipitation by the simulation. The figures consist of slices that show horizontal snapshots of the tropical Indian Ocean to the western Pacific Ocean (10°S–10°N, 40°E–160°W). The resolution of the simulated OLR is lowered to 2.5° mesh to equal the resolution of the NOAA-OLR data set. The resolutions of the TRMM 3B42 data set and the simulated precipitation are lowered to 1° mesh. After Fig. 3 of Miyakawa et al. (2014)

6. BOREAL SUMMER INTRA-SEASONAL OSCILLATION AND TROPICAL CYCLOGENESIS

6.1 Background

Tropical cyclones cause huge socio-economic impacts, so more accurate TC forecasts are demanded. Because forecasting tropical cyclogenesis (TC genesis) is still a challenge in numerical weather forecasting, the predictability of TCs at sub-seasonal to seasonal time scales (2 weeks to 2 months) is promoted worldwide (Vitart et al. 2012, 2017). TC genesis is affected by large-scale environmental conditions, such as low-level cyclonic vorticity, small vertical wind shear, the convective instability, and mid-level humidity (Gray 1975, 1979). The boreal summer intra-seasonal oscillation (BSISO; Wang and Rui 1990; Wang and Xie 1997) modulates the large-scale environment and TC activity. For example, Nakazawa (2006) pointed out that the active phase of BSISO in the 2004 summer season successively generated TCs, and persistent steering flow led to a record-breaking number of TC landfalls in Japan. Therefore, the accurate prediction of BSISO and the modulation of large-scale environmental conditions associated with BSISO are believed to lead to accurate forecasting of TC genesis.

Studies using NICAM show that the model is a promising tool for accurate BSISO and TC genesis forecasting. For example, Oouchi et al. (2009a) demonstrated that NICAM successfully simulated modulations of BSISO and that TC genesis was captured 3 weeks in advance. However, the authors examined only a limited number of TC genesis cases using a single run because of the limitations in computational resources. The K computer enables us to examine the predictability of the BSISO and TC genesis using the ensemble approach. Here, we examine the predictability of eight cases of TC genesis that occurred successively in August 2004.

6.2 Experimental setup

Nakano et al. (2015) performed 30-day simulations for each day of August 2004; that is, a total of 31 ensemble 30-day simulations were performed. The initial conditions of the atmosphere and SST were derived from the ERA-Interim dataset (Dee et al. 2011). The resolution and the physical schemes of NICAM were the same as those described in the "Asian summer monsoon" section. This setting is "forecast-mode" because the model never knows the observed SST during the time integration.

Table 1 Simulation performance for forecasting TC genesis

	Malou	Meranti	Rananim	Malakas	Megi	Chaba	Aere	Songda
Central pressure (hPa)	996	960	950	990	970	910	955	925
Lifetime (days)	0.875	4.75	4.5	2.75	4.125	11.75	6.25	11
Week 1	0	100	75	43	86	71	71	86
Week 2	N/A	N/A	N/A	N/A	80	43	57	57
Week 3	N/A	N/A	N/A	N/A	N/A	N/A	N/A	29

Each of the eight TCs occurring in August 2004 is listed with their observed central pressure and lifetime, and the simulation's hitting ratio (%) in weeks 1, 2, and 3. Adapted from Table 1 of Nakano et al. (2015).

A candidate vortex that corresponds to an observed TC should pass within 10° of the location of the observed TC genesis within 1 day of the observed TC genesis time. The candidate vortex should meet TC criteria (e.g., surface wind speed >17.5 m s^{-1}) within 5 days of the observed TC genesis time. Detailed descriptions of the experimental setup and method of TC detection are provided in Nakano et al. (2015).

6.3 Results

Table 1 summarizes the results of the TC genesis forecast. The model captured six out of the eight cases of TC genesis 1 week in advance with a low missing rate (all were above 70%). Moreover, four cases of TC genesis in late August were captured 2 weeks in advance with high probability (all were above 40% and 3 of 4 were above 50%). The genesis of TC Songda was well captured 2 weeks in advance, though not 3 weeks in advance. Note that two TC genesis events (Malou and Malakas) were not captured, probably because they were very weak (minimum central pressure >990 hPa) and had short lifespan (less than 3 days).

Figure 9 shows the zonal wind at 850 hPa and the OLR, which is related to convective activities. In the analyzed field, the shear line between easterly wind in the tropics and monsoonal westerly wind extended eastward, accompanied by active convection in mid-August. This eastward extension was maintained throughout late August. TC Songda was generated on the shear line. The experiments initialized between August 5 and 11 (3 weeks before Songda's genesis) captured the eastward extension of the shear line in mid-August but it retreated in late August. The experiments initialized between August 12 and 18 (2 weeks before Songda's genesis) captured the eastward extension of the shear line in late August. Therefore, we conclude that the eastward extension of the shear line is closely related to Songda's genesis, and if models reproduce the longitudinal evolution of the shear line, the associated TC genesis will be simulated effectively.

6.4 Future directions

Once we have achieved more reliable predictability of TC genesis with a longer lead time, our next targets are the probabilities of TC tracks, intensities, and landfall. Such information could be calculated from the large samples of ensemble simulations. Since the NICAM-LETKF system (Terasaki et al. 2015) is an ensemble-based data assimilation system, it can be used to make the large number of initial atmospheric fields fit to the model's dynamics and physics. It is expected that the probability forecast of TC tracks and landfall probability would be affected by the mid-latitude atmospheric conditions including the structure and evolution of the jet stream and the subtropical high. Given that the mid-latitude jet stream and the subtropical high are affected by the weather systems in the tropics such as MJO (Kawamura et al. 1996; Moon et al. 2013; Molinari and Vollaro 2012), we should further examine the relationship between mid-latitude model biases and those in the tropics to improve probability forecasting of TC tracks. For forecasts of TC intensity, current atmospheric models have typical biases: they are not able to reproduce rapid intensification, and TCs in the models continuously intensify even in the mid-latitude, where TCs begin to weaken, then resulting in overintensification. A higher resolution model is needed to improve the representation of the structure of the TC's inner core, which results in accurate simulation of rapid intensification (Rogers et al. 2013; Wang and Wang, 2014). Interestingly, a recent study shows that high-resolution global models reduce forecast error in TC intensity and tracking forecasts (Nakano et al. 2017). The atmosphere-ocean interacting processes should also be incorporated in the numerical model to avoid overintensification (Wada et al. 2014; Zarzycki 2016). Simulations of TC intensities are sensitive to details of physical schemes such as the boundary layer scheme and the cloud microphysics scheme (Kanada et al. 2012). More improvement and development of NICAM together with evaluations using available observational data will be necessary to pursue the aforementioned future directions in TC predictability studies.

6.5 Atmospheric Model Intercomparison Project-type experiments

Atmospheric Model Intercomparison Project (AMIP) is a standard experimental protocol for atmospheric GCMs to test models' climatology with multi-decadal simulations. Such climate simulations are now achievable using a high-resolution global non-hydrostatic model, and a new era of climate-system research has begun. For example, the simulated intensities of TCs strongly depend on horizontal resolution (Camargo 2013) and convection scheme (Murakami et al. 2012). Thus, by taking advantage of high-resolution global non-hydrostatic simulations without deep convection schemes, the climatology of TCs is expected to be simulated more informatively, thereby facilitating further investigations of the probability and geographical distributions of TC intensities and structures and their projected changes under a warmer climate (Oouchi et al. 2006; Murakami et al. 2012; Knutson et al. 2015). The global non-hydrostatic climate model is also a powerful tool for simultaneously investigating phenomena at various spatiotemporal scales of atmospheric disturbances, such as planetary-scale atmospheric general circulations, MJO, tropical waves, diurnal meso-scale convective events, severe rainfall, and atmospheric gravity waves and their interactions (Holt et al. 2016). In addition, better representation of clouds, convection, and circulations is key to reducing uncertainties of climate sensitivity (Bony et al. 2015; Stevens and Bony 2013).

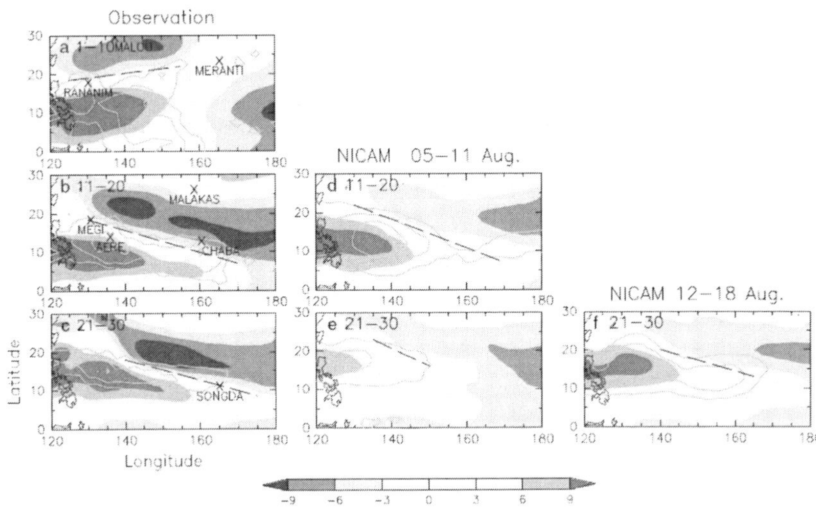

Figure 9 Comparison of 10-day mean OLR and zonal wind at 850 hPa between observation and simulation. Observed and ensemble mean of 10-day mean OLR (*green contour*) and zonal wind at 850 hPa (*color shading*) averaged during **a** 1–10 August, **b**, **d** 11–20 August, and **c**, **e**, **f** 21–30 August 2004. **d**, **e** show NICAM simulation initialized from 5 to 11 August; **f** shows 12–18 August. **a** *Black crosses* indicate the locations of tropical cyclogenesis. OLR values less than 200 W m^{-2} are shown with contour intervals of 20 W m^{-2}. *Broken lines* depict the shear line. After Fig. 1 of Nakano et al. (2015)

Motivated by the aforementioned demands for AMIP-type experiments with high-resolution global models, we performed present (Kodama et al. 2015) and future (Satoh et al. 2015) climate simulations using the 14 km grid spacing NICAM. The model settings were the same as those described in the "Asian summer monsoon" section. Specifically, we should note that instead of following the strict AMIP protocol, the slab ocean model and nudging technique were employed instead of imposing a fixed SST condition to obtain better performance of the distribution of tropical precipitation, as shown in Kodama et al. (2015). The simulations were performed to reproduce the climates of 1979–2008 (Kodama et al. 2015) and to project the climates of 2074–2100 (Satoh et al. 2015).

Kodama et al. (2015) reported the simulated climatology of basic state, TCs, MJO, Asian monsoon, diurnal precipitation cycle, and quasi-biennial oscillation (QBO). In-depth analysis of the MJO and mean and intense precipitation around Japan were presented in Kikuchi et al. (2016) and Fujita et al. (2017, personal communication), respectively. Fukutomi et al. (2015) evaluated the activity and structure of the simulated tropical synoptic scale waves over the western Pacific.

Here, we extend the analysis of Fukutomi et al. (2015) to the various types of convectively coupled equatorial waves (Kiladis et al. 2009), which govern the tropical day-to-day weather. Spatiotemporal spectral and filter analyses were applied to OLR, precipitation, and wind fields to diagnose OLR variances contributed from equatorial Kelvin waves, equatorial Rossby waves, mixed Rossby gravity waves, and tropical depression waves. The National Oceanic and Atmospheric Administration (NOAA) OLR (Liebmann and Smith 1996) and the Japanese 55-year Reanalysis (JRA-55; Kobayashi et al. 2015b) datasets were used to evaluate the results. Here, the equatorial wave-filtered OLR variances are plotted (Fig. 10). Geographical distributions of tropical waves over the Indian and Pacific Oceans were qualitatively simulated, though their overall activities tended to be weaker than the observed ones (Fig. 10). Significant biases were found over Central and South America, the Atlantic, and Africa in all the wave components. Kelvin waves were very weakly simulated, particularly over the Atlantic, whereas simulated equatorial Rossby waves were stronger over the Amazonian region and weaker over the Atlantic than those of the observation. Distribution of the simulated mixed Rossby gravity waves was biased over Central and South America and the Atlantic. Tropical depression waves were weaker over Central and South America and the western Atlantic, whereas the African easterly waves in the simulation were more active over the eastern Atlantic compared with those in the observation. The reasoning behind the model biases in convectively coupled equatorial waves will be further investigated from the perspective of reproducibility of the seasonal scale background fields.

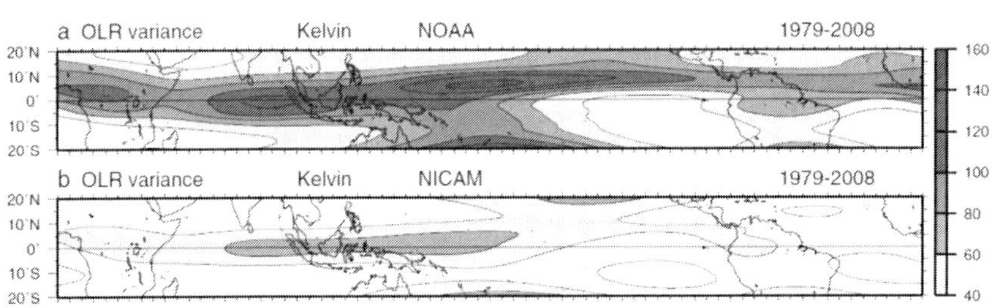

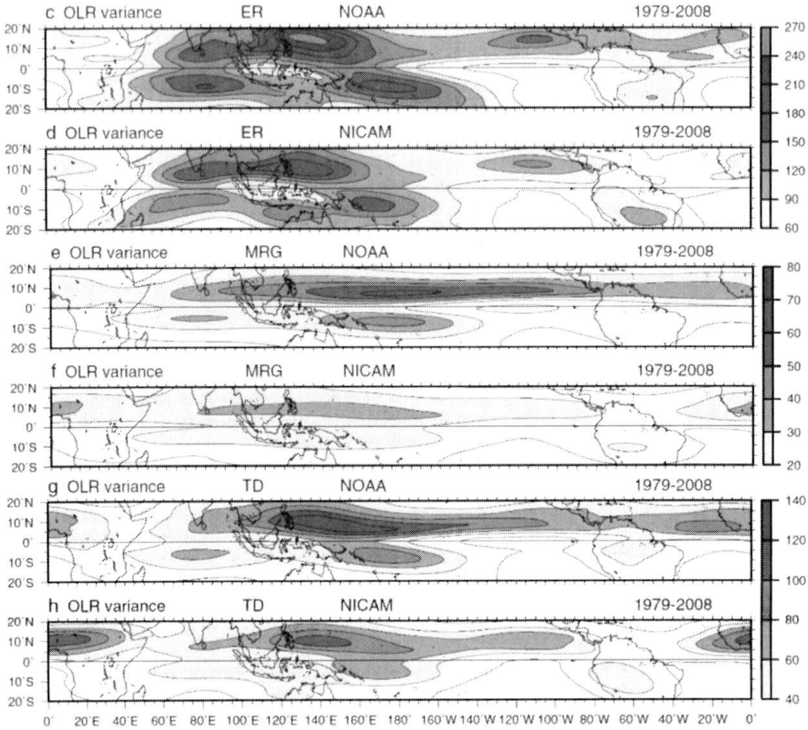

Figure 10 Comparison of climatology of the OLR variances of equatorial waves between observation and simulation. Annual mean climatology (1979–2008) of the OLR variances in $(W\ m^{-2})^2$ for Kelvin wave (Kelvin) (**a**, **b**), equatorial Rossby waves (ER) (**c**, **d**), mixed Rossby gravity waves (MRG) (**e**, **f**), and tropical depression waves (TD). The observed NOAA OLR variances are shown by (**a**, **c**, **e**, **g**), whereas the NICAM simulated OLR variances are shown by (**b**, **d**, **f**, **h**)

Future projected climate simulations were analyzed by Satoh et al. (2015), in which the response of TCs to global warming was investigated using present and future climate simulation data. Yamada et al. (2017, personal communication) further investigated structural changes in TCs. In addition to TCs, how the precipitation around Japan (Fujita et al., 2017, personal communication) and the precipitation associated with extratropical cyclones respond to global warming (Kodama et al. 2017, personal communication) has been investigated. Before the K computer was available, projection studies using NICAM were limited to short time-slice simulations such as one year or season (Yamada et al. 2010; Satoh et al. 2012a; Tsushima et al. 2014; Noda et al. 2014, 2015), and the insufficient integration period made it difficult to achieve statistical significance of the global warming response. It is now possible to confirm the results derived from shorter-integration NICAM data. For example, Chen et al. (2016) performed seasonal scale sensitivity experiments and showed a large longwave cloud radiative feedback, irrespective of the cloud microphysics schemes that were employed. This result was further supported by an additional analysis of the inter-annual variability found in the NICAM AMIP-type data. Figure 11 shows another example depicting the distribution of high-cloud size that was evaluated using the method described by Noda et al. (2014). Qualitatively, the size distributions under the present and the future conditions show a result similar to those evaluated in a 1-year NICAM simulation data in Noda et al. (2014). Responding to global warming, the number of smaller clouds (<40 km radius) notably increases, while changes to the number of larger clouds (≥40 km) are much smaller and not significant. We found that the difference in the smaller clouds is statistically significant. The result implies less organization of tropical cloud systems in the warmer world. As analyzed by Chen et al. (2016), NICAM shows larger positive high cloud feedback than the CMIP5 models due to the increase of high cloud cover and more positive longwave feedback.

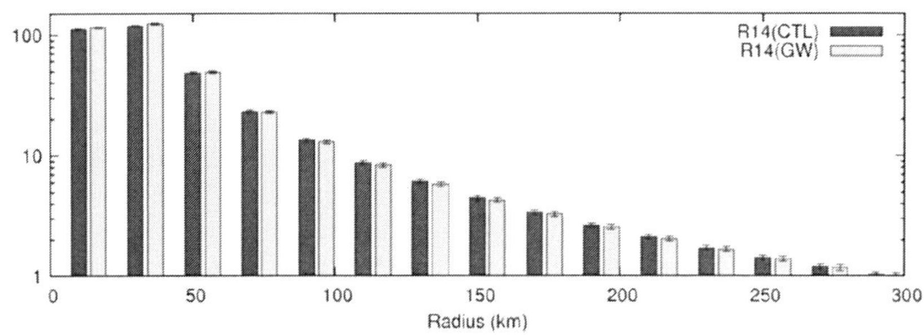

Figure 11 Size distribution of high clouds showing its changes between present and future climate simulations. Size distribution of high clouds in present (*dark blue*) and future (*light blue*) climate simulations using the NICAM AMIP-type simulation are analyzed by following the method described by Noda et al. (2014). Standard variations are designated at each bin

The NICAM AMIP-type simulations described above are the first step toward global cloud-resolving climate simulation. Our next plan, the High Resolution Model Intercomparison Project (HighResMIP), is briefly described in the "Future perspectives" section.

7. PROJECTION OF TROPICAL CYCLONES

7.1 Background

A primary energy source of TCs is latent heat supplied from the warm tropical ocean (Emanuel 2003). The change in TC activities in a warmer climate is one of the important issues of climate change (Walsh et al. 2016). Broccoli and Manabe (1990) used a GCM to project the responses of TCs to global warming for the first time. They used a horizontal grid spacing of a few hundred kilometers, which was too coarse to resolve TC structures such as a warm core. However, their simulation showed the global distribution of simulated TCs in general agreement with the observation. They indicated that climate models have the potential to predict future changes in TCs and advocated reducing the horizontal grid interval in order to realistically reproduce TC structures.

Many GCMs have used finer resolutions than that of Broccoli and Manabe (1990) to assess the response of TCs to global warming, indicating some consensus in the change of TC statistics in a warmer world: decrease in TC genesis frequency, increase in intense TCs, mean lifetime maximum intensity, and rainfall around TCs (Knutson et al. 2010; Christensen et al. 2013; Walsh et al. 2016). However, Emanuel (2013) showed an increase in global annual frequency of TCs under warmer climate conditions using a downscaling method in which incipient vortices are embedded into large-scale climate conditions projected by CMIP5 models. To achieve a better understanding of the relationships between TC genesis and climate change, the Hurricane Working Group (HWG) established by the US Climate and Ocean: Variability, Predictability and Change (CLIVAR) group coordinated common experiments (Walsh et al. 2015). Although the downscaling methodology of Emanuel (2013) projected an increase in TC genesis under the warmer climate condition projected by the HWG models, there were no models that generated a substantial increase in global TC frequency. The reduction of global TC numbers is associated with a decrease in upward mass flux (Sugi et al. 2002, 2012), and an increase in saturation deficit of the mid-troposphere (Emanuel et al. 2008). The mechanism underpinning the decrease in TC numbers, however, remains controversial.

Due to recent advances in supercomputing and GCMs, fine-resolution models with grid spacing less than 25 km have been used to investigate future change in TC activities (Murakami et al. 2012; Manganello et al. 2014; Knutson et al. 2015; Roberts et al. 2015; Wehner et al. 2015). These models reproduced TC structure more realistically. The advent of the K computer enabled researchers to perform long-term simulations using a high-resolution global non-hydrostatic model (Kodama et al. 2015; Satoh et al. 2015).

In this section, future changes in TC activities are investigated using the outputs of the aforementioned NICAM AMIP-type simulation (referred to as the present day (PD) simulation; Kodama et al. 2015) and a warmer climate simulation (referred to as the global warming (GW) simulation; Satoh et al. 2015). Kodama et al. (2015) documented the model settings and experimental design of the PD simulation. The model settings of the GW simulation were identical to those of the PD simulation, and the experimental design is described in Satoh et al. (2015).

7.2 Results

Figure 12 shows geographical distributions of TC genesis density for the PD and GW simulations and the future change (GW -- PD). The PD simulation reproduces the observed distribution of TC genesis fairly well; although, the model has clear bias particularly over the north Atlantic for less TC genesis (See Fig. 10 of Kodama et al. 2015). The GW simulation showed a salient decrease in TC genesis over the eastern North Pacific (Fig. 12c). Satoh et al. (2015) documented the annual global TC number of 104.0 for the PD simulation and 81.8 for the GW simulation. The decrement in the TC number due to global warming is 21.3%, which is within the range of previous studies (Knutson et al. 2010; Christensen et al. 2013). Figure 12c shows that TCs are mainly decreased over the north eastern Pacific.

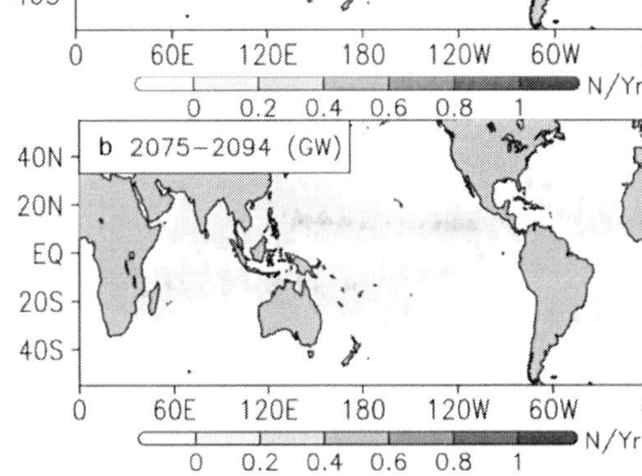

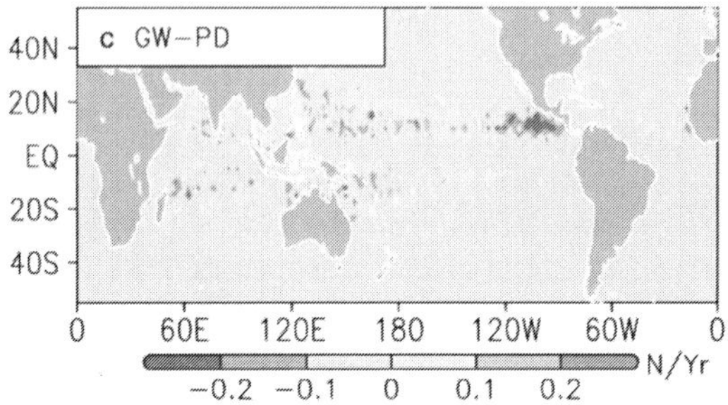

Figure 12 Geographical distributions of annual TC genesis density. **a** for the PD simulation for 1979–1998; **b** for the GW simulation of 2074–2093; **c** shows the difference between the GW and PD simulations. TC genesis density is defined as the number of TCs generated in each 2.5° × 2.5° grid box per year

As the mechanism of the decrement was not completely understood, Satoh et al. (2015) used outputs of the PD and GW simulations to propose a new concept based on the relationship between convective mass flux and TCs. They extracted a convective mass flux associated with TCs and defined its contribution rate to convective mass flux in the tropics (R) as follows:

$$R = \frac{M_T}{T_{\text{int}} M},$$

where M_T is convective mass flux associated with TCs (defined as integration over the area with a radius smaller than 500 km from the center of TCs), T_{int} is the integration period, and M is the total convective mass flux over the tropical domain (30°N–30°S). Since $M_T = NTm_T$, where N is the number of global TCs, T is the average lifetime of TCs, and m_T is the average convective mass flux per single TC, we can rewrite this as (Eq. 15 of Satoh et al. 2015)

$$N = R\frac{T_{\text{int}} M}{T m_T}.$$

Satoh et al. (2015) found that the contribution of TC (R) in GW is almost unchanged or less than that in PD. The convective mass flux in the tropics (M) generally decreases in GW (Vecchi and Soden 2007), and our results confirm this change. In the analysis of the TC convective mass flux, Satoh et al. (2015) also found that the convective mass flux associated with TCs and its fractional area become greater as TCs become more intense (Fig. 13a and b, respectively). They also found that the TC convective mass flux and its frictional area in GW are greater than those in PD for the same intensity of TC (Fig. 13a and b, respectively). These results are contrasted by Fig. 13c, where the area-averaged convective mass flux, or vertical velocity, does not show a robust change with warming. Intense TCs increase under a warmer climate, leading to greater convective mass flux associated with TCs. In these simulations, the incidence of intense TCs that develop to less than 945 hPa increases by 38.6% due to global warming (Fig. 14). Thus, Satoh et al. (2015) showed that m_T increases as warming. Assuming that R is unchanged between PD and GW, this constraint implies that TC number decreases in order to intensify TCs under a warmer climate.

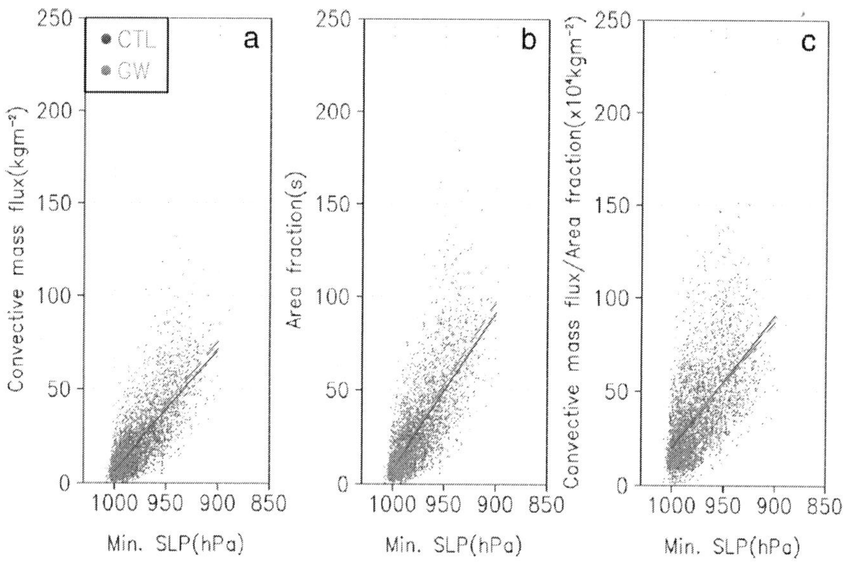

Figure 13 Scatter plots of the TC properties versus the minimum sea-level pressure. Scatter plot represents **a** the relationship between the minimum sea-level pressure (Min. SLP) and the convective mass flux over the TC domain [*A*], **b** the relationship between the min. SLP and the fractional area of the strong updrafts [*B*], and (**c**) the relationship between the min. SLP and the area-averaged convective mass flux [*A*/*B*] for all the simulated TCs. The PD simulation is shown by *blue dots* and the GW simulation by *red dots*; *blue* and *red lines* show regressions for the PD and GW simulations, respectively. The convective mass flux and the fractional area of the strong updrafts are accumulated for the lifetime of each TC, so that these have units of kg m^{-2} and s, respectively. Only the vertical velocities exceeding the threshold value of 0.5 m s^{-1} were used. The values at an altitude of 4 km are shown. After Fig. 5 of Satoh et al. (2015)

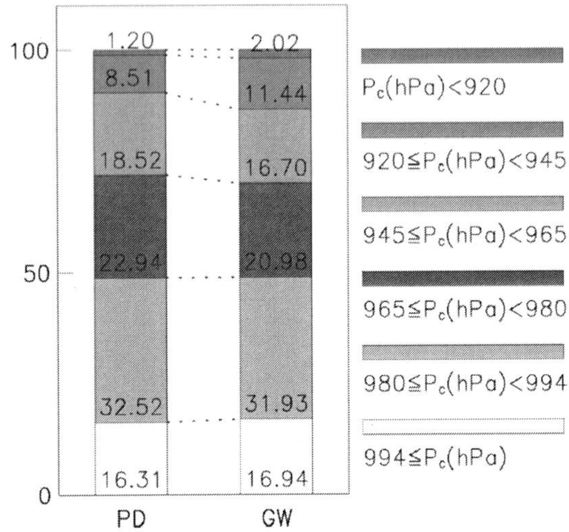

Figure 14 Normalized frequencies (%) for each TC intensity from the PD and GW simulations. The *two boxes* represent the number of TCs categorized into six intensities using the lifetime minimum central sea-level pressure (Pc). The number in each colored box indicates the number of TCs categorized in each intensity level

7.3 Future directions

Walsh et al. (2016) highlighted the importance of the following topics for future research on TCs and climate change: detection methods as a source of uncertainty, lack of climate theory of TC formation, confidence in detection and attribution of observed changes in TCs to date, impact of uncertainty in projections of large-scale climate fields, and correct simulations of TC frequency response in high-resolution climate models. It is expected that high-resolution models will be useful tools for investigating these topics. Thanks to the K computer, we completed the first-ever climate simulation with a non-hydrostatic GCM using 14 km grid spacing (Kodama et al. 2015; Satoh et al. 2015). The outputs from this work will contribute important new findings on TC climatology.

Higher-resolution models have simulated more realistic TC climatology (Walsh et al. 2016). However, TC distributions simulated in high-resolution models still show inconsistency with observed TC activities such as fewer TCs over the north Atlantic (Kodama et al. 2015; Murakami et al. 2015; Roberts et al. 2015), more TCs over the western North Pacific (Manganello et al. 2012; Murakami et al. 2015) and the eastern North Pacific (Kodama et al. 2015; Roberts et al. 2015), weak maximum wind speed compared with low central sea-level pressure (Roberts et al. 2015), and an overestimation of mean TC intensities at higher latitudes (Murakami et al. 2012; Kodama et al. 2015). Reducing these biases will enhance confidence in GCM simulations of future change. For example, Lloyd and Vecchi (2011) investigated the observation that the passage of TCs causes SST cooling, indicating that this oceanic feedback suppresses intensification of TCs. Atmosphere-ocean coupled models will improve the overestimation of TC intensities at higher latitudes.

In addition, high-resolution models simulate TC structures more realistically. An increase in the convective mass flux associated with TCs due to global warming is not caused by an enhancement of updraft velocities (Fig. 13c), but by an increase in the area of the updraft region around TCs (Fig. 13b). This suggests that global warming influences the dynamical structure of TCs. The scale of damage caused by a disaster associated with TCs is measured by TC gale force wind radii as well as its frequency, intensity, and track. Recent studies have focused on future changes in the size of TCs (Kanada et al. 2013; Manganello et al. 2014; Knutson et al. 2015). High-resolution global non-hydrostatic models should also make it possible to statistically evaluate future changes of TC structure.

Ohno and Satoh (2015) revealed that the flow from the lower stratosphere near the TC eyewall plays a key role in constituting the upper-level warm core which develops TC intensity prior to reaching the mature stage. To increase confidence in this theory, it must be compared with an authentic observation with high temporal and spatial resolution. Today, geostationary satellites can observe atmospheric wind vectors associated with TCs in the upper-troposphere with high temporal and spatial resolution (Bessho et al. 2016). Outputs of high-resolution models should be evaluated with such new satellite data, promoting our understanding of TCs.

8. FUTURE PERSPECTIVES

8.1 Post-K supercomputer project

Among the outcomes of Field 3 of SPIRE summarized in this review, the series of studies using the sub-kilometer global atmospheric simulation (Miyamoto et al. 2013, 2015; Kajikawa et al. 2016) are typical examples of "capability computing" using the K computer; that is, a single complex problem solved in the shortest time possible using the maximum computing power available. Although this kind of simulation does not always lead to the maximum output or outcome, the knowledge gained has great potential to bolster next-generation research. In much larger computer system such as the Post-K supercomputer, problems of this size will be easily handled with various configurations. We can expect to get much more information from these supercomputers in terms of "capacity computing," or the solution of smaller or less complex problems using more efficient computing power.

A research strategy similar to that mentioned above has already been adopted. In the era of the Earth Simulator, Miura et al. (2007a) take a capability computing approach to demonstrate the ability to represent the MJO with 3.5 and 7 km grid spacing using NICAM. Using a large number of ensemble simulations, Miyakawa et al. (2014) statistically confirmed the MJO predictability suggested by Miura et al. (2007a) and suggested that high-resolution global non-hydrostatic models like NICAM generally have better skill in MJO predictability than GCMs. Thus, combining the two strategies of capability computing and capacity computing may be an effective approach for future research using the Post-K supercomputer.

The project to develop the next-generation flagship supercomputer is now underway. The Post-K supercomputer is designed by using homogeneous CPU architecture. The CPU will be a many-core processor, and accelerators are not included in the system. The system will consist of a six-dimensional mesh/torus interconnected network (Tofu; Ajima et al. 2009) and a three-level hierarchical storage system. The co-design is based on collaboration between computational scientists and computer scientists, because a system constructed using commodity computers along current computer trends cannot satisfy the socio-scientific demands of all research areas, including earth sciences. NICAM has been selected as a co-design application from the viewpoint of a structured stencil calculation. In addition, LETKF has also been selected in this co-design project as an application requiring the dense-matrix calculation. NICAM-LETKF (Terasaki et al. 2015) is one of the most important target applications for the development of the new computer system. The current target problem for the NICAM-LETKF is as follows: in the data assimilation cycle, 1000 ensembles by NICAM with 3.5 km grid spacing and 100 vertical levels are conducted every 6 h. The prediction using 440 m horizontal grid spacing is integrated for several months. In terms of the computational performance, the workflow between the simulation code and the data assimilation system, including I/O and communication in a massively parallel supercomputer becomes a serious problem. A framework proposed by Yashiro et al. (2016c), which was designed to reduce data movement and to utilize parallel I/O, showed a reduction in the workflow's total elapse time. Based on the scientific outcomes summarized in the present review, the big data assimilation by NICAM-LETKF using tremendous volumes of observational data, such as satellite data, is expected to further contribute to the improvement of mid-range predictions around the globe.

8.2 Post-K supercomputer studies with NICAM

Following the SPIRE project, the FLAGSHIP2020 project began in 2014 to examine possible maximum use of the Post-K supercomputer. Among the research subjects defined for the weather and climate simulation studies, seamless simulations using NICAM have been proposed. Seamless means beyond a week to seasonal prediction, especially for the predictability of TC statistics such as TC genesis, tracks, and intensities. Furthermore, seamless studies cover a year to decadal scales, and even to the centennial scale, as aimed for by the HighResMIP study, which is introduced below. Successful extended-range forecasts of MJO and BSISO (Miyakawa et al. 2014; Nakano et al. 2015) suggest greater predictability of TCs beyond a week to a few weeks. Through ensemble simulations, many aspects of TCs such as TC genesis, tracks, and intensities will be obtained. In addition, together with the development of the high-resolution atmosphere and ocean coupled model (NICAM-COCO), the seasonal scale predictability of ISO and associated TC characteristics will be explored. For further multi-year simulation, we will examine the potential predictability of the near future from a year to a decadal scale, particularly focusing on TC statistics (as described in the "Projection of tropical cyclones" section). These longer time-scale simulations have also been examined in the "Atmospheric Model Intercomparison Project-type experiments" section and are further pursued in the next subsection, "High Resolution Model Intercomparison Project."

8.3 High Resolution Model Intercomparison Project

As a part of Coupled Model Intercomparison Project phase 6 (CMIP6), the High Resolution Model Intercomparison Project (HighResMIP; Haarsma et al. 2016) is now under way. In this project, climate modeling with a horizontal resolution of 25–50 km will be performed for more than half a century. The aim is that such fine-resolution climate model data will be available for everyone within a few years as CMIP6 continues. Following the HighResMIP protocol, we plan to perform 56 and 28 km grid spacing NICAM for 1950–2014 and 14 km grid spacing NICAM for ten years. Currently, a series of sensitivity experiments are being performed to improve climatology in the simulation. Our aim in participating in the HighResMIP is to investigate multi-scale phenomena including tropical and extratropical cyclones, MJO, tropical waves, convection, and clouds in a seamless manner (see Section 7.2 of Haarsma et al. 2016). Process-based assessment using satellite datasets will also be key to evaluating and constraining the high-resolution climate model.

9. CONCLUSIONS

NICAM-related studies using the K computer under Field 3 of SPIRE have produced a wealth of valuable weather and climate modeling research and present clear challenges that may be solved with the Post-K supercomputer. Key achievements of these studies include high-resolution sub-kilometer global simulations, ensemble simulations for MJO and BSISO, and simulations of TC genesis. The longer multi-decadal simulation by NICAM was conducted for the first time with the global non-hydrostatic model. The climatology of the 14 km grid spacing NICAM shows comparable results with other climate models; although, biases similar to those of other models exist. The results suggest that, even though it has been arguable whether deep convective parameterization is required for the 14 km grid spacing model, the simulation without the parameterization reproduces climatology consistent with observation. Thanks to the K computer, we obtained considerable evidence of NICAM's usability for weather and climate simulations. Further research using the Post-K supercomputer in the next decade is expected to contribute even more crucial knowledge to this exciting research field.

REFERENCES

1. Ajima Y, Sumimoto S, Shimizu T (2009) Tofu: a 6D mesh/torus interconnect for exascale computers. Computer 42:36–40. doi:10.1109/MC.2009.370
2. Amdahl GM (1967) Validity of the single processor approach to achieving large scale computing capabilities. In: AFIPS '67 (Spring) Proceedings of the April 18-20, 1967, spring joint computer conference. ACM, New York, pp 483–485. doi:10.1145/1465482.1465560
3. Ando K, Hyodo M, Baba T, Hori T, Kato T, Watanabe M, Ichikawa S, Kitahara H, Uehara H, Inoue H (2016) Parallel-algorithm extension for tsunami and earthquake-cycle simulators for massively parallel execution on the K computer. Int J High Perform Comput Appl 30:454–468. doi:10.1177/1094342016636670
4. Arakawa A, Schubert WH (1974) Interaction of a cumulus cloud ensemble with large-scale environment, Part 1. J Atmos Sci 31:674–701
5. Arakawa T, Inoue T, Sato M (2014) Performance evaluation and case study of a coupling software ppOpen-MATH/MP. Procedia Comp Sci 29:924–935. doi:10.1016/j. procs.2014.05.083
6. Bessho K, Date K, Hayashi M, Ikeda A, Imai T, Inoue H, Kumagai Y, Miyakawa T, Murata H, Ohno T, Okuyama A, Oyama R, Sasaki Y, Shimazu Y, Shimoji K, Sumida Y, Suzuki M, Taniguchi H, Tsuchiyama H, Uesawa D, Yokota H, Yoshida R (2016) An introduction to Himawari-8/9—Japan's new-generation geostationary meteorological satellites. J Meteorol Soc Jpn 94:151–183. doi:10.2151/jmsj.2016-009
7. Bony S, Stevens B, Frierson DMW, Jakob C, Kageyama M, Pincus R, Shepherd TG, Sherwood SC, Siebesma AP, Sobel AH, Watanabe M, Webb MJ (2015) Clouds, circulation and climate sensitivity. Nat Geosci 8:261–268. doi:10.1038/ngeo2398

8. Broccoli AJ, Manabe S (1990) Can existing climate models be used to study anthropogenic changes in tropical cyclone climate? Geophys Res Lett 17:1917–1920. doi:10.1029/GL017i011p01917
9. Bryan GH, Wyngaard JC, Fritsch JM (2003) Resolution requirements for the simulation of deep moist convection. Mon Weather Rev 131:2394–2416. doi:10.1175/1520-0493(2003)131<2394:RRFTSO>2.0.CO;2
10. Camargo SJ (2013) Global and regional aspects of tropical cyclone activity in the CMIP5 models. J Clim 26:9880–9902. doi:10.1175/JCLI-D-12-00549.1
11. Chen G, Zhu X, Sha W, Iwasaki T, Seko H, Saito K, Iwai H, Ishii S (2015a) Toward improved forecasts of sea-breeze horizontal convective rolls at super high resolutions. Part I: configuration and verification of a down-scaling simulation system (DS3). Mon Weather Rev 143:1849–1872. doi:10.1175/MWR-D-14-00212.1
12. Chen G, Zhu X, Sha W, Iwasaki T, Seko H, Saito K, Iwai H, Ishii S (2015b) Toward improved forecasts of sea-breeze horizontal convective rolls at super high resolutions. Part II: the impacts of land use and buildings. Mon Weather Rev 143:1873–1894. doi:10.1175/MWR-D-14-00230.1
13. Chen Y-W, Seiki T, Kodama C, Satoh M, Noda AT, Yamada Y (2016) High cloud responses to global warming simulated by two different cloud microphysics schemes implemented in the Nonhydrostatic Icosahedral Atmospheric Model (NICAM). J Clim 29:5949–5964. doi:10.1175/JCLI-D-15-0668.1
14. Christensen JH, Krishna Kumar K, Aldrian E, An S-I, Cavalcanti IFA, de Castro M, Dong W, Goswami P, Hall A, Kanyanga JK, Kitoh A, Kossin J, Lau N-C, Renwick J, Stephenson DB, Xie S-P, Zhou T (2013) Climate phenomena and their relevance for future regional climate change. In: Stocker TF, Qin D, Plattner G-K, Tignor M, Allen SK, Boschung J, Nauels A, Xia Y, Bex V, Midgley PM (eds) Climate Change 2013: The Physical Science Basis. Contribution of Working Group I to the Fifth Assessment Report of the Intergovernmental Panel on Climate Change. Cambridge University Press, Cambridge
15. Dee DP, Uppala SM, Simmons AJ, Berrisford P, Poli P, Kobayashi S, Andrae U, Balmaseda MA, Balsamo G, Bauer P, Bechtold P, Beljaars ACM, van de Berg L, Bidlot J, Bormann N, Delsol C, Dragani R, Fuentes M, Geer AJ, Haimberger L, Healy SB, Hersbach H, Hólm EV, Isaksen L, Kållberg PK, Köhler M, Matricardi M, McNally AP, Monge-Sanz BM, Morcrette J-J et al (2011) The ERA-Interim reanalysis: configuration and performance of the data assimilation system. Q J R Meteorol Soc 137:553–597. doi:10.1002/qj.828
16. Duc L, Kuroda T, Saito K, Fujita T (2015) Ensemble Kalman Filter data assimilation and storm surge experiments of tropical cyclone Nargis. Tellus A 67:25941. doi:10.3402/tellusa.v67.25941
17. Emanuel KA (2003) Tropical cyclones. Annu Rev Earth Planet Sci 31:75–104. doi:10.1146/annurev.earth.31.100901.141259
18. Emanuel KA (2013) Downscaling CMIP5 climate models shows increased tropical cyclone activity over the 21st century. Proc Natl Acad Sci U S A 110:12219–12224. doi:10.1073/pnas.1301293110
19. Emanuel KA, Sundararajan R, Williams H (2008) Hurricanes and global warming: results from downscaling IPCC AR4 simulations. Bull Am Meteorol Soc 89:347–367. doi:10.1175/BAMS-89-3-347
20. Fudeyasu H, Wang YQ, Satoh M, Nasuno T, Miura H, Yanase W (2008) Global cloud-system-resolving model NICAM successfully simulated the lifecycles of two real tropical cyclones. Geophys Res Lett 35:L22808. doi:10.1029/2008gl036003
21. Fukutomi Y, Kodama C, Yamada Y, Noda AT, Satoh M (2015) Tropical synoptic scale wave disturbances over the western Pacific simulated by a global cloud-system resolving model. Theor Appl Climatol 124:737–755. doi:10.1007/s00704-015-1456-4
22. Gottschalck J, Wheeler M, Weickmann K, Vitart F, Savage N, Lin H, Hendon H, Waliser D, Sperber K, Nakagawa M, Prestrelo C, Flatau M, Higgins W (2010) A framework for assessing operational model MJO forecasts: a project of the CLIVAR Madden-Julian Oscillation working group. Bull Am Meteorol Soc 91:1247–1258
23. Graph 500 (2014) The Graph 500 List., http://www.graph500.org. Accessed 10 Mar 2017

24. Gray WM (1975) Tropical cyclone genesis. Atmospheric Science Paper, No. 234. Colorado State University, Fort Collins
25. Gray WM (1979) Hurricanes: their formation, structure and likely role in the tropical circulation. In: Shaw DB (ed) Meteorology over the tropical oceans. Royal Meteorological Society, Reading, pp 155–218
26. Haarsma RH, Roberts MJ, Vidale PL, Senior CA, Bellucci A, Bao Q, Chang P, Corti S, Fučkar NS, Guemas V, von Hardenberg J, Hazeleger W, Kodama C, Koenigk T, Leung LR, Lu J, Luo J-J, Mao J, Mizielinski MS, Mizuta R, Nobre P, Satoh M, Scoccimarro E, Semmler T, Small J, von Storch J-S (2016) High Resolution Model Intercomparison Project (HighResMIP v1.0) for CMIP6. Geosci Model Dev 9:4185–4208. doi:10.5194/gmd-9-4185-2016
27. Hasegawa Y, Iwata JI, Tsuji M, Takahashi D, Oshiyama A, Minami K, Boku T, Shoji F, Uno A, Kurokawa M, Inoue H, Miyoshi I, Yokokawa M (2011) First-principles calculations of electron states of a silicon nanowire with 100,000 atoms on the K computer. In: Proceedings of 2011 International Conference for High Performance Computing, Networking, Storage and Analysis. ACM Press, New York. doi:10.1145/2063384.2063386
28. Holt LA, Alexander MJ, Coy L, Molod A, Putman W, Pawson S (2016) Tropical waves and the quasi-biennial oscillation in a 7-km global climate simulation. J Atmos Sci 73:3771–3783. doi:10.1175/JAS-D-15-0350.1
29. Huffman GJ, Adler RF, Morrissey MM, Bolvin DT, Curtis S, Joyce R, McGavock B, Susskind J (2001) Global precipitation at one-degree daily resolution from multi-satellite observations. J Hydrometeorol 2:36–50. doi:10.1175/1525-7541(2001)002<0036:GPAODD>2.0.CO;2
30. Huffman GJ, Bolvin DT, Nelkin EJ, Wolff DB, Adler RF, Gu G, Hong Y, Bowman KP, Stocker EF (2007) The TRMM multi-satellite precipitation analysis (TMPA): quasi-global, multiyear, combined-sensor precipitation estimates at fine scale. J Hydrometeorol 8:38–55. doi:10.1175/JHM560.1
31. Hung M-P, Lin J-L, Wang W, Kim D, Shinoda T, Weaver SJ (2013) MJO and convectively coupled equatorial waves simulated by 20 CMIP5 models. J Clim 26:6185–6214. doi:10.1175/JCLI-D-12-00541.1
32. Hyodo M, Hori T, Kaneda Y (2016) A possible scenario for earlier occurrence of the nest Nankai earthquake due to triggering by an earthquake at Hyuga-nada, off southwest Japan. Earth Planets Space 68:6. doi:10.1186/s40623-016-0384-6
33. Ichimura T, Fujita K, Tanaka S, Hori M, Lalith M, Shizawa Y, Kobayashi H (2014) Physics-based urban earthquake simulation enhanced by 10.7 BlnDOF x 30 K time-step unstructured FE non-linear seismic wave simulation. In: SC14: International Conference for High Performance Computing, Networking, Storage and Analysis (2014). IEEE Press, Piscataway, pp 15–26. doi:10.1109/SC.2014.7
34. Ichimura T, Fujita K, Quinay PEB, Maddegedara L, Hori M, Tanaka S, Shizawa Y, Kobayashi H, Minami K (2015) Implicit nonlinear wave simulation with 1.08T DOF and 0.270T unstructured finite elements to enhance comprehensive earthquake simulation. In: Proceedings of the International Conference for High Performance Computing, Networking, Storage and Analysis. ACM Press, New York. doi:10.1145/2807591.2807674
35. Ishiyama T, Nitadori K, Makino J (2012) 4.45 Pflops astrophysical N-body simulation on K computer: the gravitational trillion-body problem. In: SC '12 Proceedings of the International Conference on High Performance Computing, Networking. Storage and Analysis. IEEE CS Press, Los Alamitos, pp 1–10
36. Ito K, Kuroda T, Saito K, Wada A (2015) A large number of tropical cyclone intensity forecasts around Japan using a coupled high-resolution model. Weather Forecast 30:793–808
37. Kain JS, Fritsch JM (1990) A one-dimensional entraining detraining plume model and its application in convective parameterization. J Atmos Sci 47:2784–2802
38. Kajikawa Y, Yamaura T, Tomita H, Satoh M (2015) Impact of tropical disturbance on the Indian summer monsoon onset simulated by a global cloud-system-resolving model. SOLA 11:80–84. doi:10.2151/sola.2015-020

39. Kajikawa Y, Miyamoto Y, Yoshida R, Yamaura T, Yashiro H, Tomita H (2016) Resolution dependence of deep convections in a global simulation from over 10-kilometer to sub-kilometer grid spacing. Prog Earth Planet Sci 3:16. doi:10.1186/s40645-016-0094-5
40. Kanada S, Wada A, Nakano M, Kato T (2012) Effect of planetary boundary layer schemes on the development of intense tropical cyclones using a cloud-resolving model. J Geophys Res 117:D03107. doi:10.1029/2011JD016582
41. Kanada S, Wada A, Sugi M (2013) Future changes in structures of extremely intense tropical cyclones using a 2-km mesh nonhydrostatic model. J Clim 26:9986–10005. doi:10.1175/JCLI-D-12-00477.1
42. Kawamura R, Murakami T, Wang B (1996) Tropical and midlatitude 45-day perturbations over the western Pacific during the northern summer. J Meteorol Soc Jpn 74:867–890
43. Kikuchi K, Kodama C, Nasuno T, Miura H, Satoh M, Noda AT, Yamada Y (2016) Tropical intraseasonal oscillation simulated in an AMIP-type experiment by NICAM. Climate Dynam 48:2507-2528. doi:10.1007/s00382-016-3219-z
44. Kiladis GN, Wheeler MC, Haertel PT, Straub KH, Roundy PE (2009) Convectively coupled equatorial waves. Rev Geophys 47. doi:10.1029/2008RG000266
45. Kinter JL, Cash B, Achuthavarier D, Adams J, Altshuler E, Dirmeyer P, Doty B, Huang B, Marx L, Manganello J, Stan C, Wakefield T, Jin E, Palmer T, Hamrud M, Jung T, Miller M, Towers P, Wedi N, Satoh M, Tomita H, Kodama C, Nasuno T, Oouchi K, Yamada Y, Taniguchi H, Andrews P, Baer T, Ezell M, Halloy C et al (2013) Revolutionizing climate modeling with Project Athena: a multi-institutional, international collaboration. Bull Am Meteorol Soc 94:231–245. doi:10.1175/Bams-D-11-00043.1
46. Knutson TR, Mcbride JL, Chan J, Emanuel K, Holland G, Landsea C, Held I, Kossin JP, Srivastava AK, Sugi M (2010) Tropical cyclones and climate change. Nat Geosci 3:157–163. doi:10.1038/NGEO779
47. Knutson TR, Sirutis JJ, Zhao M, Tuleya RE, Bender M, Vecchi GA, Villarini G, Chavas D (2015) Global projections of intense tropical cyclone activity for the late twenty-first century from dynamical downscaling of CMIP5/RCP4.5 scenarios. J Clim 28:7203–7224. doi:10.1175/JCLI-D-15-0129.1
48. Kobayashi K, Kitamura D, Ando K, Ohi N (2015a) Parallel computing for high-resolution/large-scale flood simulation using the K supercomputer. Hydrol Res Lett 9:61–68
49. Kobayashi S, Ota Y, Harada Y, Ebita A, Moriya M, Onoda H, Onogi K, Kamahori H, Kobayashi C, Endo H, Miyaoka K, Takahashi K (2015b) The JRA-55 reanalysis: general specifications and basic characteristics. J Meteorol Soc Jpn 93:5–48. doi:10.2151/jmsj.2015-001
50. Kodama C, Yamada Y, Noda AT, Kikuchi K, Kajikawa Y, Nasuno T, Tomita T, Yamaura T, Takahashi HG, Hara M, Kawatani Y, Satoh M, Sugi M (2015) A 20-year climatology of a NICAM AMIP-type simulation. J Meteorol Soc Jpn 93:393–424. doi:10.2151/jmsj.2015-024
51. Krishnamurthy V, Stan C, Randall DA, Shukla RP, Kinter JL (2014) Simulation of the south Asian monsoon in a coupled model with an embedded cloud-resolving model. J Clim 27:1121–1142
52. Kumahata K, Minami K, Maruyama N (2016) High-performance conjugate gradient performance improvement on the K computer. Int J High Perform Comput Appl 30(1):55–70. doi:10.1177/1094342015607950
53. Kunii M (2014a) Mesoscale data assimilation for a local severe rainfall event with the NHM-LETKF system. Weather Forecast 29:1093–1105
54. Kunii M (2014b) The 1000-member ensemble Kalman filtering with the JMA nonhydrostatic mesoscale model on the K computer. J Meteorol Soc Jpn 91:623–633. doi:10.2151/jmsj.2014-607
55. Kusunoki S, Arakawa O (2015) Are CMIP5 models better than CMIP3 models in simulating precipitation over East Asia? J Clim 28:5601–5621
56. Lee JY, Wang B (2014) Future change of global monsoon in the CMIP5. Climate Dynam 42:101–119

57. Liebmann B, Smith CA (1996) Description of a complete (interpolated) outgoing longwave radiation dataset. Bull Am Meteorol Soc 77:1275–1277
58. Lloyd ID, Vecchi GA (2011) Observational evidence for oceanic controls on hurricane intensity. J Clim 24:1138–1153. doi:10.1175/2010JCLI3763.1
59. Madden R, Julian P (1972) Description of global-scale circulation cells in the tropics with a 40-50 day period. J Atmos Sci 29:1109–1123
60. Manganello JV, Hodges KI, Kinter JL III, Cash BA, Marx L, Jung T, Achuthavarier D, Adams JM, Altshuler EL, Huang B, Jin EK, Stan C, Towers P, Wedi N (2012) Tropical cyclone climatology in a 10-km global atmospheric GCM: toward weather-resolving climate modeling. J Clim 25:3867–3893. doi:10.1175/JCLI-D-11-00346.1
61. Manganello JV, Hodges KI, Dirmeyer B, Kinter JL III, Cash BA, Marx L, Jung T, Achuthavarier D, Adams JM, Altshuler EL, Huang B, Jin EK, Towers P, Wedi N (2014) Future changes in the western North Pacific tropical cyclone activity projected by a multidecadal simulation with a 16-km global atmospheric GCM. J Clim 27:7622–7646. doi:10.1175/JCLI-D-13-00678.1
62. Miura H, Satoh M, Nasuno T, Noda AT, Oouchi K (2007a) A Madden-Julian oscillation event realistically simulated by a global cloud-resolving model. Science 318:1763–1765. doi:10.1126/science.1148443
63. Miura H, Satoh M, Tomita H, Nasuno T, Iga S, Noda AT (2007b) A short-duration global cloud-resolving simulation with a realistic land and sea distribution. Geophys Res Lett 34:L02804. doi:10.1029/2006GL027448
64. Miyakawa T, Satoh M, Miura H, Tomita H, Yashiro H, Noda AT, Yamada Y, Kodama C, Kimoto M, Yoneyama K (2014) Madden-Julian oscillation prediction skill of a new-generation global model demonstrated using a supercomputer. Nat Commun 5:3769. doi:10.1038/Ncomms4769
65. Miyamoto Y, Kajikawa Y, Yoshida R, Yamaura T, Yashiro H, Tomita H (2013) Deep moist atmospheric convection in a subkilometer global simulation. Geophys Res Lett 40(18):4922–4926. doi:10.1002/grl.50944
66. Miyamoto Y, Yoshida R, Yamaura T, Yashiro H, Tomita H, Kajikawa Y (2015) Does convection vary in different cloud disturbances? Atmos Sci Lett 16(3):305–309. doi:10.1002/asl2.558
67. Molinari J, Vollaro D (2012) A subtropical cyclonic gyre associated with interactions of the MJO and the midlatitude jet. Mon Weather Rev 140:343–357. doi:10.1175/MWR-D-11-00049.1
68. Moon JY, Wang B, Ha KJ, Lee JY (2013) Teleconnections associated with Northern Hemisphere summer monsoon intraseasonal oscillation. Climate Dynam 40:2761–2774. doi:10.1007/s00382-012-1394-0
69. Murakami T, Matsumoto J (1994) Summer monsoon over the Asian continent and western North Pacific. J Meteorol Soc Jpn 72:719–745
70. Murakami H, Wang Y, Yoshimura H, Mizuta R, Sugi M, Shindo E, Adachi T, Yukimoto S, Hosaka M, Kusunoki S, Ose T, Kitoh A (2012) Future changes in tropical cyclone activity projected by the new high-resolution MRI-AGCM. J Clim 25:3237–3260. doi:10.1175/JCLI-D-11-00415.1
71. Murakami H, Vecchi GA, Underwood S, Delworth TL, Wittenberg AT, Anderson WG, Chen JH, Gudgel RG, Harris LM, Lin SJ, Zeng F (2015) Simulation and prediction of category 4 and 5 hurricanes in the high-resolution GFDL HiFLOR coupled climate model. J Clim 28:9058–9079. doi:10.1175/JCLI-D-15-0216.1
72. Nakano M, Sawada M, Nasuno T, Satoh M (2015) Intraseasonal variability and tropical cyclogenesis in the western North Pacific simulated by a global nonhydrostatic atmospheric model. Geophys Res Lett 42:565–571. doi:10.1002/2014gl062479
73. Nakano M, Wada A, Sawada M, Yoshimura H, Onishi R, Kawahara S, Sasaki W, Nasuno T, Yamaguchi M, Iriguchi T, Sugi M, Takeuchi Y (2017) Global 7-km mesh nonhydrostatic Model Intercomparison Project for improving TYphoon forecast (TYMIP-G7): Experimental design and preliminary results. Geosci Model Dev 10:1363-1381. doi:10.5194/gmd-2016-184

74. Nakazawa T (2006) Madden-Julian oscillation activity and typhoon landfall on Japan 2004. SOLA 2:136–139. doi:10.2151/sola.2006-035
75. Nasuno T, Satoh M (2011) Properties of precipitation and in-cloud vertical motion in a global nonhydrostatic aquaplanet experiment. J Meteorol Soc Jpn 89:413–439. doi:10.2151/jmsj.2011-502
76. Noda AT, Satoh M, Yamada Y, Kodama C, Seiki T (2014) Responses of tropical and subtropical high-cloud statistics to global warming. J Clim 27:7753–7768. doi:10.1175/JCLI-D-14-00179.1
77. Noda AT, Yamada Y, Kodama C, Miyakawa T, Seiki T, Satoh M (2015) Cold and warm rain simulated using a global nonhydrostatic model without cumulus parameterization, and their responses to a warmer atmospheric condition. J Meteorol Soc Jpn 93:181–197. doi:10.2151/jmsj.2015-010
78. Ogata T, Ueda H, Inoue T, Hayasaki M, Yoshida A, Watanabe S, Kira M, Ooshiro M, Kumai A (2014) Projected future changes in the Asian monsoon: a comparison of CMIP3 and CMIP5 model results. J Meteorol Soc Jpn 92:207–225. doi:10.2151/jmsj.2014-302
79. Ohno T, Satoh M (2015) On the warm core of a tropical cyclone formed near the tropopause. J Atmos Sci 72:551–571. doi:10.1175/JAS-D-14-0078.1
80. Oishi Y, Imamura F, Sugawara D (2015) Near-field tsunami inundation forecast using the parallel TUNAMI-N2 model: application to the 2011 Tohoku-Oki earthquake combined with source inversions. Geophys Res Lett 42:1083–1091. doi:10.1002/2014GL062577
81. Oouchi K, Yoshimura J, Yoshimura H, Mizuta R, Kusunoki S, Noda A (2006) Tropical cyclone climatology in a global-warming climate as simulated in a 20 km-mesh global atmospheric model: frequency and wind intensity analyses. J Meteorol Soc Jpn 84:259–276
82. Oouchi K, Noda AT, Satoh M, Miura H, Tomita H, Nasuno T, Iga S (2009a) A simulated preconditioning of typhoon genesis controlled by a boreal summer Madden-Julian oscillation event in a global cloud-system-resolving model. SOLA 5:65–68. doi:10.2151/sola.2009-017
83. Oouchi K, Noda AT, Satoh M, Wang B, Xie SP, Takahashi HG, Yasunari T (2009b) Asian summer monsoon simulated by a global cloud-system-resolving model: Diurnal to intra-seasonal variability. Geophys Res Lett 36:L11815. doi:10.1029/2009gl038271
84. Palmer T (2014) Build high-resolution global climate models. Nature 515:338–339
85. Petch JC, Brown AR, Gray MEB (2002) The impact of horizontal resolution on the simulations of convective development over land. Q J R Meteorol Soc 128:2031–2044. doi:10.1256/003590002320603511
86. Prein AF, Langhans W, Fosser G, Ferrone A, Ban N, Goergen K, Keller M, Tölle M, Gutjahr O, Feser F, Brisson E, Kollet S, Schmidli J, van Lipzig NPM, Leung R (2015) A review on regional convection-permitting climate modeling: Demonstrations, prospects, and challenges. Rev Geophys 53:323–361
87. Roberts MJ, Vidale PL, Mizielinski MS, Strachan J, Hodges K, Bell R, Camp J (2015) Tropical cyclone in the UPSCALE ensemble of high resolution global climate models. J Clim 28:574–596. doi:10.1175/JCLI-D-14-00131.1
88. Rogers RF, Reasor P, Lorsolo S (2013) Airborne Doppler observations of the inner-core structural differences between intensifying and steady-state tropical cyclones. Mon Weather Rev 141:2970–2991
89. Saito K, Tsuyuki T, Seko H, Kimura F, Tokioka T, Kuroda T, Duc L, Ito K, Oizumi T, Chen G, Ito J, SPIRE Field 3 Mesoscale NWP group (2013) Super high-resolution mesoscale weather prediction. J Phys Conf Ser 454:012073. doi:10.1088/1742-6596/454/1/012073
90. Sato T, Miura H, Satoh M, Takayabu YN, Wang Y (2009) Diurnal cycle of precipitation in the tropics simulated in a global cloud-resolving model. J Clim 22:4809–4826. doi:10.1175/2009jcli2890.1
91. Satoh M, Matsuno T, Tomita H, Miura H, Nasuno T, Iga S (2008) Nonhydrostatic Icosahedral Atmospheric Model (NICAM) for global cloud resolving simulations. J Comp Phys 227:3486–3514. doi:10.1016/j.jcp.2007.02.006

92. Satoh M, Iga S, Tomita H, Tsushima Y, Noda AT (2012a) Response of upper clouds due to global warming tested by a global atmospheric model with explicit cloud processes. J Clim 25:2178–2191. doi:10.1175/JCLI-D-11-00152.1
93. Satoh M, Oouchi K, Nasuno T, Taniguchi H, Yamada Y, Tomita H, Kodama C, Kinter J, Achuthavarier D, Manganello J, Cash B, Jung T, Palmer T, Wedi N (2012b) The intraseasonal oscillation and its control of tropical cyclones simulated by high-resolution global atmospheric models. Clim Dyn 39:2185–2206. doi:10.1007/s00382-011-1235-6
94. Satoh M, Tomita H, Yashiro H, Miura H, Kodama C, Seiki T, Noda AT, Yamada Y, Goto D, Sawada M, Miyoshi T, Niwa Y, Hara M, Ohno Y, Iga S, Arakawa T, Inoue T, Kubokawa H (2014) The Non-hydrostatic Icosahedral Atmospheric Model: Description and development. Prog Earth Planet Sci 1:18. doi:10.1186/s40645-014-0018-1
95. Satoh M, Yamada Y, Sugi M, Kodama C, Noda AT (2015) Constraint on future change in global frequency of tropical cyclones due to global warming. J Meteorol Soc Jpn 93:489–500. doi:10.2151/jmsj.2015-025
96. Seko H, Kunii M, Yokota S, Tsuyuki T, Miyoshi T (2015) Ensemble experiments using a nested LETKF system to reproduce intense vortices associated with tornadoes of 6 May 2012 in Japan. Prog Earth Planet Sci 2:42. doi:10.1186/s40645-015-0072-3
97. Sperber K, Annamalai H, Kang IS, Kitoh A, Moise A, Turner A, Wang B, Zhou T (2013) The Asian summer monsoon: an intercomparison of CMIP5 vs. CMIP3 simulations of the late 20th century. Clim Dyn 41:2711–2744. doi:10.1007/s00382-012-1607-6
98. Stevens B, Bony S (2013) What are climate models missing? Science 340:1053–1054. doi:10.1126/science.1237554
99. Sugi M, Noda A, Sato N (2002) Influence of the global warming on tropical cyclone climatology: An experiment with the JMA global model. J Meteorol Soc Jpn 80:249–272. doi:10.2151/jmsj.80.249
100. Sugi M, Murakami H, Yoshimura J (2012) On the mechanism of tropical cyclone frequency changes due to global warming. J Meteorol Soc Jpn 90A:397–408. doi:10.2151/jmsj.2012-A24
101. Taniguchi H, Yanase W, Satoh M (2010) Ensemble simulation of cyclone Nargis by a global cloud-system-resolving model - Modulation of cyclogenesis by the Madden-Julian oscillation. J Meteorol Soc Jpn 88:571–591
102. Terai M, Yashiro H, Sakamoto K, Iga S, Tomita H, Satoh M, Minami K (2014) Performance optimization and evaluation of a global climate application using a 440 m horizontal mesh on the K computer, Abstract presented at the 2014 International Conference for High Performance Computing. Networking, Storage and Analysis, Ernest N. Morial Convention Center, New Orleans, 16-21 November 2014
103. Terasaki K, Sawada M, Miyoshi T (2015) Local ensemble transform Kalman Filter experiments with the Nonhydrostatic Icosahedral Atmospheric Model NICAM. SOLA 11:23–26. doi:10.2151/sola.2015-006
104. Tomita H (2008) New microphysical schemes with five and six categories by diagnostic generation of cloud ice. J Meteorol Soc Jpn 86:121–142
105. Tomita H, Satoh M (2004) A new dynamical framework of nonhydrostatic global model using the icosahedral grid. Fluid Dyn Res 34:357–400
106. Tomita H, Miura H, Iga S, Nasuno T, Satoh M (2005) A global cloud-resolving simulation: Preliminary results from an aqua planet experiment. Geophys Res Lett 32:L08805. doi:10.1029/2005GL022459
107. Top 500 (2011a) Top 500 List June 2011., https://www.top500.org/lists/2011/06/. Accessed 10 Mar 2017
108. Top 500 (2011b) Top 500 List November 2011., https://www.top500.org/lists/ 2011/ 11/. Accessed 10 Mar 2017
109. Tsuboi S, Ando K, Miyoshi T, Peter D, Komatitsch D, Tromp J (2016) A 1.8 trillion degrees-of-freedom, 1.24 petaflops global seismic wave simulation on the K computer. Int J High Perform Comput Appl 30(4):411–422. doi:10.1177/1094342016632596
110. Tsushima Y, Iga S, Tomita H, Satoh M, Noda AT, Webb M (2014) High cloud increase in a perturbed SST experiment with a global nonhydrostatic model including explicit convective processes. J Adv Model Earth Syst 6:571–585. doi:10.1002/2013MS000301

111. Vecchi GA, Soden BJ (2007) Global warming and the weakening of the tropical circulation. J Clim 20:4316–4340. doi:10.1175/JCLI4258.1
112. Vitart F, Robertson AW, Anderson DLT (2012) Subseasonal to Seasonal Prediction Project: Bridging the gap between weather and climate. WMO Bull 61:23–28
113. Vitart F, Ardilouze C, Bonet A, Brookshaw A, Chen M, Codorean C, Déqué M, Ferranti L, Fucile E, Fuentes M, Hendon H, Hodgson J, Kang H, Kumar A, Lin H, Liu G, Liu X, Malguzzi P, Mallas I, Manoussakis M, Mastrangelo D, MacLachlan C, McLean P, Minami A, Mladek R, Nakazawa T, Najm S, Nie Y, Rixen M, Robertson A et al (2017) The Subseasonal to Seasonal (S2S) Prediction Project Database. Bull Am Meteorol Soc 98:163–173. doi:10.1175/BAMS-D-16-0017.1
114. Wada A, Uehara T, Ishizaki S (2014) Typhoon-induced sea surface cooling during the 2011 and 2012 typhoon seasons: observational evidence and numerical investigations of the sea surface cooling effect using typhoon simulations. Prog Earth Planet Sci 1:11. doi:10.1186/2197-4284-1-11
115. Walsh KJE, Camargo SJ, Vecchi GA, Daloz AS, Elsner J, Emanuel K, Horn M, Lim Y-K, Roberts M, Patricola C, Scoccimarro E, Sobel AH, Strazzo S, Villarini G, Wehner M, Zhao M, Kossin JP, LaRow T, Oouchi K, Schubert S, Wang H, Bacmeister J, Chang P, Chauvin F, Jablonowski C, Kumar A, Murakami H, Ose T, Reed KA, Saravanan R et al (2015) Hurricanes and climate: the U.S. CLIVAR working group on hurricanes. Bull Am Meteorol Soc 96:997–1017. doi:10.1175/BAMS-D-13-00242.1
116. Walsh KJE, Mcbride JL, Klotzbach PJ, Balachandran S, Camargo SJ, Holland G, Knutson TR, Kossin JP, Lee T, Sobel A, Sugi M (2016) Tropical cyclone and climate change. WIREs Clim Change 7:65–89. doi:10.1002/wcc.371
117. Wang B, Rui H (1990) Synoptic climatology of transient tropical intraseasonal convection anomalies: 1975-1985. Meteorol Atmos Phys 44:43–61
118. Wang B, Xie X (1997) A model for the boreal summer intraseasonal oscillation. J Atmos Sci 54:72–86
119. Wang H, Wang Y (2014) A numerical study of typhoon Megi (2010). Part I: rapid intensification. Mon Weather Rev 142:29–48. doi:10.1175/MWR-D-13-00070.1
120. Wang B, Ding QH, Fu XH, Kang IS, Jin K, Shukla J, Doblas-Reyes F (2005) Fundamental challenge in simulation and prediction of summer monsoon rainfall. Geophys Res Lett 32:L15711. doi:10.1029/2005gl022734
121. Wehner M, Reed KA, Stone D, Collins WD, Bacmeister J (2015) Resolution dependence of future tropical cyclone projections of CAM5.1 in the US CLIVAR Hurricane Working Group idealized configurations. J Clim 28:3905–3925. doi:10.1175/JCLI-D-14-00311.1
122. Wheeler MC, Hendon HH (2004) An all-season real-time multivariate MJO index: development of an index for monitoring and prediction. Mon Weather Rev 132:1917–1932
123. Yamada Y, Oouchi K, Satoh M, Tomita H, Yanase W (2010) Projection of changes in tropical cyclone activity and cloud height due to greenhouse warming: global cloud-system-resolving approach. Geophys Res Lett 37:L07709. doi:10.1029/2010GL042518
124. Yamamoto K, Uno A, Murai H, Tsukamoto T, Shoji F, Matsui S, Sekizawa R, Sueyasu F, Uchiyama H, Okamoto M, Ohgushi N, Takashina K, Wakabayashi D, Taguchi Y, Yokokawa M (2014) The K computer operations: experiences and statistics. Procedia Comput Sci 29:576–585. doi:10.1016/j.procs.2014.05.052
125. Yamaura T, Kajikawa Y, Tomita H, Satoh M (2013) Possible impact of a tropical cyclone on the northward migration of the Baiu frontal zone. SOLA 9:89–93. doi:10.2151/sola.2013-020
126. Yanase W, Taniguchi H, Satoh M (2010) The genesis of tropical cyclone Nargis (2008): environmental modulation and numerical predictability. J Meteorol Soc Jpn 88:497–519. doi:10.2151/jmsj.2010-314
127. Yashiro H, Kajikawa Y, Miyamoto Y, Yoshida R, Yamaura R, Tomita H (2016a) Resolution dependency of diurnal precipitation cycle simulated by global cloud resolving model. SOLA 12:272–276
128. Yashiro H, Terai M, Yoshida R, Iga S, Minami K, Tomita H (2016b) Performance analysis and optimization of Nonhydrostatic ICosahedral Atmospheric Model (NICAM) on the K

Computer and TSUBAME2.5. In: Proceedings of the Platform for Advanced Scientific Computing Conference (PASC'16). ACM, New York, p Article 3. doi:10.1145/2929908.2929911

129. Yashiro H, Terasaki K, Miyoshi T, Tomita H (2016c) Performance evaluation of a throughput-aware framework for ensemble data assimilation: the case of NICAM-LETKF. Geosci Model Dev 9:2293–2300. doi:10.5194/gmd-9-2293-2016
130. Yokokawa M, Shoji F, Uno A, Kurokawa M, Watanabe T (2011) The K computer: Japanese next-generation supercomputer development project. In: IEEE/ACM International Symposium on Low Power Electronics and Design (ISLPED) 2011. IEEE, New York, pp 371–372. doi:10.1109/ISLPED.2011.5993668
131. Yokota S, Seko H, Kunii M, Yamauchi H, Niino H (2016) The tornadic supercell on the Kanto Plain on 6 May 2012: polarimetric radar and surface data assimilation with EnKF and ensemble-based sensitivity analysis. Mon Weather Rev 144:3133–3157. doi:10.1175/MWR-D-15-0365.1
132. Yoneyama K, Zhang C, Long CN (2013) Tracking pulses of the Madden-Julian oscillation. Bull Am Meteorol Soc 94:1871–1891
133. Zarzycki CM (2016) Tropical cyclone intensity errors associated with lack of two-way ocean coupling in high-resolution global simulations. J Clim 29:8589–8610
134. Zou LW, Zhou TJ (2015) Asian summer monsoon onset in simulations and CMIP5 projections using four Chinese climate models. Adv Atmos Sci 32:794–806

CHAPTER 11

Selection of a representative subset of global climate models that captures the profile of regional changes for integrated climate impacts assessment

Alex C. Ruane[1,2]*Sonali P. McDermid[3,1,2]

[1]*NASA Goddard Institute for Space Studies, 2880 Broadway, New York, NY10025, USA*

[2]*Columbia University Center for Climate Systems Research, 2880 Broadway, New York, NY 10025, USA*

AUTHORS' CONTRIBUTIONS

ACR developed the idea, conducted work and led manuscript writing. SM helped develop the idea and contributed to the manuscript writing. Both authors read and approved the final manuscript.

ABSTRACT

We present the Representative Temperature and Precipitation (T&P) GCM Subsetting Approach developed within the Agricultural Model

Intercomparison and Improvement Project (AgMIP) to select a practical subset of global climate models (GCMs) for regional integrated assessment of climate impacts when resource limitations do not permit the full ensemble of GCMs to be evaluated given the need to also focus on impacts sector and economics models. Subsetting inherently leads to a loss of information but can free up resources to explore important uncertainties in the integrated assessment that would otherwise be prohibitive. The Representative T&P GCM Subsetting Approach identifies five individual GCMs that capture a profile of the full ensemble of temperature and precipitation change within the growing season while maintaining information about the probability that basic classes of climate changes (relatively cool/wet, cool/dry, middle, hot/wet, and hot/dry) are projected in the full GCM ensemble. We demonstrate the selection methodology for maize impacts in Ames, Iowa, and discuss limitations and situations when additional information may be required to select representative GCMs. We then classify 29 GCMs over all land areas to identify regions and seasons with characteristic diagonal skewness related to surface moisture as well as extreme skewness connected to snow-albedo feedbacks and GCM uncertainty. Finally, we employ this basic approach to recognize that GCM projections demonstrate coherence across space, time, and greenhouse gas concentration pathway. The Representative T&P GCM Subsetting Approach provides a quantitative basis for the determination of useful GCM subsets, provides a practical and coherent approach where previous assessments selected solely on availability of scenarios, and may be extended for application to a range of scales and sectoral impacts.

Keywords: Climate change, GCMs. Climate model. Subset. Representative T&P GCM subset approach. Climate impacts. AgMIP Integrated assessment

1. INTRODUCTION

Societal applications of climate change information are driven by the needs of stakeholders deciding upon particular adaptation, mitigation, policy, or risk management strategies. Model-based projections of climate changes and related uncertainties form a core component of these climate impact applications, however uncertainty may also be introduced by sector models (e.g., crop, livestock, health, ecosystem, fisheries, energy, or water resource models), biophysical or socioeconomic scenarios (e.g., greenhouse gas concentrations; regional development pathways), and the economic models upon which many decisions are evaluated. Connecting climate models, sector impact models, and economics models results in an integrated assessment framework capable of exploring the societal ramifications of climate impacts as well as opportunities to build resilience through interventions under a risk management framework. The Agricultural Model Intercomparison and Improvement Project (AgMIP) was developed to facilitate best practices and collaborative development of agricultural models with the aim of informing stakeholder decisions across a variety of regions, scales, models, and scenarios (Rosenzweig et al. 2013, 2015, 2016). One goal of AgMIP is to provide guidance on the creation of climate change scenarios for agricultural experts unfamiliar with the climate modeling community.

Impact assessments across sectors have too often been influenced by selection bias, leading to inconsistencies across studies and confusion among policymakers. White et al. (2011) revealed large differences across crop modeling studies in the number and types of global climate models (GCMs) used, with many using a small number of models and GCM selection often influenced by the availability of output. The first phase of the Inter-Sectoral Impacts Model Intercomparison Project (ISI-MIP; Warszawski et al. 2014) utilized a common set of 5 GCMs for all sectoral impacts assessments owing to the need for consistency across regions and sectors. The ramifications of choosing this subset, based in part upon which GCMs available at the time, continues to be explored (McSweeney and Jones 2016). Early AgMIP regional integrated assessments also used 5 GCMs selected according to prominence in publications, length of participation in CMIP, spatial resolution, and historical monsoon patterns (Ruane et al. 2015b).

A common limitation in AgMIP and related studies stems from the overwhelming number of possible combinations of individual elements within an integrated assessment framework, leading to a prohibitive number of simulations. For example, ongoing AgMIP work across sub-Saharan Africa and South Asia seeks to utilize information from a set of 29 GCMs, 3 time periods, 2 greenhouse gas concentration pathways, 2 climate scenario generation methodologies, 3 adaptation scenarios, 2 agricultural development pathways, at least 2 crop models, and 40 or more households representing the distribution of impacts for a given region. Together, this would be an impractical number of simulations to conduct and analyze (at least $29x3x2x2x3x2x2x40 = 167{,}040$) for each crop species examined, so representative subsets are required wherever possible.

Subsetting GCMs necessarily leads to a loss of information, which begs the question: what types of information are most important to maintain? In the end this is a subjective decision that is best decided in discussion between researchers and stakeholders, however it must be justified (Knutti et al. 2010a) and physical and statistical relationships provide useful guidance.

The most prominent information for assessing sectoral climate impacts in a given region remain projected temperature change and precipitation change (the concentration of greenhouse gases such as carbon dioxide can be taken directly from driving scenarios). These quantities are indicative of large-scale changes in energy and water cycles in a region, and many other climate variables are closely correlated with these quantities. Even an increasing emphasis on extreme events requires the context of long-term shifts in mean temperature and precipitation, which also form the basis of intuitive climate scenario generation methodologies such as the "delta approach" wherein mean monthly temperature and precipitation changes are imposed upon historical observations (Wilby et al. 2004). The delta approach was the most common climate scenario generation methodology used in the White et al. (2011) review of agricultural impact models. Many impact sector models also respond strongly to mean temperature and rainfall shifts, allowing for the development of simple but effective emulators (e.g., Ruane et al. 2014; McDermid et al. 2015a; Howden and Crimp 2005).

Characteristics of the overall distribution of projections from the full GCM ensemble are priority information to maintain, as multi-model ensembles generally outperform individual models when validated across many observations (though not necessarily for any single observation). This has been illustrated in many studies within the climate modeling community (Reichler and Kim 2008; IPCC 2010; Flato et al. 2013) and for impacts models in various sectors including agriculture (Asseng et al. 2013; Bassu et al. 2014; Martre et al. 2015; Li et al. 2015). The overall spread and skewness of projections is also a worthwhile indication of model-based uncertainty, confidence, and the range of plausible outcomes.

This study describes the Representative Temperature and Precipitation (T&P) GCM Subsetting Approach that may be used to select a practical and coherent subset of GCMs for use in regional integrated assessments that conserves resources and captures the general physical and statistical uncertainty in projections. Here we aim for 5 GCMs given typical resource requirements within recent AgMIP activites, although it is important to resist the temptation to boil this down further as the information loss gradient is increasingly steep at lower numbers. The goals of the Representative T&P GCM Subsetting Approach are fivefold:

1) To reduce the number of GCMs required to sample climate change uncertainty and thereby free up resources for other elements of a regional integrated assessment,

2) To focus the GCM selection and assessment process on the season of interest,

3) To ensure that the assessment framework exposes the system to major classes of change without averaging to an extent that temporal and spatial patterns are not physical,

4) To avoid the selection of extreme outliers that may skew results, and

5) To maintain information about overall uncertainty in GCM projections.

The Representative T&P GCM Subsetting Approach builds upon the work of Semenov and Stratonovich (2015), who suggested the use of "climate sensitivity indices" to subset GCMs based upon annual-mean temperature and precipitation changes in aggregated regions around the world. This methodology was developed independently from a similar approach recently described by Lutz et al. (2016) that focuses on describing the limits of an envelope of potential temperature and precipitation changes and the largest shifts in extremes. As described below, we add additional guidance on classes of regional climate change, the maintenance of uncertainty information from the broader ensemble, sector specific seasonal focus, a connection to anticipated scenario generation methods, and coherence across space, time, and greenhouse gas scenario. We also provide a regional demonstration, show how this framework may be applied at global scales, and explore how the perspective of GCM subset priorities enables statistical and physical analysis of the ensemble of GCM projections.

It is important to emphasize that selecting a subset of GCMs to conserve computational and analytic resources is distinct from efforts to develop unequal weights for GCMs in order to better capture the signal of future climate change (e.g., Giorgi and Mearns 2003; Tebaldi et al. 2005; see reviews in Knutti 2010; IPCC 2010; Flato et al. 2013). The development of unequal weights is particularly appealing given that CMIP represents an ensemble of opportunity rather than an ensemble designed to systematically capture uncertainty around a true projection. Weighting approaches allow models with substantial biases to be lowered in emphasis, although biases in the historical period do not necessarily reflect biases in climate response to anthropogenic forcings and it can be difficult to determine which metrics should form the basis of weighting. Knutti et al. (2010b) called for the use of multi-model ensembles but indicate that equal weighting is preferred until weighting methods can be more fully developed and validated.

2. MATERIALS AND METHODS

2.1 Climate datasets

The Representative T&P GCM Subsetting Appraoch is built upon climate change projections provided by state-of-the-art climate and earth system models such as those submitted to the Coupled Model Intercomparison Project (CMIP; Taylor et al. 2012; Eyring et al. 2016) in support of the

Intergovernmental Panel on Climate Change (IPCC). These projections are then combined with a gridded historical climate dataset in order to overcome common biases in GCM precipitation climatology. Meteorological observations for the Ames, Iowa, example utilized below were drawn from the Iowa Environmental Mesonet maintained by Iowa State University.

2.2 Future climate projections

This study utilizes 29 GCMs (Table1) commonly used within AgMIP for climate change projections from the Fifth Coupled Model Intercomparison Project (CMIP5; Taylor et al.2012). This represents the set of GCMs that: 1) were available (as of May, 2015) on the Earth Systems Grid; 2) contained daily data from 1980 to 2100; and 3) included historical and future outputs for both the moderate and high representative concentration pathways (RCP4.5 and RCP8.5, respectively; Moss et al.2010; Knutti2014). Over 40 modeling groups have now contributed results to the Earth Systems Grid. These GCMs represent a range of institutions, horizontal and vertical resolutions, components, and climate sensitivities. For the purposes of this study each GCM is considered to be equally likely and exchangeable as there is no clear method to evaluate GCM performance in a climate that has not yet occurred (Gleckler et al.2008; IPCC2010), and the approach would likely not make sense for perturbed physics ensembles of a single GCM. The specific ensemble members for which GCM groups archived daily outputs is also likely to affect regional trends (Sriver et al.2015), which suggests the continuing need to analyze many-member ensembles for probabilistic impacts research.

Table 1 Summary of 29 CMIP5 GCMs that form the ensemble of climate projections used in this study

GCM	Institution	Horizontal resolution	2x [CO_2] Eq. climate Sens. (°C)
ACCESS1-0	Commonwealth Scientific and Industrial Research Organization (CSIRO) and Bureau of Meteorology (BOM), Australia	1.25° × 1.875°	3.8
BCC-CSM1-1	Beijing Climate Center, China Meteorological Administration	~2.8° × 2.8°	2.8
BNU-ESM	College of Global Change and Earth Systems Science, Beijing Normal University (BNU)	~2.8° × 2.8°	4.1

CanESM2	Canadian Centre for Climate Modelling & Analysis	~2.8° × 2.8°	3.7
CCSM4	US National Center for Atmospheric Research (NCAR)	~0.9° × 1.25°	2.9
CESM1-BGC	US National Science Foundation (NSF), US Department of Energy (DOE), and the US National Centre for Atmospheric Research (NCAR)	~0.9° × 1.25°	n.a.
CMCC-CM	Euro-Mediterranean Center on Climate Change	~0.75° × 0.75°	n.a.
CMCC-CMS	Euro-Mediterranean Center on Climate Change	~1.9° × 1.875°	n.a.
CNRM-CM5	France National Centre for Meteorological Research	~1.4° × 1.4°	3.3
CSIRO-Mk3-6-0	Queensland Climate Change Centre of Excellence and Commonwealth Scientific and Industrial Research Organization (CSIRO)	~1.9° × 1.875°	4.1
FGOALS-g2	Chinese Academy of Sciences	~2.8° × 2.8°	n.a.
GFDL-CM3	NOAA/Geophysical Fluid Dynamic Laboratory (GFDL)	2.0° × 2.5°	4.0
GFDL-ESM2G	NOAA/Geophysical Fluid Dynamic Laboratory (GFDL)	~2.0° × 2.5°	2.4
GFDL-ESM2M	NOAA/Geophysical Fluid Dynamic Laboratory (GFDL)	~2.0° × 2.5°	2.4
GISS-E2-H	National Aeronautics and Space Association Goddard Institute for Space Studies (NASA GISS)	2° × 2.5°	2.3
GISS-E2-R	National Aeronautics and Space Association Goddard Institute for Space Studies (NASA GISS)	2° × 2.5°	2.1
HadGEM2-AO	UK Meteorological Office - Hadley Centre	1.25° × 1.875°	n.a.
HadGEM2-CC	UK Meteorological Office - Hadley Centre	1.25° × 1.875°	n.a.
HadGEM2-ES	UK Meteorological Office - Hadley Centre	1.25° × 1.875°	4.6
INMCM4.0	Russian Institute for Numerical	1.5° × 2°	2.1

	Mathematics (INM)		
IPSL-CM5A-LR	Institute Pierre Simon Laplace (IPSL)	~1.9° × 3.75°	4.1
IPSL-CM5A-MR	Institute Pierre Simon Laplace (IPSL)	~1.3° × 2.5°	n.a.
IPSL-CM5B-LR	Institute Pierre Simon Laplace (IPSL)	~1.9° × 3.75°	2.6
MIROC5	University of Tokyo, Japanese National Institute for Environmental Studies (NIES), and Japan Agency for Marine-Earth Science and Technology (JAMSTEC)	~1.4° × ~1.4°	2.7
MIROC-ESM	University of Tokyo, Japanese National Institute for Environmental Studies (NIES), and Japan Agency for Marine-Earth Science and Technology (JAMSTEC)	~2.8° × ~2.8°	4.7
MPI-ESM-LR	Max Planck Institute (MPI) for Meteorology (low resolution)	~1.9° × 1.875°	3.6
MPI-ESM-MR	Max Planck Institute (MPI) for Meteorology (mixed resolution)	~1.9° × 1.875°	n.a.
MRI-CGCM3	Japanese Meteorological Research Institute (MRI)	~1.1° × 1.125°	2.6
NorESM1-M	Norwegian Climate Centre	~1.9° × 2.5°	2.8

Climate sensitivities represent an equilibrium global mean surface temperature increase resulting from a doubling of pre-industrial CO_2concentrations (from Table 9.5 of Flato et al.2013; n.a. = not available). References for these models are in Flato et al.2013, and Ruane et al.2015b

GCM projections of temperature change are calculated as absolute differences (future-current) while projections of future precipitation are calculated as percentages of the current period (future/current * 100%). When averaging GCM projections of seasonal precipitation changes over multiple months, historical climate information over the "current" period (1980–2010) is needed to recognize that percentage changes in wet months impact total season precipitation more than changes in dry months. To demonstrate similar characteristics of GCM projections, we create simple climate scenarios by imposing the temperature changes (an additive factor) and precipitation changes (a multiplicative factor) upon historical observations as is done in the "Delta" approach among many other statistical methods (Wilby et al. 2004; Ruane et al. 2015b). Future time periods are defined as Near-term (2010–2039), Mid-Century (2040–2069), and "End-of-Century" (2070–2099) as in Ruane et al. (2015b).

2.3 Historical climate data

Local observations of meteorological variables are the ideal source of historical climate information, however these are missing or incomplete in many regions and gridded applications require spatial consistency. This study therefore uses AgMERRA (Ruane et al.2015a), a hybrid of the Modern-Era Retrospective-analysis for Research and Applications (MERRA; Rienecker et al.2011) and various gridded and satellite products in order to fill in gaps and remove common biases needed for global agricultural modeling. AgMERRA compares favorably with similar climate forcing datasets and reanalyses, although substantial uncertainty remains in regions with limited observational coverage (much of the tropics and many developing countries; Ruane et al.2015a). AgMERRA provides a global, daily, 0.25 × 0.25 gridded climate dataset spanning 1980–2010 containing maximum and minimum temperatures, precipitation, solar radiation, wind speed, and relative humidity at the maximum temperature time of day. AgMERRA has been used in numerous AgMIP activities (Elliott et al.2015; Ruane et al.2015b).

2.4 The representative T&P GCM subsetting approach

To demonstrate the Representative T&P GCM Subsetting Approach, we consider an integrated assessment using crop and economic models to assess the impacts of climate change on maize production systems in Ames, Iowa, USA (93°45′W, 42°1′N). The framework of this hypothetical study includes the use of multiple climate and crop models, as well as a range of climate, agricultural development, and adaptation scenarios utilizing a regional economic model. This example is typical of many applications within AgMIP and similar projects which aim to provide stakeholders with information about potential challenges and vulnerabilities in order to aid in investment and risk management decisions (Antle et al.2015).

2.5 Defining growing seasons

As crop model simulations that drive this analysis consider only the maize growing season, GCMs must also be selected based upon their projections of the relevant months. This seasonal focus highlights the differences between GCMs that would affect the overall outcome of the integrated assessment while avoiding the influence of potential biases in months that are never simulated. Here, we utilize local growing season as defined in the harmonized simulations of AgMIP's Global Gridded Crop Model Intercomparison (GGCMI; Elliott et al. 2015), which derived planting windows from Sacks et al. (2010) and Portmann et al. (2008, 2010). We round planting and harvesting dates to whole months, with months included only if crops were in the ground for the majority of days (this would prevent over-representation of a month if planting occurred on its last day, for example). For Ames, Iowa, the average maize season corresponds to planting on May 13th and harvesting on September 27th, so GCMs will be selected based upon their projections of the May–September period (MJJAS). In other cases it may be desirable to utilize only a subset of the growing season, particularly if anomalies in a given month are shown to have a particularly strong impact on crop development. As maize is often grown in the summertime for mid-latitude farms and in rainy seasons within the tropics, maize season results presented in the Results section are consistent in representing the challenges faced by maize farmers despite including results from different months in different locations on the same map. Soil moisture in crop models is often initialized shortly before planting, so we do not include preceding months.

Integrated assessments that link together the economics of multiple growing seasons in a given farming system often benefit from the use of a consistent set of GCMs for all crops. In these situations it is likely desirable to analyze a growing season that covers all months when crops are in the ground, as detailed in the Discussion section below.

2.6 Delineating climate change quadrants

To identify fundamental classes of projected climate change for a region during a specific time period and RCP, we characterize an individual model's projected, location-specific temperature and precipitation changes in terms of its deviation from the ensemble median. A GCM's projections will therefore be relatively cool or hot and relatively wet or dry. In this demonstration we use the climate information from the GCM output's native grid boxes, although a similar procedure could subset an ensemble of downscaled products and may be advisable where there is substantial disagreement between native projections and downscaled analyses (Christensen et al. 2007). It is important to underscore that these classifications are relative, as the ensemble median is likely different than today's conditions. This is particularly important in the climate change context, as nearly all GCMs project long-term warming trends across the world (alternatively, "relatively cool" and "relatively hot" may be classified as "warm" and "hot"). A given GCM with low levels of overall warming may be classified as "relatively cool" over a given region if it is cooler than the median of all GCMs. Likewise, a given GCM exhibiting a slight long-term drying trend may still be "relatively wet" if the full GCM ensemble median is projecting a more extreme drying trend. With a GCM's projection classified as either relatively hot/cool or relatively wet/dry, this makes four basic quadrants of change: "cool/wet", "cool/dry", "hot/wet", "hot/dry".

A fifth, "Middle", classification is also introduced in order to capture the models that represent the nexus of these quadrants around the median of the ensemble. Using the ensemble standard deviation (σ) of growing season temperature and precipitation changes as a representation of this model spread, we capture these models in a fifth quadrant including all models whose temperature and precipitation changes are within $\pm(\rho^*\sigma)$of the median, where ρ is a standard deviation factor designed as a simple measure of spread in order to capture approximately $1/5^{th}$(20%) of all GCM projections. This targets each quadrant to contain approximately the same number of models to minimize the number of GCMs that each quadrant's selected model has to represent (the selection of this GCM is described in the next section). The use of standard deviations to define the bounds of the "Middle" quadrant does not hinge on an assumption that GCM projections are characterized by a Gaussian distribution of individual projections for both temperature and precipitation changes. Although a Gaussian form is most common, bimodality and distributions more precisely represented by other statistical forms do occur (Tebaldi et al. 2005). The middle quadrants in in this study are defined using $\rho = 0.5$ as it is a close fit to the Mid-Century optimal values for RCP8.5 and RCP4.5, as described in the Patterns of quadrant weights section below. When communicating the relevance of this quadrant to stakeholders it is important to emphasize that this middle quadrant is not an invitation to select only a single GCM. While this quadrant captures the center of the distribution, on average it is not substantially more likely than any of the other quadrants.

Each GCM may thus be classified as falling within a particular quadrant of relative temperature and precipitation change. As an example, the dashed lines in Figureure. 1 show the delineation of the five quadrants of relative temperature and precipitation change for the maize-growing season in Ames, Iowa, under Mid-Century RCP8.5 conditions. The color of the triangles shows the classification of each GCM according to its quadrant of relative temperature and precipitation change.

2.7 Choosing a representative model for each quadrant

It is not necessarily desirable that the model representing a given quadrant be centered within the quadrant, rather our desire is to pick a model that is in the center of the GCMs that fall in this quadrant as its role is to represent the types of change projected by these models. The selection of a representative model is therefore aided by calculating the quadrant center of mass for temperature and precipitation changes; achieved by marking the mean temperature and precipitation change for all of the GCMs that fall into a given quadrant (dots in Figureure. 1). The models that fall into the middle quadrant of Figureure. 1, for example, are centered slightly cooler and wetter than the median of the entire GCM ensemble. This should be reflected in the selection of a representative GCM for the middle quadrant.

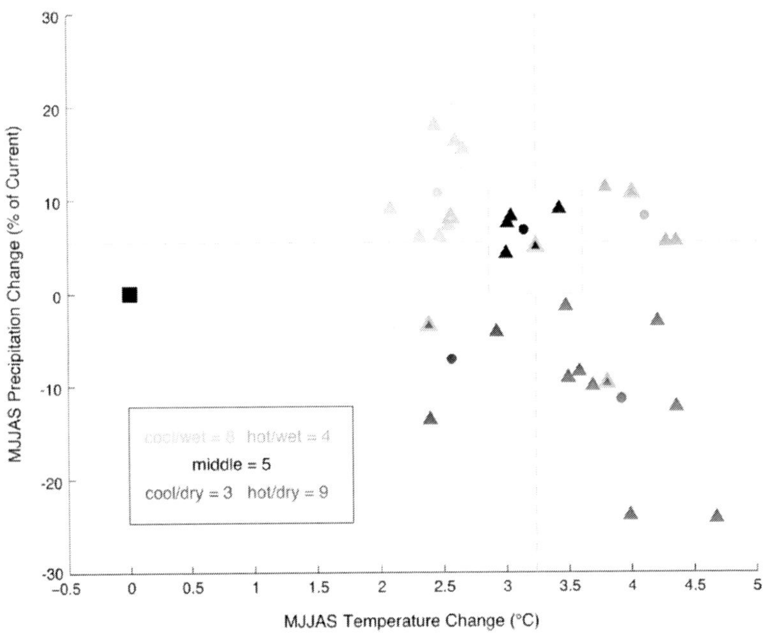

Figure. 1 Basic definition of the quadrants and Ames, Iowa, Maize season examples from 29 GCMs for Mid-Century RCP8.5. Each triangle is colored by its quadrant and represents one of the 29 GCMs, and the square represents the baseline condition (no change from observed average of 20.2 °C and 3.7 mm/day). *Dots* represent the mean of GCMs within any given quadrant, and the selected representative GCMs are denoted with a gray outline

Although the quadrant centers of mass are the best representations of GCMs within a given quadrant, these growing season averages do not contain sufficient information for many impacts assessments. To build the types of scenarios needed for more complex assessments, information is needed on the daily time series, frequency of extreme events, number of rainy days, covariation of meteorological variables, sub-seasonal variability, and other factors that cannot be gleaned from this simple averaging. Multi-model average projections can also wash out spatial and temporal patterns to a point where they are not physically plausible. By selecting a single model we end up with physically-consistent climate information from a simulation that resulted in the type of climate changes projected by GCMs in this quadrant. Drawbacks of this reliance upon a single model's physics are discussed in the Discussion section.

The first guess at a representative model for a given quadrant is therefore the GCM whose temperature change and precipitation projection fall closest to the center of mass of GCMs within that quadrant. A degree of subjectivity is possible at the stage of representative model selection, however, depending on three factors that are worthy of extended consideration below:

1) Uncertainty in model projections,
2) GCM biases related to major patterns of atmospheric circulation
3) The sensitivity of a given system or assessment.

Model uncertainty may justify selecting a model other than that which is closest to the quadrant center of mass if an examination of the GCM spread reveals that the center of mass was drawn far away from most GCMs by an extreme outlier. For example, if a single model increases precipitation by 300% while the precipitation changes of the other models in a quadrant are limited to 20% gains or less, the model closest to the center of mass will likely be the wettest of the non-outliers. In this case it may make more sense to drop the outlier as a candidate and select a model close to the center of the remaining models.

Particular care must be taken in assessment regions governed by major atmospheric circulation patterns or seasonal cycles (IPCC 2010; Knutti et al.2010a). The most prominent examples are monsoons (e.g., in South Asia, East Asia, Southwest North America, or West Africa), the migration of the Inter-Tropical Convergence Zone (ITCZ), and the seasonal cycle of sea-ice extent; each of which can dominate seasonal precipitation and temperature patterns for a given area. In these regions it is possible that the GCM that is closest to the quadrant center of mass misses the onset or exit of a rainy season (as captured in observational products such as AgMERRA), leading to climate changes for a given month that are based upon dry conditions when observations reveal periodic rainfall.

To illustrate a potential motivation for subjective selection based upon major atmospheric circulation patterns, imagine a quadrant where GCM A and GCM B both project nearly identical seasonal temperature and precipitation changes but GCM A is slightly closer to the quadrant center of mass. If GCM B has a climatology that generally matches the observed seasonal cycle of rainfall but GCM A reveals a delayed rainy season and thus a dry month when observations contain substantial rain, two likely biases may manifest in GCM A that would not be found in GCM B. First, rainfall changes in excess of ±50% are possible with even a small change in rainfall, which could have dramatic effects when imposed upon wetter conditions. Second, temperature changes are likely biased toward greater increase as excess energy that should drive latent heat exchanges is instead pushed into sensible heat due to moisture limitations (resulting in a higher Bowen Ratio in GCM A than GCM B). A comparison between observed rainfall seasonality and GCM rainfall seasonality is thus required before selecting the representative GCM for a given quadrant, with results potentially justifying selecting GCM B despite GCM A being the closest GCM to the quadrant center of mass. Within a given quadrant the approach is therefore similar to the Reliability Ensemble Approach (Giorgi and Mearns2003; Xu et al.2010) in its reliance on a historical period comparison, although the focus is on selecting a leading representative GCM rather than on reducing the impact of a GCM through weighting.

A final subjective consideration may be necessary in situations where the two GCMs closest to the quadrant center of mass are likely to produce substantially different results owing to the sensitivity of the system being assessed. To illustrate this point, imagine that the hot and dry quadrant contained GCM C and GCM D, each of which is approximately the same temperature and precipitation change away from the center of mass but in opposite directions (e.g., GCM C is 1 °C warmer and 10% drier than the quadrant center of mass; GCM D is 1 °C cooler and 10% wetter than the quadrant center of mass). In this example either GCM could be selected, but GCM C is more highly recommended due to its greater exploration of the sensitivity of the system to the hot and dry changes that its quadrant represents. This necessarily leads to slightly more extreme responses (particularly where systems are non-linear), but this information may be conveyed to stakeholders who can respond accordingly.

2.8 Quantifying the weights of representative models for each quadrant

Uncertainty in the ensemble of GCM projections is quantified using the probability that projections fall into any given quadrant. This follows an implicit assumption that models are exchangeable and equally likely manifestations of future climate, thus individual models are not differentially weighted (Knutti et al.2010a). A quadrant weighting factor ($W_{quadrant}$) may therefore be calculated by dividing the number of models that falls into a given quadrant and by the total number of models in the ensemble ($W_{quadrant} = N_{quadrant}/N_{Total}$). These factors may then be passed on to later phases of the integrated assessment to represent model-based probability and confidence, allowing eventual results to be aggregated using a set of quadrant weight in factors that sum to 1. As the goal is to represent the entire ensemble of GCMs through selected GCMs, studies examining multiple time period/RCP combinations do not require the same subset of GCMs to be used for each combination. Rather, selecting the most representative GCMs and noting corresponding weights maintains more information about the overall ensemble.

2.9 Evaluating skewness

In addition to carrying information about the probability of particular classes of climate change, quadrant weights also convey information about physical mechanisms prevalent in projected climate changes for a given region and season. The latter is recognizable as strong deviations from the expected weights if the 29 members of the GCM projection ensemble were distributed evenly across all five quadrants (20% of models in each quadrant).

2.10 Diagonal skewness

"Diagonal skewness" is defined as existing when there are substantially more GCMs falling into a particular diagonal orientation of three quadrants than the 17.4 (60%) GCMs that would be expected. Specifically, site projections are considered to exhibit "hot/wet vs. cool/dry skewness" when at least 22 (75%) of the GCMs fall in either the hot/wet, middle, or cool/dry quadrants. Conversely, site projections feature "hot/dry vs. cool/wet skewness" when at least 22 (75%) of the GCMs fall in either the hot/dry, middle, or cool/wet quadrants.

2.11 Extreme skewness

"Extreme skewness" is defined as existing when there are substantially more (or fewer) GCMs falling into the middle quadrant than the 5.8 (20%) GCMs that would be expected. Specifically, site projections are considered "very extreme" when 3 (10.3%) or less of the 29 GCMs fall within the middle quadrant. This is approximately half of the expected value the middle quadrant, and also means that 26 (89.7%) or more of the GCMs fall into the cool/wet, cool/dry, hot/wet, or hot/dry quadrants. Site projections are considered "Non-extreme" when 9 (31%) or more models fall within the middle quadrant, which is 1.5 times the number expected.

3. RESULTS

3.1 Point-based sub-setting

The Ames, Iowa, maize-growing season example demonstrates the approach and reveals probabilistic information about climate projections for the area (Figureure.1). During the May-September maize-growing season the 29 CMIP5 GCMs project median temperature (T) rises of 3.24 C and 5% precipitation (P) increases. These changes form the criteria for defining GCM projections as being relatively hot ($\Delta T > +3.24$ C), cool ($\Delta T < +3.24$ C), wet ($P > +5\%$ change), or dry ($P < +5\%$ change). A GCM projecting increases in rainfall of 0–5% would therefore still be considered "relatively dry" in comparison to the 29-member GCM ensemble despite not being drier than the historical conditions (there is only one model projecting an increase in Ames precipitation that would still be considered "relatively dry"). All GCMs (including both the "relatively hot" and "relatively cool") project temperatures warmer than the current period.

The middle quadrant is defined using a range centered upon the median of one standard deviation of projected temperature (0.75 °C) and precipitation changes (11%). GCMs that are within one half standard deviation in both the temperature and precipitation change dimensions are therefore considered to be in the "middle" of the projection ensemble.

With the five quadrants defined using ensemble statistics, we find that there are 8 GCMs classified as "cool/wet" for the Ames maize-growing season (green triangles in Figureure.1), 3 are "cool/dry" (blue), 4 are "hot/wet" (yellow), 9 are "hot/dry" (red), and 5 are "middle" (black). Using the center of mass for projected changes in each quadrant (noted as a dot in Figureure.1) as a target temperature and precipitation change that best represents each quadrant, we can select five individual models to represent the broader ensemble at this location and growing season (triangles highlighted with gray edges in Figureure.1). None of these models demonstrate mean biases in the historical period that would justify elimination, and biases in the baseline period are not associated with projected changes (Additional file1: Figureureure S1).

These five selected models may therefore be used to drive further crop, livestock, and economic model analysis at Ames as part of an integrated assessment. The number of models within each quadrant may also provide useful probabilistic insight that may be incorporated into final analyses. To illustrate the potential of this information, imagine that the economic impacts of both the hot/dry and hot/wet scenarios were particularly worrying to a stakeholder. That stakeholder's risk management may consider that 9 of the 29 GCMs (31%) fell into the hot/dry category while only 4 of the 29 GCMs (14%) projected hot/wet conditions. This probabilistic information could also be used as weights to estimate the expected value of an outcome (E(C)E(C)) across the full 29-member GCM ensemble based upon the outcome (*C*) and quadrant weighting factor ($W_{quadrant}$) of cool/wet (*cw*), cool/dry (*cd*), middle (*m*), hot/wet (*hw*), and hot/dry (*hd*):

$$E(C) = (C_{cw} * W_{cw}) + (C_{cd} * W_{cd}) + (C_m * W_m) + (C_{hw} * W_{hw}) + (C_{hd} * W_{hd}) \quad (1)$$

In this way the Representative T&P GCM Subsetting Approach encourages the system to be tested against the types of changes that could happen while also providing guidance about the changes that are more likely to happen according to the ensemble of GCM projections.

To elucidate the Representative T&P GCM Subsetting Approach's effect on GCM selection in Ames, Figureure. 2 shows simple future climate scenarios based upon projected monthly mean temperature (Figureure. 2a) and precipitation (Figureure. 2b) for all 29 GCMs compared against the historical observations. While all GCMs project increasing temperatures, the "relatively warm" models are on the upper end of the distribution throughout the year, while the "relatively cool" models lie on the lower end and the "middle" models fall in between. Future precipitation scenarios also separate upon the relatively wet/middle/dry classification as expected, however there is much more variation across months than was seen for temperature given the larger coefficient of variation for GCM projections of precipitation change. This leads to examples where a "relatively wet" model may actually be drier than a "relatively dry" model (e.g., where green lines are below red lines in July) and also reveals that "middle" models may actually be wetter than the ensemble average in the beginning of the growing season and drier than the ensemble average at the end of the season. As GCM projections were classified according to the May–September maize-growing season in Ames, it is not surprising that the quarterly average temperature and rainfall projections that overlap this season are neatly sorted into cool/wet, cool/dry, hot/wet, hot/dry, and middle classifications. The annual, JFM, and DJF seasons, on the other hand, show the danger of selecting a GCM subset based upon seasons that may have different GCM characteristics than the season of application. Although some of this sub-seasonal information is lost in the selection of a GCM subset, for Ames the Representative T&P GCM Subsetting Approach is able to capture the basic behaviors of the GCM ensemble.

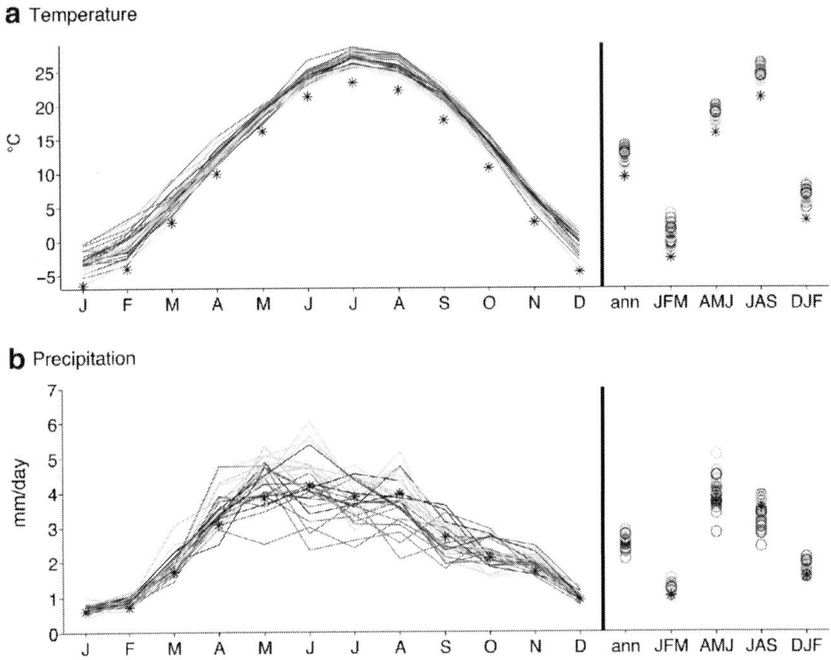

Figure. 2 Ames, Iowa, scenarios of future, monthly, annual, and quarterly temperature (*top*) and precipitation rates*(bottom)* from each of the 29 CMIP5 GCMs for Mid-Century RCP8.5, along with current period values (black stars). Colors indicate the quadrant classification for each GCM (*green* = cool/wet, *blue* = cool/dry, *gray* = middle, *dark yellow* = hot/wet, *red* = hot/dry)

3.2 Patterns of change

The process by which GCMs are classified into five quadrants using the median and standard deviation of temperature and precipitation change projections also yields interesting information about the GCMs and the simulated climate system. While there is an element of subjectivity in the guided selection of representative GCMs for each quadrant, the delineation of quadrants is an objective process which can be repeated for all regions. Below, we analyze the classification (described in the Delineating climate change quadrants section) of all 29 CMIP5 GCMs listed in Table 1 for each 0.25° × 0.25° grid box of AgMERRA over both the entire year and over the maize-growing season as defined in the AgMIP GGCMI (grid boxes without a planting date are omitted).

3.3 Median and standard deviations of projections

Quadrants are defined based upon the median and standard deviation across 29 GCMs of temperature and precipitation changes within a given season. Figureureure 3 shows these projections for the 2040–2069 ("Mid-Century") period under RCP8.5 for both the full year and the maize growing season. Annual mean change patterns were a focus of discussion by Flato et al. (2013), revealing prominently increased rates of warming at high-latitudes where the snow-albedo feedback is in full effect and longwave radiation forms a larger proportion of the energy budget (Figureure.3a). Warming is also stronger away from the coastal buffering provided by the oceans' higher heat capacity. Precipitation patterns (Figureure.3c) tend to follow the "rich get richer" rule of thumb described by Trenberth (2011) whereby areas and seasons that are currently wet tend to get wetter while areas and seasons that are currently dry become drier. Median temperature change patterns are similar for the full year and the maize growing seasons (Figureure.3b), with only slightly less warming during the maize season over India and the Amazon and slightly higher warming over Western Europe. Increased wintertime precipitation that fed the annual increases in many mid-latitude regions are absent in the spring and summer maize growing seasons (Figureure. 3d), leading to slightly more drying or reductions in the wet signal in comparison to the full year median projections (Figureure.3c).

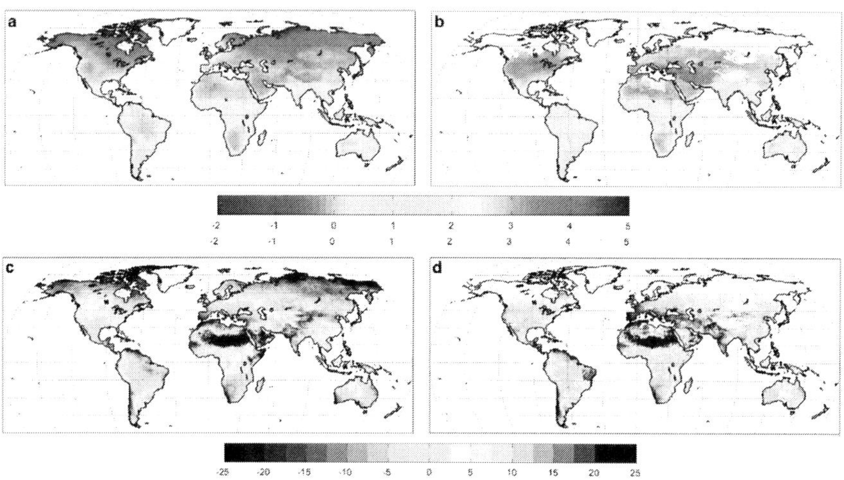

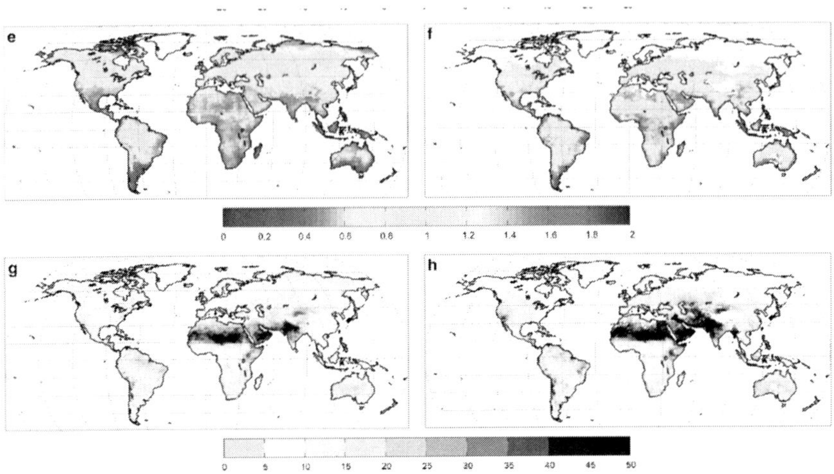

Figure. **3** Annual (*left*) and maize season (*right*) median change in (a, b) temperature (°C) and (c,d) precipitation (%); standard deviation (across 29 GCMs) of median change in (e, f) temperature (C) and (g,h) precipitation (%). 2040–2069 RCP8.5 compared to 1980–2009 current period

The standard deviation of temperature changes across the 29 GCMs (Figureure. 3e-f) show the largest values at high latitudes, over the Himalaya Mountains, and in the Amazon Basin. These are due in large part to model differences in ice-albedo feedback, the resolution of complex topography, and atmosphere-biosphere interactions, respectively (Flato et al. 2013). The standard deviation of temperature also tends to scale with median temperature increase. Precipitation change is remarkably consistent (Figureure. 3g) except over the arid portions of North Africa and Western Asia where even small changes can lead to large percentage shifts in some GCMs. These patterns are largely replicated in the maize growing season (Figureure. 3h), although standard deviations are slightly higher when fewer months are averaged together as individual models' monthly biases tend to be larger than their annual bias.

3.4 Patterns of quadrant weights

Figure4presents the percentage of GCMs that fall within each of the five quadrants for annual mean and maize season RCP8.5 climate change projections. Analysis across time slices and RCPs suggests that the optimum standard deviation factor ρ (that captures ~20% of GCMs within the middle quadrant) increases slightly with climate change as GCM uncertainty increases. The optimal ρ rises from 0.453 in the Near-Term RCP8.5 to 0.560 in the End-of-Century RCP8.5. The optimal ρ in the Mid-Century RCP4.5

(0.503) and RCP8.5 (5.18) are quite close to 0.5, which is an appealingly simple factor to convey to stakeholders. Using $\rho = 0.5$ for the Mid-Century RCP8.5 results in the middle quadrant containing 19.1% of models for the annual period and 18.9% of models for the maize-growing season. Despite an ability to optimize the standard deviation factor ρ for various time periods, the importance of this optimization is small compared to the benefit of a simple definition of the middle quadrant that is intuitive (the range of the quadrant is one standard deviation in each direction).

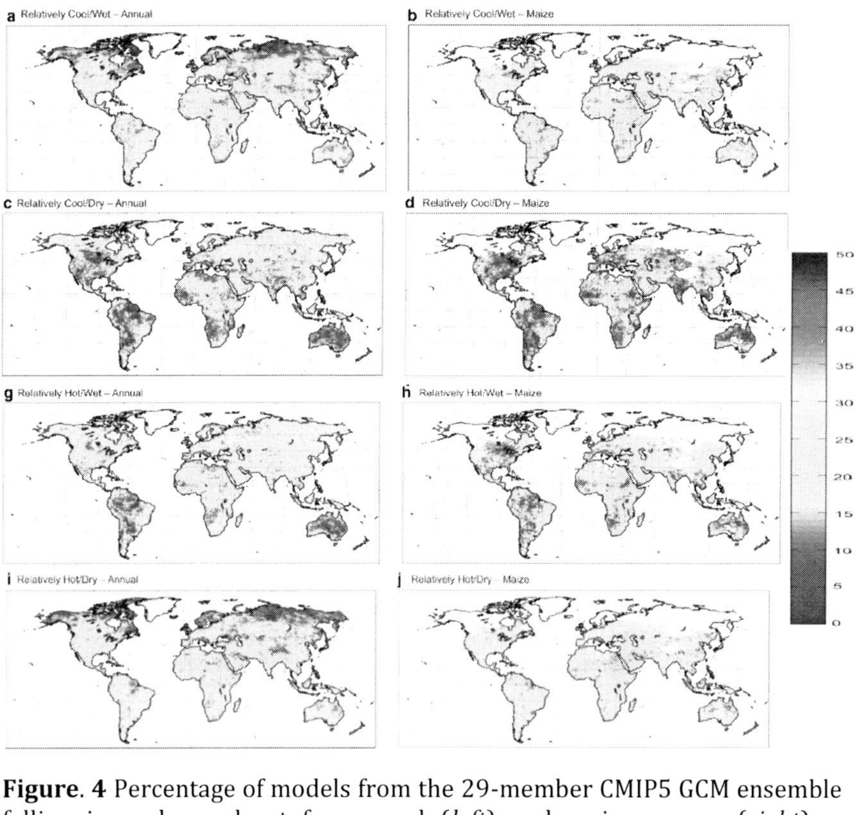

Figure. **4** Percentage of models from the 29-member CMIP5 GCM ensemble falling in each quadrant for annual (*left*) and maize season (*right*). a Relatively cool/wet - annual. b Relatively cool/wet - maize. c Relatively cool/dry - annual. d Relatively cool/dry - maize. e Middle - annual. f Middle - maize. g Relatively hot/wet - annual. h Relatively hot/wet - maize. i Relatively hot/dry - annual. j Relatively hot/dry – maize

Horizontal variation in quadrant weights reveals that small-scale deviations (within a couple degrees latitude and longitude) are typically small in comparison to emergent large-scale patterns that reveal a disproportionate number of GCMs in a given quadrant. Large-scale patterns highlight meridional variation (e.g., a reduction in hot and dry models at high latitudes for annual changes; Figureure. 4i), major mountain chains and tundra (e.g., an increase in the number of middle quadrant GCMs over the Himalayas for annual changes; Figureure. 4e), and semi-arid zones (e.g., common deviations over Australia and Southwestern Africa in many quadrants). This indicates that the probabilities captured by quadrant weights are not likely to vary tremendously over a region less than a couple degrees latitude or longitude across (which is typical of many impact studies) unless there is a major shift in aridity or orography. The physical mechanisms behind large scale patterns and resulting skewness are highlighted in the next section.

The large-scale patterns of quadrant weights are quite similar between the annual and maize season. The most striking differences are a result of the maize-growing area not including high-latitude regions that stand out in the annual quadrant weight maps as described above. More subtle differences are also apparent over the maize-growing regions, which tend to be characterized by slightly higher cool/wet and hot/dry weights at the expense of cool/dry and hot/wet weights (over the Midwestern United States and India, for example; Figureure. 4b, d, h, j). These subtle changes between the annual and maize distributions are substantial in the aggregate, as noted in the ~3% shift from cool/dry to cool/wet quadrants and corresponding ~3% shift from hot/wet to hot/dry quadrants (Table 2). It is also not surprising that more GCMs fall into the wet quadrants than the dry quadrants, as the percentage change metric used to define quadrants is not limited in its increase but cannot decrease by more than 100%.

Table 2 Percentage of 29 CMIP5 GCMs that fall into relative temperature and precipitation change quadrants for Mid-Century (2040–2069) RCP8.5

Quadrant	Annual changes over all land areas	Maize-season changes over all maize-growing areas
Relatively Cool and Wet	20.3%	23.0%
Relatively Cool and Dry	18.0%	15.3%
Middle	19.1%	18.8%
Relatively Hot and Wet	21.8%	19.1%

Relatively Hot and Dry	20.8%	23.8%

Quadrant weights averaged accounting for diminishing area of grid boxes with higher latitude

3.5 Patterns of skewness

The shifts toward hot/dry and cool/wet quadrant weights in the maize areas and season is a manifestation of increasing hot/dry vs. cool/wet skewness. Although the ~3% shifts are small compared to what is possible at a single point, deviations to that extent on the global average indicate a powerful signal. Figureureures 5a, b provide a useful illustration of the skewness represented in Table 2, showing more maize-season weight in the hot/dry vs. cool/wet diagonal (65.6% of all GCMs) than in the hot/wet vs. cool/dry diagonal (53.2% of all GCMs). In contrast, the annual average of all land areas represented by AgMERRA shows a lack of noteworthy diagonal skewness (60.2 and 58.9% of GCMs along the respective diagonals). While a precise determination of the physical causes for these shifts requires analysis beyond the scope of this study, patterns of skewness highlight mechanisms worthy of further study.

An understanding of the geographical patterns of diagonal skewness (Figureure. 6a, b) suggests an explanation for the overall increase in cool/wet vs. hot/dry diagonal skew for the maize-season. While climate change projections for most of the world are not substantially skewed, pockets of hot/dry vs. cool/wet skewness appear over many semi-arid regions as well as over the Amazon Basin. To understand this skewness in climate change projections it is useful to consider that energy added to a given land surface is forced primarily into either sensible (Q_h) or latent (Q_e) heat fluxes, which are commonly related via the Bowen Ratio (*B*; defined as $B = Q_h / Q_e$). Over semi-arid areas *B* is high (there is very little surface moisture for evapotranspiration), so excess energy drives mostly sensible heat flux and therefore the models projecting drier conditions also tend to project the largest temperature increases. Skewness over the Amazon reflects the tendency of some GCMs to dry out the Basin considerably (Collins et al. 2013). This leads to a local warming enhancement through a dramatic increase in sensible heat flux (and thus an elevated *B*). The annual maps also show a tendency toward the opposite skew (hot/wet vs. cool/dry) over the highest latitudes and elevations. These regions feature a low *B* with precipitation typically limited by moisture recycling (as warm air evaporates surface moisture) and moisture flux convergence (Ruane and Roads 2008); both of which increase with warming temperatures.

GCMs that project higher temperatures in these regions therefore typically also project larger increases in precipitation.

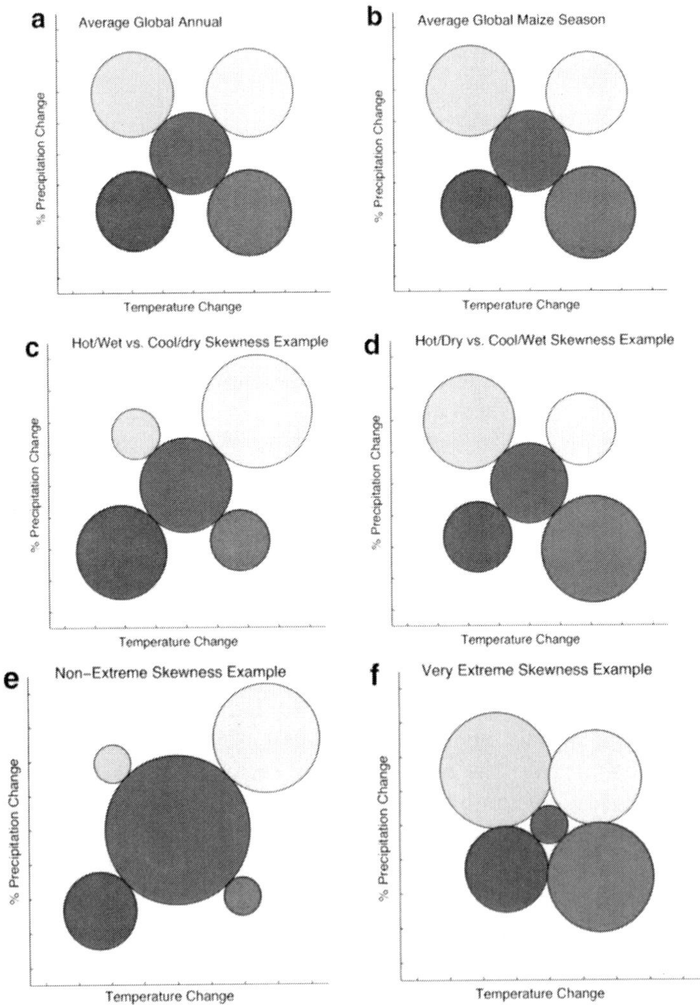

Figure. **5** Representation of quadrant weights (denoted as radius of circle in each quadrant) for (a, b) global average and (c-f) examples of characteristic types of skewness. Examples drawn from (c) Massachusetts, USA (annual); (d) Ames, USA (maize season); (e) Taymyr Peninsula, Russia (annual); and (f) Southern France (maize season). Note that panel d differs from Figure. 1 due to differences between in situ observations and AgMERRA as well as the location of the observational site vs. the AgMERRA grid box

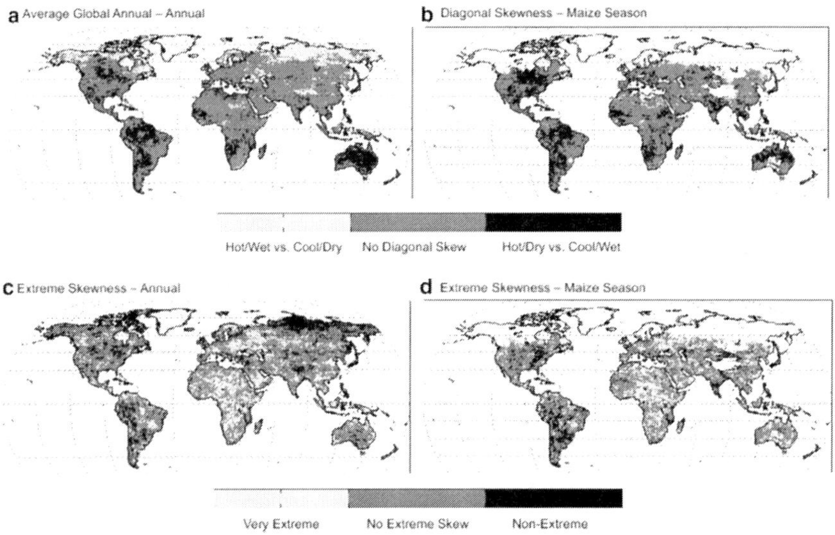

Figure. **6** Geographical pattern of (a, b) diagonal skewness and (c, d) extreme skewness for the (a, c) annual average and (b, d) maize growing season

The increase in maize skewness likely comes from the co-location of maize-growing areas with frost-free climates and seasons containing moderate amounts of rainfall. Due to temperature constraints maize is typically not grown at high-latitudes and is not in the ground during the wintertime storms that deliver a large portion of seasonal rainfall to the dry, primarily mid-latitude regions. Maize is therefore concentrated where soil moisture is variable and prevailing weather patterns associate rainfall with cool and wet air masses (typical of a mid-latitude summer) rather than warm and wet air masses (as are often found in wintertime or polar cyclones). Increased temperatures and energy are projected to raise global evapotranspiration demand, however some maize-growing regions will not be able to keep up with the elevated demand posed by the warmest GCMs and will therefore shift toward drier conditions. This in turn raises *B*, increasing sensible heat (and local temperatures) and leading to hot/dry vs. cool/wet skew such as that we saw over the Amazon for annual changes. The maize season maps do not include the winter season and high-latitude areas constrained by surface moisture and energy, leading to more instances where the warmest models increase evapotranspiration and drive wetter conditions (hot/wet vs. cool/dry skew). Figureureure 5c illustrates hot/wet vs. cool/dry skewness using an annual grid box in Massachusetts that is dominated by wintertime precipitation, while Figureure. 5d uses the

Ames maize-growing season from the gridded analysis to demonstrate hot/dry vs. cool/wet skewness.

The middle quadrant of both the globally-averaged annual and maize change distributions contain similar weights (19.1 and 18.8%, respectively for the Mid-Century RCP8.5), but some regions do exhibit extreme skew (Figureures. 6c, d). Non-extreme skew is common in northern Siberia, where median temperature and precipitation changes are both among the highest on the planet. Temperature changes, in particular, have a very high standard deviation owing to model differences in climate sensitivity and local factors like the snow-ice albedo feedback. Non-extreme skew therefore is a description of the spread of GCM projections rather than the extremity of the GCM projections themselves. In fact non-extreme skew is found in places where standard deviation of GCM projections is so large due to outliers that the middle quadrant expands to capture many of the other GCMs. Figureureure 5e displays an example of this non-extreme skew near Lake Tamyr, Siberia, where the middle quadrant contains a disproportionate number of GCMs. Non-extreme skew is also seen where variations in the GCMs' resolution of orography cause slightly larger standard deviations of temperature and precipitation change, including the highly productive plains and foothills just south of the Himalayas. Regions displaying very extreme skew (e.g., Figureure. 5f) tend to be patchy but are most prevalent in Africa and Eastern Europe. Very extreme skew occurs when the GCMs tend to separate evenly into each of the outer quadrants while leaving very few models in the center of the distribution. In these regions GCMs show patterns of temperature and rainfall responses that are bimodal or disjointed in another way (potentially due to the acceleration of temperature or precipitation shifts beyond a given threshold or associated with a particular regime shift).

3.6 Re-constructing a GCM's change patterns using quadrants

By classifying each GCM into a relative change quadrant for the global 0.25*x*0.25 grid, we can illustrate climate change patterns for each GCM relative to the 29 member CMIP5 ensemble. Figure 7 shows the GISS-E2-R and UKMO HadGEM2-ES annual change quadrant classifications for the mid-century RCP8.5, mid-century RCP4.5, and near-term RCP8.5. It is immediately evident that HadGEM2-ES is among the warmer models for nearly all locations, while GISS-E2-R tends to be relatively cool over much of the land area. This is a reflection of HadGEM2-ES's climate sensitivity (4.6 °C) being among the highest in Table 1 while GISS-E2-R's (2.1 °C) is among the lowest. Both models are characterized by patterns of quadrant classifications, for example HadGEM2-ES's relatively wet designation over the highest latitudes in comparison to relatively dry designation over Eastern Europe, as well as the GISS-E2-R's relatively hot/dry projections over much of Africa and Latin America that stand out from the overall cooler projections. The spatial coherence and general consistency of these patterns across time periods and RCPs reveals distinguishing tendencies of the models and serve as an additional verification of the basic stability of the quadrant designation approach. It is not surprising that the patterns are not perfectly consistent, as shifts in any model's circulation can potentially change the median or standard deviation of projections for a given region. [Mid-century annual RCP8.5 classifications for each GCM are provided in Additional file 1: Figure S2, revealing that no GCM overwhelmingly falls into the middle quadrant].

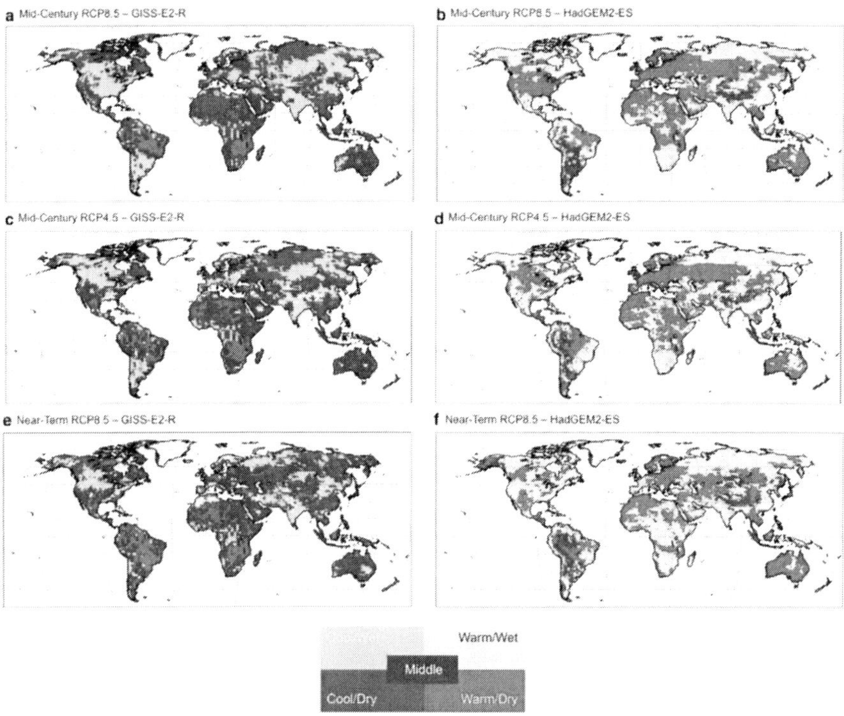

Figure. **7** Quadrant designation for GISS-E2-R (*left*) and HadGEM2-ES (*right*), over the time period of the (a, b) mid-century RCP8.5; (c, d) mid-century RCP4.5; and (e, f) near-term RCP8.5

The global maps of GCM quadrant classifications also enable large-scale analysis through a recombination of regional integrated assessment results that were conducted using representative GCMs. For example, the first phase of AgMIP regional integrated assessments in sub-Saharan Africa and South Asia focused upon distributed regions linked together through the use of 5 common GCMs for all sites (Ruane et al. 2015b; Kihara et al. 2015; McDermid et al., 2015b). This guaranteed consistency in continent-wide analysis but also incorrectly assumed that the 5 selected GCMs adequately sampled projected climate changes for each region. In future phases of this work the Representative GCM approach will be used to capture the key classes of climate change within each region, providing more useful regional information. Results from these analyses will then enable continent-wide analyses linked either by quadrant (e.g., impact of relatively cool/wet climate scenarios on Africa) or by GCM (e.g., impact of HadGEM2-ES over Africa). For the latter analysis the impact pattern of the HadGEM2-ES can be constructed by identifying the type of change projected by that GCM over a given region and then utilizing the corresponding representative GCM from the regional integrated analysis. Through this relatively inexpensive approach one could examine the larger-scale patterns of change (which affect international trade, for example) across many GCMs while maintaining practically representative subsets for analysis within each region.

4. DISCUSSION

4.1 Benefits of representative T&P GCM subsetting approach

The Representative T&P GCM Subsetting Approach provides criteria for the selection of a GCM subset that is:

- Practical in the number of resources required (reducing the CMIP5 GCM ensemble by ~1/6th to free up resources for other elements of a regional integrated assessment);
- Defined according to changes in the season and variables that most affect the system of interest;
- Characteristic of the major classes of climate change (relatively cool/wet, cool/dry, hot/wet, hot/dry, and middle) projected by the GCM ensemble;
- Cognizant of potential outliers that may be evidence of substantial biases;

- Connected to weights that denote the ensemble-based probability of GCM projections that are similar to each GCM within the subset;
- Generalizable through standard definitions;
- Capable of utilizing local observations or being built upon gridded climate products (such as AgMERRA);
- Transferable from region to region and sector to sector;
- Able to be recombined for larger-scale analyses;
- Generally stable across alike geographical areas, time periods, and RCPs; and
- Relatively simple to communicate to stakeholders.

5. LIMITATIONS AND AREAS FOR CONTINUING DEVELOPMENT

By definition, any subset of GCMs is a reduction in information compared to the full ensemble and therefore subsetting should only be undertaken when limited resources are required elsewhere in an integrated assessment process. A clear understanding of assumptions and limitations is therefore necessary when subsetting is necessary.

The Representative T&P GCM Subsetting Approach's reliance on mean temperature and precipitation change projections will not account for differences in other climatic properties that may affect the system of interest (climate models have many degrees of freedom). Agricultural outcomes, for example, may be particularly sensitive to changes in the number of rainy days or the sub-seasonal breakdown of temperature increases (heat waves are particularly damaging during key plant development stages). Ruane et al. (2013) found that scenarios featuring only growing season mean temperature and precipitation changes slightly reduce the range of simulated yield changes compared to scenarios containing additional information about monthly temperature and rainfall changes. That representative models are similar to other models within their quadrant on variables other than growing season mean temperature and precipitation is implicitly assumed but worthy of further study. It is likely that close relationships exist (e.g., mean rainfall with relative humidity, cloud cover, or the number of rainy days) but that other variables may be more independent (e.g., interannual variability, the frequency of extreme events). The subset approach only selects GCMs for further analysis, and therefore the overall importance of variable changes other than the seasonal mean depends on the choice of a scenario generation approach that accounts for these changes as well as the sensitivity of the integrated assessment models to these changes.

Although the Representative T&P GCM Subsetting Approach can be generalized and applied objectively, subjective considerations related to the selection of specific representative GCMs (described in the Choosing a representative model for each quadrant section) may be very important in some regions. As an illustration, consider a case in which GCM E falls nicely near the center of mass as defined by other GCMs within the hot/dry quadrant but each of the other GCMs lies quite far from that center of mass. If GCM E contains substantial bias in comparison to historical observations of seasonal temperature or precipitation, an additional objective criterion is needed to decide at which point we accept GCM E rather than selecting the next-best GCM and therefore relying on a GCM that is less representative of the mean growing season climate changes.

Systems with multiple crops require special attention in defining a growing season in which to evaluate projected changes in temperature and precipitation and select representative GCMs. The Rice-Wheat systems of the Punjab in Pakistan, for example, are famously productive in utilizing monsoon rains during the wet (*Kharif;* May-October) season and harnessing the power of irrigation in the dry (Rabi; November-April) season (Ahmad et al. 2015). The economics of this rice-wheat system is better assessed using a single driving climate model rather than attempting to stitch together rice impacts simulated with one GCM and wheat impacts simulated by another. In this situation it is better to define the growing season as encompassing all months in which rice and wheat are cultivated, however investigation of the individual seasons is informative. Projections of temperature and precipitation change for the individual Kharif and Rabi seasons may distinguish models that are consistently within a given quadrants from those with different behaviors from season to season; information which could be incorporated into the selection of representative GCMs or passed on to the final risk assessment.

Objective rules for dealing with outliers would also benefit those seeking to automate the GCM selection process. These are most prevalent in the precipitation change dimension, and therefore several approaches merit further study. A first approach would be to cap precipitation changes (below 25% or above 200% of present day rainfall for any given month, for example) in order to limit their influence on the quadrant center of mass and the overall standard deviation of the GCM distribution. An alternate approach would be to fully eliminate GCMs where precipitation changes exceeded given thresholds, utilizing the remaining models within a quadrant to form a center of mass and select a representative GCM. This is often necessary in arid and monsoon-driven regions, as mismatches between the GCM and observed rainfall can lead to implausibly large percentage increases for months at the beginning or end of distinct rainy seasons (Additional file 1: Figureureure S3 presents the number of models that exceed a high threshold for each region). In both approaches the outlier could still be considered in the weighting of the quadrant, although examples are possible where large biases in a wet month that is dry within a GCM actually shift the growing season from a dry to a wet classification. Strong connections between a GCM's baseline temperature and precipitation biases with that GCM's eventual relative change quadrant may also suggest that dynamical or thermodynamical biases are affecting projections for a region (potentially due to sea ice/snow cover, monsoon circulations, or soil moisture anomalies), indicating the need to disqualify particularly egregious models (Additional file 1: Figureureure S1 shows that this is not the case for the rainfed maize season in Ames).

Any objective approach must also be careful to recognize that some extreme regional changes are quite plausible and grounded in strong model physics. These may be of utmost interest to stakeholders and should not be too quickly dismissed. For example, a maize farmer in Iowa may be more interested in the probability that his crop fails (an event at the distributional tail) than the expected future yield (a value in the center of the distribution). A stakeholder that is interested in simply stress-testing a system may therefore be more interested in selecting representative GCMs that bound the projected impacts than in establishing a probability of occurrence.

Very rarely (<2.5% of grid boxes for annual; <1% for maize) diagonal or extreme skew in a given location is so dramatic that an entire quadrant will be devoid of any GCM. For these locations it would be consistent to either ignore this quadrant (if its weight is 0% than it is not worth simulating) or define an additional quadrant to separate GCMs within the most heavily weighted quadrant.

6. EXTENSIONS

The methods and analyses above were based upon an effort to select a 5-GCM subset from a broader 29-GCM CMIP5 ensemble, but different resource levels or systems of focus may call for larger or smaller subsets. In terms of selecting a larger number of GCMs, the next logical step would be to select 9 GCMs representing a 3x3 matrix of temperature and precipitation changes (allowing a quadrant for "hot/middle" changes, for example). Alternatively, a third dimension of analysis could be added, for example the standard deviation of temperatures (to emphasize extreme events) or end-of-season mean rainfall (to emphasize the important grain-filling stage of crops). This would also select 9 GCMs, as combinations of relatively high or low values for each of the three dimensions defines 8 quadrants in addition to the middle quadrant. While there is an appeal in using more complex statistics such as 99th percentile rainfall levels or the number of consecutive dry days (see, e.g., Peterson 2005), these values are not well simulated in many GCMs and there is resulting risk of determining selection on areas of GCM weakness. If an even-numbered subset is desired to avoid focus on a central scenario, the next reasonable level would be 8 GCMs, forming outer (cool/wet, cool/dry, hot/wet, hot/dry) and inner (middle/cool, middle/hot, middle/wet, middle/dry) quadrants. Alternatively, assessments of farm systems that are carefully irrigated to manage water stress may only be interested in selecting 3 representative GCMs for high, middle, and low temperature changes (likely closely related to their climate sensitivities). Irrigated farms may also prefer to use solar radiation change as an alternative dimension for quadrant analysis.

Other forms of selecting GCMs to represent classes of change are also possible, including the definition of alternative quadrants or more complex cluster methodologies potentially incorporating more variables. As subsetting always loses information contained in the complete ensemble, many of the same challenges addressed above remain, including the potential that individual GCMs do not fall nicely into clusters or quadrants in some regions. More complex methods may become additionally problematic, however, if they are too complex for dissemination among partners from the variety of disciplines often included in integrated assessments. The temperature and precipitation quadrant approach described in this study is bolstered by its ease of generalization and communication with stakeholders.

On the global scale a quadrant approach defined by mean temperature and precipitation change is hindered by the globally-closed hydrologic cycle, which tends to respond to increasing temperatures by increasing the overall levels of both evapotranspiration and precipitation. As integrated assessments are dependent on the resolution of regional impacts, mean precipitation changes are not a sufficient metric as GCMs balance precipitation increases in one region with compensating drying in other areas. A better metric to gauge hydrologic impacts at the global scale is therefore the mean absolute percentage change of precipitation, which is indicative of the way that climate change strengthens and alters the geographical patterns within the water cycle. Global average temperature changes are closely related to the GCMs' climate sensitivities listed in Table 1. In the process of selecting GCMs for a global study it would also be useful to examine the annual quadrant classifications in important regions of interest, as presented in Additional file 1: Figureureure S2. Further validation exercises that explore the extent and ramifications of regional temperature and precipitation biases within the GCMs would also be quite useful for a wide number of climate applications (Ruane et al. 2016).

Representative GCM selection for a national or regional-level integrated assessment is challenging in that it is unlikely that a single set of GCMs can represent the classes of change for all points within a large domain. Sub-regional assessments may be recombined for larger-scale evaluation as described in the Re-constructing a GCM's change patterns using quadrants section, but in cases where economic analyses span the broader region it is important that inputs are consistent. In this situation an initial first step would be to perform quadrant analysis at a distributed network of sites within the domain and then look for particular GCMs that are suitably consistent representatives of a given quadrant. The final selection of GCMs may prioritize accurate representation of the GCM ensemble over key sub-regions (e.g., a country's breadbasket in an agricultural study) and can also note which sub-regions may be missing a representative for a quadrant within the selected GCMs.

7. SUMMARY AND NEXT STEPS

The Representative T&P GCM Subsetting Approach provides a practical way to reduce computational and analytical resources in integrated assessments of climate change impacts. Although information is lost in any subsetting of GCMs, this efficient approach captures the basic combinations of important climate change factors and their relative probabilities in order to enable stakeholder risk management. The core of the approach involves the analysis of major types of climate changes likely to affect a given sector (illustrated above for agriculture), with the goal of selecting GCMs that represent each major type of change and are associated with probabilistic information related to the broader ensemble. In some cases this analysis may lead to further stakeholder inquiry as to the extremes possible within a given quadrant, which could form the basis for continuing study.

The process of classifying GCMs relative to the wider ensemble of projected temperature and precipitation changes for a given region also provides useful insight into the sensitivity of these variables and the coherence of regional patterns across space, time, and greenhouse gas scenario. GCMs generally demonstrate noteworthy consistency, with some regions also demonstrating various forms of skewness in the full ensemble that are indicative of climatic processes or model uncertainty. As public attention following accords at the 21st Conference of Parties in Paris shifts increasingly toward the implementation of mitigation and adaptation strategies rooted in climate applications research, this type of analysis may help in the selection of GCM subsets covering the range of regional changes needed to increase resilience.

ACKNOWLEDGEMENTS

The authors benefited from multiple conversations with members of the AgMIP community who helped communicate the need for consistent and informative climate subsets. Dr. Ruane's research was funded under the NASA Earth Sciences Division (WBS: 509496.02.08.04.24 and 281945.02.03.03.96). Dr. McDermid's contributions were supported by the United Kingdom Department for International Development (DFID; Program Code 202108). We thank Olivier Crespo, Jack Simmons, and Shari Lifson for their support and advice. We acknowledge the World Climate Research Programme's Working Group on Coupled Modelling, which is responsible for CMIP, and we thank the climate modeling groups (listed in Table 1 of this paper) for producing and making available their model output. For CMIP the U.S. Department of Energy's Program for Climate Model Diagnosis and Intercomparison provides coordinating support and led development of software infrastructure in partnership with the Global Organization for Earth System Science Portals.

REFERENCES

1. Ahmad A, et al (2015) Impact of Climate Change on the Rice–Wheat Cropping System of Pakistan. In: Rosenzweig C, Hillel, D (eds) Handbook of Climate Change and Agroecosystems: The Agricultural Model Intercomparison and Improvement Project (AgMIP). C. Rosenzweig, and D. Hillel, Eds., ICP Series on Climate Change Impacts, Adaptation, and Mitigation Vol. 3 Part 2. Imperial College Press, 219–258, doi:10.1142/9781783265640_0019
2. Antle JM, Valdivia RO, Boote K, Janssen S, Jones JW, Porter CH, Rosenzweig C, Ruane AC, Thorburn PJ (2015) AgMIP's transdisciplinary agricultural systems approach to regional integrated assessment of climate impacts, vulnerability, and adaptation. In: Rosenzweig C, Hillel, D (eds) Handbook of Climate Change and Agroecosystems: The Agricultural Model Intercomparison and Improvement Project (AgMIP). ICP Series on Climate Change Impacts, Adaptation, and Mitigation Vol. 3. Part 1, Imperial College Press, 27–44, doi:10.1142/9781783265640_0002
3. Asseng S, Ewert F, Rosenzweig C, Jones JW, Hatfield JL, Ruane AC, Boote KJ, Thorburn PJ, Rötter RP, Cammarano D, Brisson N, Basso B, Martre P, Aggarwal PK, Angulo C, Bertuzzi P, Biernath C, Challinor AJ, Doltra J, Gayler S, Goldberg R, Grant R, Heng L, Hooker J, Hunt LA, Ingwersen J, Izaurralde RC, Kersebaum KC, Müller C, Naresh Kumar S, Nendel C, O'Leary G, Olesen JE, Osborne TM, Palosuo T, Priesack E, Ripoche D, Semenov MA, Shcherbak I, Steduto P, Stöckle C, Stratonovitch P, Streck T, Supit I, Tao F, Travasso M, Waha K, Wallach D, White JW, Williams JR, Wolf K (2013) Uncertainty in simulating wheat yields under climate change. Nature Clim Change 3:827–832. doi:10.1038/nclimate1916

4. Bassu S, Brisson N, Durand J-L, Boote K, Lizaso J, Jones JW, Rosenzweig C, Ruane ACR, Adam M, Baron C, Basso B, Biernath C, Boogaard H, Conijn S, Corbeels M, Deryng D, De Sanctis G, Gayler S, Grassini P, Hatfield J, Hoek S, Izaurralde C, Jongschaap R, Kemanian AR, Kersebaum KC, Kim SH, Kumar NS, Makowski D, Müller C, Nendel C, Priesack E, Pravia MV, Sau F, Shcherbak I, Tao F, Teixeira E, Timlin D, Waha K (2014) How do various maize crop models vary in their responses to climate change factors? Glob. Change Biol 20(7):2301–2320. doi:10.1111/gcb.12520
5. Christensen JH, Hewitson B, Busuioc A, Chen A, Gao X, Held I, Jones R, Kolli RK, Kwon W-T, Laprise R, Magana Rueda V, Mearns L, Menendez CG, Raisanen J, Rinke A, Sarr A, Whetton P (2007) Regional climate projections. In: Solomon S, Qin D, Manning M, Chen Z, Marquis M, Averyt KB, Tignor M, Miller HL (eds) Climate change 2007: the physical science basis. Contribution of Working Group I to the fourth assessment report of the intergovernmental panel on climate change. Cambridge University Press, Cambridge, United Kingdom, and New York, pp 847–845
6. Collins M, Knutti R, Arblaster J, Dufresne J-L, Fichefet T, Friedlingstein P, Gao X, Gutowski WJ, Johns T, Krinner G, Shongwe M, Tebaldi C, Weaver AJ, Wehner M (2013) Long-term climate change: projections, commitments and irreversibility. In: Stocker TF, Qin D, Plattner GK, Tignor M, Allen SK, Boschung J, Nauels A, Xia Y, Bex V, Midgley PM (eds) Climate change 2013: the physical science basis. Contribution of Working Group I to the fifth assessment report of the intergovernmental panel on climate change. Cambridge University Press, Cambridge, United Kingdom and New York
7. Elliott J, Müller C, Deryng D, Chryssanthacopoulos J, Boote KJ, Büchner M, Foster I, Glotter M, Heinke J, Iizumi T, Izaurralde RC, Mueller ND, Ray DK, Rosenzweig C, Ruane AC, Sheffield J (2015) The global gridded crop model intercomparison: data and modeling protocols for phase 1 (v1.0). Geosci Model Dev 8:261–277. doi:10.5194/gmd-8-261-2015View ArticleGoogle Scholar
8. Eyring V, Bony S, Meehl GA, Senior CA, Stevens B, Stouffer RJ, Taylor KE (2016) Overview of the Coupled Model Intercomparison Project Phase 6 (CMIP6) experimental design and organization. Geosci Model Dev 9:1937–1958. doi:10.5194/gmd-9-1937-2016
9. Flato G, Marotzke J, Abiodun B, Braconnot P, Chou SC, Collins W, Cox P, Driouech F, Emori S, Eyring V, Forest C, Gleckler P, Guilyardi E, Jakob C, Kattsov V, Reason C, Rummukainen M (2013) Evaluation of climate models. In: Stocker TF, Qin D, Plattner GK, Tignor M, Allen SK, Boschung J, Nauels A, Xia Y, Bex V, Midgley PM (eds) Climate Change 2013: The Physical Science Basis. Contribution of Working Group I to the Fifth Assessment Report of the Intergovernmental Panel on Climate Change. Cambridge University Press, 741–882, doi:10.1017/CBO9781107415324.020
10. Giorgi F, Mearns LO (2003) Probability of regional climate change based on the Reliability Ensemble Averaging (REA) method. Geophys Res Lett 30(12):1629. doi:10.1029/2003GL017130
11. Gleckler PJ, Taylor KE, Doutriaux C (2008) Performance metrics for climate models. J Geophys Res 113(D6):D06104, doi:10.1029/2007JD008972
12. Howden M, Crimp S (2005) Assessing dangerous climate change impacts on Australia's wheat industry. MODSIM 2005 International Congress on Modeling and Simulation. Modelling and Simulation Society of Australia and New Zealand., pp 170–176. ISBN 0-9758400-2-9
13. IPCC (2010) Meeting report of the intergovernmental panel on climate change expert meeting on assessing and combining multi model climate projections. In: Stocker TF, Qin D, Plattner G-K, Tignor M, Midgley PM (eds) IPCC Working Group I Technical Support Unit. University of Bern, Bern, p 117. ISBN 978-92-9169-129-6
14. Kihara J, MacCarthy DS, Bationo A, Koala S, Hickman J, Koo J, Vanya C, Adiku S, Beletse Y, Masikate P, Rao KPC, Mutter CZ, Rosenzweig C, Jones JW (2015) Perspectives on Climate Effects on Agriculture: The International Efforts of AgMIP in Sub-Saharan Africa. In: Rosenzweig C, Hillel, D (eds) Handbook of Climate Change and Agroecosystems: The Agricultural Model Intercomparison and Improvement Project (AgMIP). ICP Series on Climate Change Impacts, Adaptation, and Mitigation Vol. 3 Part 2. Imperial College Press, p. 3–23. doi:10.1142/9781783265640_0013

15. Knutti R (2010) The end of model democracy? Clim Change 102(3):395–404. doi:10.1007/s10584-010-9800-2View ArticleGoogle Scholar
16. Knutti R (2014) IPCC Working Group I AR5 snapshot: the historical, rcp4.5, and rcp8.5 experiments. WDCC at DKRZ., doi:10.1594/WDCC/ETHhi; doi:10.1594/WDCC/ETHr4; doi:10.1594/WDCC/ETHr8
17. Knutti R, Abramowitz G, Collins M, Eyring V, Gleckler PJ, Hewitson B, Mearns L (2010a) Good practice guidance paper on assessing and combining multi model climate projections. In: Stocker TF, Qin D, Plattner G-K, Tignor M, Midgley PM (eds) Meeting report of the intergovernmental panel on climate change expert meeting on assessing and combining multi model climate projections. IPCC Working Group I Technical Support Unit. University of Bern, Bern
18. Knutti R, Furrer R, Tebaldi C, Cermak K, Meehl GA (2010b) Challenges in combining projections from multiple models. J Clim 23:2739–2756. doi:10.1175/2009JCLI3361.1
19. Li T, Hasegawa T, Yin X, Zhu Y, Boote K, Adam M, Bregaglio S, Buis S, Confalonieri R, Fumoto T, Gaydon D, Marcaida M III, Nakagawa H, Oriol P, Ruane AC, Ruget F, Singh B, Singh U, Tang L, Tao F, Wilkens P, Yoshida H, Zhang Z, Bouman B (2015) Uncertainties in predicting rice yield by current crop models under a wide range of climatic conditions. Glob Change Biol 21(3):1328–1341. doi:10.1111/gcb.12758
20. Lutz AF, ter Maat HW, Biemans H, Shresth AB, Wester P, Immerzeel WW (2016) Selecting representative climate models for climate change impact studies: an advanced envelope-based selection approach. Int J Climatol 36:3988–4005. doi:10.1002/joc.4608
21. Martre P, Wallach D, Asseng S, Ewert F, Jones JW, Rötter RP, Boote KJ, Ruane AC, Thorburn PJ, Cammarano D, Hatfield JL, Rosenzweig C, Aggarwal PK, Angulo C, Basso B, Bertuzzi P, Biernath C, Brisson N, Challinor AJ, Doltra J, Gayler S, Goldberg R, GrantRF HL, Hooker J, Hunt LA, Ingwersen J, Izaurralde RC, Kersebaum KC, Müller C, Kumar SN, Nendel C, O'Leary G, Olesen JE, Osborne TM, Palosuo T, Priesack E, Ripoche D, Semenov MA, Shcherbak I, Steduto P, Stöckle CO, Stratonovitch P, Streck T, Supit I, Tao F, Travasso M, Waha K, White JW, Wolf J (2015) Multimodel ensembles of wheat growth: many models are better than one. Glob Change Biol 21(2):911–925. doi:10.1111/gcb.12768
22. McDermid, SP, Ruane AC, Rosenzweig C, Hudson NI, Morales MD, Agalawatte P, Ahmad S, Ahuja LR, Amien I, Anapalli SS, Anothai J, Asseng S, Biggs J, Bert F, Bertuzzi P, Bhatia V, Bindi M, Broad I, Cammarano D, Carretero R, Chattha AA, Chung U, Debats S, Deligios P, De Sanctis G, Dhliwayo T, Dumont B, Estes L, Ewert F, Ferrise R, Gaiser T, Garcia G, Gbegbelegbe S, Geethalakshmi V, Gerardeaux E, Goldberg R, Grant B, Guevara E, Hickman J, Hoffmann H, Huang H, Hussain J, Justino FB, Karunaratne AS, Koehler A-K, Kouakou PK, Kumar SN, Lakshmanan A, Lieffering M, Lin X, Luo Q, Magrin G, Mancini M, Marin FR, Marta AD, Masutomi Y, Mavromatis T, McLean G, Meira S, Mohanty M, Moriondo M, Nasim W, Negm N, Orlando F, Orlandini S, Ozturk I, Pinto HMS, PodestaG, Qi Z, Ramarohetra J, Rahman HH, Raynal H, Rodriguez G, Rötter R, Sharda V, Shuo L, Smith W, Snow V, Soltani A, Srinivas K, Sultan B, Swain DK, Tao F, Tesfaye K, Travasso MI, Trombi G, Topaj A, Vanuytrecht E, Viscarra FE, Wajid SA, Wang E, Wang H, Wang J, Wijekoon E, Byun-Woo L, Xiaoguang Y, Young BH, Yun JI, Zhao Z, Zubair L (2015a) The AgMIP Coordinated Climate-Crop Modeling Project (C3MP): Methods and protocols. In: Rosenzweig C, Hillel, D, (Ed), Handbook of Climate Change and Agroecosystems: The Agricultural Model Intercomparison and Improvement Project (AgMIP) Integrated Crop and Economic Assessments, Part 1. ICP Series on Climate Change Impacts, Adaptation, and Mitigation Vol. 3. Part 1, Imperial College Press, p 191–220, doi:10.1142/9781783265640_0008
23. McDermid SP, Dileepkumar G, Dakshina Murthy KM, Nedumaran S, Singh P , Srinivasa C, Gangwar B, Subash N, Ahmad A, Zubair L, Nissanka SP (2015b) Integrated Assessments of the Impacts of Climate Change on Agriculture: An Overview of AgMIP Regional Research in South Asia. In: Rosenzweig C, Hillel, D, (Eds), Handbook of Climate Change and Agroecosystems: The Agricultural Model Intercomparison and Improvement Project (AgMIP). ICP Series on Climate Change Impacts, Adaptation, and Mitigation Vol. 3. The Agriculutral Model Intercomparison and Improvement Project Integrated Crop and Economic Assessments, Part 2 (pp. 201–217). Imperial College Press. doi:10.1142/9781783265640_0018

24. McSweeney CF, Jones RG (2016) How representative is the spread of climate projections from the 5 CMIP5 GCMs used in ISI-MIP? Climate Services 1, doi:10.1016/ j.cliser. 2016.02.001
25. Moss RH, Edmonds JA, Hibbard KA, Manning MR, Rose SK, van Vuuren DP, Carter TR, Emori S, Kainuma M, Kram T, Meehl GA, Mitchell JFB, Nakicenovic N, Riahi K, Smith SJ, Stouffer RJ, Thomson AM, Weyant JP, Wilbanks TJ (2010) The next generation of scenarios for climate change research and assessment. Nature 463(7282):747–756. doi:10.1038/nature08823
26. Peterson TC (2005) Climate change indices. WMO Bull 54(2):83–86
27. Portmann F et al (2008) Global data set of monthly growing areas of 26 irrigated crops. Institute of Physical Geography, University of Frankfurt, Frankfurt am Main, p 400
28. Portmann FT, Siebert S, Döll P (2010) MIRCA2000-Global monthly irrigated and rainfed crop areas around the year 2000: a new high-resolution data set for agricultural and hydrological modeling. Global Biogeochem Cycles 24:Gb1011
29. Reichler T, Kim J (2008) How well do coupled models simulate today's climate? Bull Am Met Soc 89:303–311
30. Rienecker MM et al (2011) MERRA: NASA's modern-era retrospective analysis for research and applications. J Climate 24:3624–3648. doi:10.1175/JCLI-D-11-00015.1
31. Rosenzweig C, Jones JW, Hatfield JL, Ruane AC, Boote KJ, Thorburn P, Antle JM, Nelson GC, Porter C, Janssen S, Asseng S, Basso B, Ewert F, Wallach D, Baigorria G, Winter JM (2013) The Agricultural Model Intercomparison and Improvement Project (AgMIP): protocols and pilot studies. Ag For Meteor 170:166–182. doi:10.1016/j.agrformet.2012.09.011
32. Rosenzweig C, Jones JW, Hatfield JL, Antle JM, Ruane AC, Mutter CZ (2015): The Agricultural Model Intercomparison and Improvement Project: Phase I activities by a global community of science. In: Rosenzweig C, Hillel, D, (Eds), Handbook of Climate Change and Agroecosystems: The Agricultural Model Intercomparison and Improvement Project (AgMIP). ICP Series on Climate Change Impacts, Adaptation, and Mitigation Vol. 3. Part 1, Imperial College Press, p 3–24. doi:10.1142/9781783265640_0001
33. Rosenzweig C, Antle J, Elliott J (2016) Assessing Impacts of Climate Change on Food Security Worldwide. Eos, 97, doi:10.1029/2016EO047387. Published on 9 March 2016
34. Ruane AC, Roads JO (2008) Dominant balances and exchanges of the atmospheric water cycle in the Reanalysis-2 at diurnal, annual, and intraseasonal time scales. J Climate 21(16):3951–3966. doi:10.1175/2007JCLI2015.1
35. Ruane AC, Cecil LD, Horton RM, Gordón R, McCollum R, Brown D, Killough B, Goldberg R, Greeley AP, Rosenzweig C (2013) Climate change impact uncertainties for maize in Panama: farm information, climate projections, and yield sensitivities. Agr Forest Meteorol 170:132–145. doi:10.1016/j.agrformet.2011.10.015
36. Ruane AC, McDermid S, Rosenzweig C, Baigorria GA, Jones JW, Romero CC, Cecil LD (2014) Carbon–Temperature–Water change analysis for peanut production under climate change: a prototype for the AgMIP Coordinated Climate-Crop Modeling Project (C3MP). Glob Chang Biol 20:394–407. doi:10.1111/gcb.12412
37. Ruane AC, Goldberg R, Chryssanthacopoulos J (2015a) Climate forcing datasets for agricultural modeling: Merged products for gap-filling and historical climate series estimation. Agr Forest Meteorol 200. doi:10.1016/j.agrformet.2014.09.016
38. Ruane AC, Winter JM, McDermid SP, Hudson NI (2015b) AgMIP climate datasets and scenarios for integrated assessment. In: Rosenzweig C, Hillel, D, (Eds), Handbook of Climate Change and Agroecosystems: The Agricultural Model Intercomparison and Improvement Project (AgMIP). ICP Series on Climate Change Impacts, Adaptation, and Mitigation Vol. 3. Part 1, Imperial College Press, p 45–78, doi:10.1142/ 9781783265640_0003
39. Ruane AC, Teichmann C, Arnell N, Carter TR, Ebi KL, Frieler K, Goodess CM, Hewitson B, Horton R, Kovats RS, Lotze HK, Mearns LO, Navarra A, Ojima DS, Riahi K, Rosenzweig C, Themessl M, Vincent K (2016) The Vulnerability, Impacts, Adaptation and Climate Services Advisory Board (VIACS AB v1.0) contribution to CMIP6. Geosci Model Dev 9:3493–3515. doi:10.5194/gmd-9-3493-2016

40. Sacks WJ et al (2010) Crop planting dates: an analysis of global patterns. Glob Ecol Biogeogr 19(5):607–620
41. Semenov MA, Stratonovich P (2015) Adapting wheat ideotypes for climate change: accounting for uncertainties in CMIP5 climate projections. Clim Res Vol 65:123–139. doi:10.3354/cr01297
42. Sriver RL, Forest CE, Keller K (2015) Effects of initial conditions uncertainty on regional climate variability: an analysis using a low-resolution CESM ensemble. Geophys Res Lett 42:5468–5476. doi:10.1002/2015GL064546
43. Taylor KE, Stouffer RJ, Meehl GA (2012) An overview of CMIP5 and the experiment design. Bull Am Meteorol Soc 93(4):485–498, doi:10.1175/BAMS-D-11-00094.1
44. Tebaldi C, Smith RL, Nychka D, Mearns LO (2005) Quantifying uncertainty in projections of regional climate change: a Bayesian approach to the analysis of multimodel ensembles. J Clim 18:1524–1540
45. Trenberth K (2011) Changes in precipitation with climate change. Clim Res 47:123–138. doi:10.3354/cr00953
46. Warszawski L, Frieler K, Huber V, Piontek F, Serdeczny O, Schewe J (2014) The Inter-Sectoral Impact Model Intercomparison Project (ISI–MIP): project framework. P Natl Acad Sci USA 111:3228–3232. doi:10.1073/pnas.1312330110
47. White JW, Hoogenboom G, Kimball BA, Wall GW (2011) Methodologies for simulating impacts of climate change on crop production. Field Crop Res 124(3):357–368. doi:10.1016/j.fcr.2011.07.001
48. Wilby RL, Charles S, Zorita E, Timbal B, Whetton P, Mearns L (2004) Guidelines for use of climate scenarios developed from statistical downscaling methods., IPCC Supporting Material, available from the DDC of IPPC TGCIA
49. Xu Y, Gao X, Giorgi F (2010) Upgrades to the reliability ensemble averaging method for producing probabilistic climate-change projections. Clim Res 1:61–81. doi: 10.3354/cr00835

CHAPTER 12

Meningitis and climate: From science to practice

Carlos Pérez García-Pando[1,2], Madeleine C Thomson[3,10], Michelle C Stanton[4], Peter J Diggle[5,6], Thomas Hopson[7], Rajul Pandya[8], Ron L Miller[1,2] and Stéphane Hugonnet

[1]NASA Goddard Institute for Space Studies, New York, USA

[2]Department of Applied Physics and Applied Math, Columbia University, New York, USA

AUTHORS' CONTRIBUTIONS

CPGP and MT conceived the paper and drafted large parts of the manuscript. MS and PD contributed to Section Potential tools and models to improve epidemic response and critically revised the manuscript. TH and RP contributed to Sections The role of climate and dust and Potential tools and models to improve epidemic response. RM contributed to Section The role of climate and dust and critically revised the manuscript. SH contributed to Section Potential tools and models to improve epidemic response and critically revised the manuscript.

ABSTRACT

Meningococcal meningitis is a climate sensitive infectious disease. The regional extent of the Meningitis Belt in Africa, where the majority of epidemics occur, was originally defined by Lapeysonnie in the 1960s. A combination of climatic and environmental conditions and biological and social factors have been associated to the spatial and temporal patterns of epidemics observed since the disease first emerged in West Africa over a century ago. However, there is still a lack of knowledge and data that would allow disentangling the relative effects of the diverse risk factors upon epidemics. The Meningitis Environmental Risk Information Technologies Initiative (MERIT), a collaborative research-to-practice consortium, seeks to inform national and regional prevention and control strategies across the African Meningitis Belt through the provision of new data and tools that better determine risk factors. In particular MERIT seeks to consolidate a body of knowledge that provides evidence of the contribution of climatic and environmental factors to seasonal and year-to-year variations in meningococcal meningitis incidence at both district and national scales. Here we review recent research and practice seeking to provide useful information for the epidemic response strategy of National Ministries of Health in the Meningitis Belt of Africa. In particular the research and derived tools described in this paper have focused at "getting science into policy and practice" by engaging with practitioner communities under the umbrella of MERIT to ensure the relevance of their work to operational decision-making. We limit our focus to that of reactive vaccination for meningococcal meningitis. Important but external to our discussion is the development and implementation of the new conjugate vaccine, which specifically targets meningococcus A.

Keywords: Meningococcal meningitis, Meningitis belt, Climate, Mineral dust, Relative humidity, Epidemics, Reactive vaccination, Forecasting, Risk factors, Africa Statistical models

1. INTRODUCTION

Meningococcal meningitis is an infection of the thin lining that surrounds the brain and spinal cord. While there are many causes of meningitis, the epidemic form of the disease is caused by bacteria *Neisseria Meningitidis.* Meningitis is one of the most feared diseases in Africa because of its rapid onset and high rates of long-term disability and fatality. Epidemics pose a serious threat to populations and place a severe burden on public health systems and on socio-economic development.

Getting science into policy and practice is a challenge, particularly in public health where evidence-based policies are expected. For diseases such as meningitis that are climate sensitive, there is the perception that climate information may be relevant to improving control measures. However, delivering apparently simple changes into an epidemic response strategy requires applied cross-disciplinary science focusing on practical solutions to problems that evolve with time.

The MERIT (Meningitis Environmental Research Information Technologies) initiative was launched in 2007 as a multi-sectoral partnership led by the World Health Organization (WHO) to provide a platform for public health specialists, epidemiologists, immunologists, microbiologists, demographers and climate and environment specialists to work together to provide innovative solutions for the control of meningococcal meningitis epidemics in the African Sahel. The history of MERIT, its membership and its processes are described in detail elsewhere (Thomson et al. 2013). The International Research Institute for Climate and Society (IRI) has played a key role in MERIT since its inception, helping to frame the initiative and Steering Committee, while IRI staff, students, adjuncts and partners followed key lines of inquiry. In 2012 the MERIT initiative was endorsed by an external panel of experts (Thomson et al. 2013).

For over a century, epidemics of meningococcal meningitis have occurred in Africa. Until recently, the main control approach has been reactive vaccination after an outbreak has reached a defined threshold. Delays in vaccine procurement and delivery limit the success of this approach, which is also unable to prevent the occurrence of new epidemics (Roberts 2008). Meningococcal conjugate vaccines, which can prevent meningococcal carriage and thus interrupt transmission, may be effective at preventing epidemics and are currently being 'rolled out' across the region (LaForce et al. 2007). Despite the rapid implementation of a new conjugate vaccine (which targets serogroup A only), a recent review indicated that there is still a need for further work to identify the host, organism and environmental factors that contribute to the geographic location, seasonality and inter-annual variability of meningococcal disease and to predict and control epidemics in Africa (Greenwood 2013).

Over the 6 years that MERIT has been active three operational areas have been identified as a priority for research - with the focus shifting over time from reactive to preventive vaccination strategies (MERIT 2012). These areas are:

1) improve the impact of reactive mass vaccination campaigns, prepare for the following epidemic season, and refine the response strategy for epidemics. After the introduction of the new conjugate vaccine, this refers to epidemics due to serogroups other than A;

2) preventive vaccination campaigns; guide the introduction of the conjugate A vaccine and estimate its impact, and

3) forecasting the location of future epidemics on 5 to 10-year time horizons in order to assess possible vaccine needs.

Broad spatial and seasonal patterns of meningitis in Africa have been related to climate and other risk factors since the eco-epidemiology of the disease was first described (Lapeyssonnie 1963). Yet, the introduction of climate information and other risk factors as part of a control strategy is relatively new (World Health Organization 2004).

This paper reviews the advances and challenges in science and practice to develop Early Warning Systems (EWSs) for meningococcal meningitis in Africa. Specifically, the paper overviews the main characteristics of epidemic meningitis in Africa, reviews the current understanding of the role of climate upon the spatial and temporal variability of the disease, discusses the conceptual and mathematical models that combine risk factors to explain the spatiotemporal epidemic patterns, and presents tools and statistical models recently developed with potential operational application in the context of epidemic response. Important but external to our discussion is the development and introduction of the new conjugate vaccine that specifically targets meningococcus A.

2. MENINGITIS EPIDEMICS IN SUB-SAHARAN AFRICA

2.1 The Meningitis Belt

The Meningitis Belt (Figure 1), comprising much of semi-arid sub-Saharan Africa, features high levels of seasonal endemicity and epidemics during the dry season. The belt was first associated to an area between latitudes 4° and 16° N receiving between 300 and 1100 mm of annual rainfall in a single rainy season (Lapeyssonnie 1963) and included parts of Burkina Faso, Niger, Nigeria, Chad and Sudan. The belt was subsequently extended to parts of Benin, Cameroon, Central Africa Republic, Uganda, Kenya, Rwanda, Ethiopia, The Gambia, Ghana, Mali and Senegal (Cheesbrough et al. 1995; Moore 1992; World Health Organization 1998). Epidemics in the Meningitis Belt are due to serogroups A, C, X, Y and W135, with serogroup A historically accounting for 80–85% of all cases. Important epidemics with serogroup W135 have emerged in the recent past, including 2002 in Burkina Faso, with 13,125 suspected cases and 1,510 deaths (World Health Organization 2003).

Figure 1 THE MENINGITIS BELT. Extracted from (World Health Organization 2003).

2.2 Spatial and temporal dynamics of meningitis

Epidemic waves following crescendo-decrescendo patterns in the Meningitis Belt have occurred every 8 to 14 years (Moore 1992; Broutin et al. 2007) (Figure 2). Periodicity of meningitis and synchrony among countries has changed through time (Broutin et al. 2007). While Burkina Faso has shown a constant cycle of around 12 years since 1939, periodicity in Sudan has increased from around 9 to 12–13 years since the 70s and in Nigeria has decreased from around 12 to 8–9 years since the mid 70s. Only after the mid 60's have Niger and Mali featured periodicities of 8–10 and 12–14 years, respectively, and Ghana, Togo and Benin showed 12-year cycles up to the 60s, with no periodicity detected thereafter. There is a significant correlation in meningitis epidemics among countries, although it alternates between perfect or out-of phase synchrony in the majority of comparisons (Broutin et al. 2007).

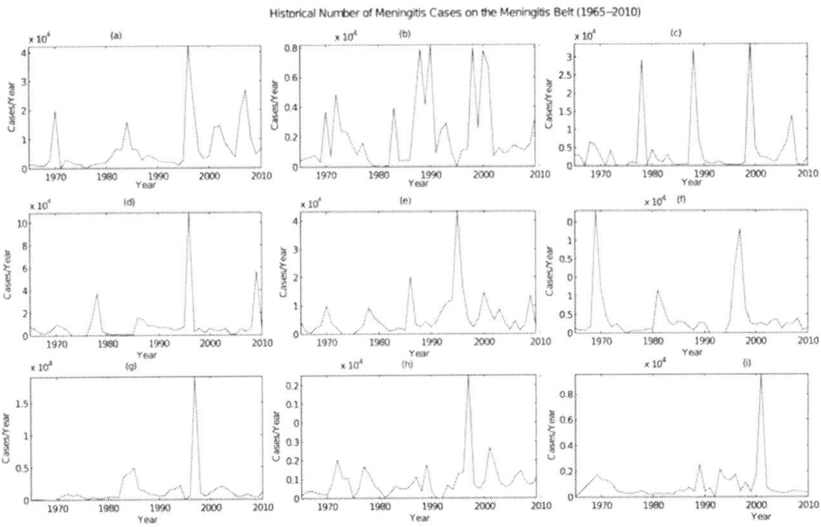

Figure 2 Times series of number of meningitis cases between 1965 to 2010 in (a) Burkina Faso, (b) Chad, (c) Sudan, (d) Nigeria, (e) Niger, (f) Mali, (g) Ghana, (h) Togo, and (i) Benin. Source of data: WHO.

From the 1980s, epidemics occurred throughout the Meningitis Belt in Benin, Burkina Faso, Chad, The Gambia, Ghana, Mali, Niger, Nigeria, Senegal and Togo. Severe epidemics occurred in Ethiopia, Sudan and Chad in 1987–1990, with more than 30,000 cases reported in Sudan in 1988 (World Health Organization 1998). In 1995–1997 a strong epidemic wave (Figure 2) affected Niger (more than 25,000 cases in 1995 and more than 16,000 cases in 1996), Northern Nigeria (more than 100,000 cases in 1996), Burkina Faso (more than 40,000 cases in 1996 and more than 20,000 in 1997) and Mali (more than 7,000 cases in 1996, more than 10,000 in 1997).

Until recently, Niger has been one of the countries in the belt with the largest disease burden and for which the longest district-level weekly records are available. Figure 3 displays the weekly number of cases at national level and the seasonal cycle box plot of incidence by week for 38 districts in Niger from 1986 through 2008. From 1986 to 2005 Niger had an average annual incidence rate of 95.9 cases per 100,000 population and case fatality rates of 8.2% (Djingarey et al. 2008). The largest epidemic occurred in 1995. In addition, 1986, 1993, 1994, 1996 and 2000 were epidemic years with incidence greater than 100 per 100,000 at the national level.

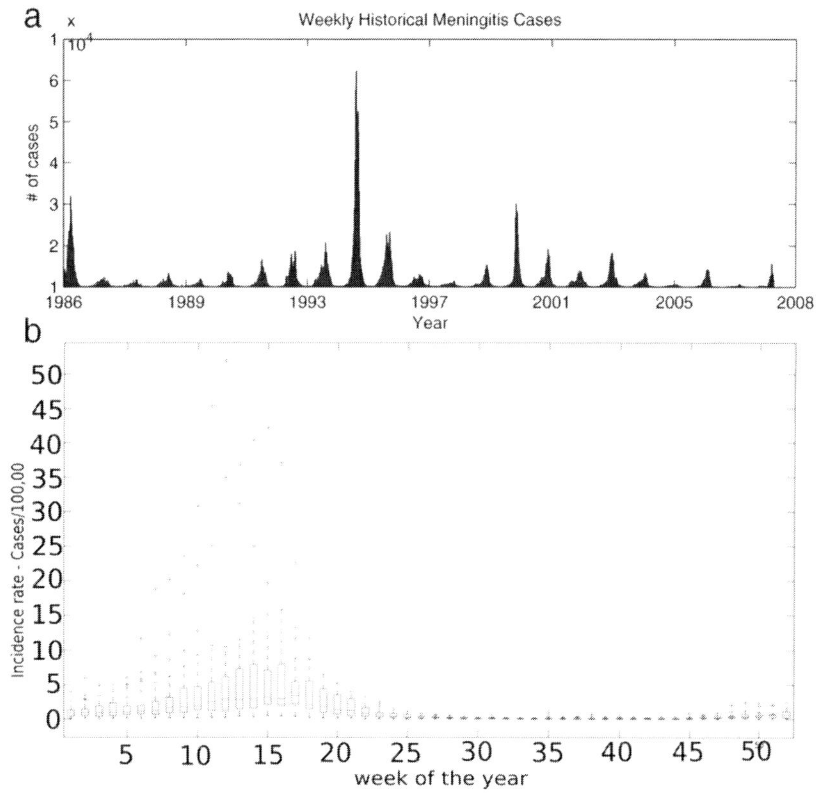

Figure 3 Number of cases per week between 1986 and 2008 in Niger (a) and seasonal cycle box plot of incidence (cases/100,000) by week for 38 districts in Niger from 1986 through 2008 (b). (b) each box shows the range in which the central 50% of the data fall. The box borders show the first and third quartiles and the crossbar shows the median. Values (outliers) exceeding the quartile values by more than factor of 1.5, the interquartile distance, are marked as crosses.

Meningitis incidence is seasonally dependent, with cases increasing at the beginning of the dry season from November/December, typically peaking in February-March-April and declining rapidly with the onset of rains in May (Figure 3). In Niger, incidence peaked at week 14 of the year on average and epidemics lasted on average 10 weeks (Djingarey et al. 2008). Incidence displays strong spatial and interannual variability at regional (Figure 2), national (Figure 3) and district levels (Paireau et al. 2012; Tall et al. 2012).

2.3 Risk factors

Risk factors for epidemics are not fully understood. The heterogeneous spatial and temporal distribution of epidemics (Figures 2 and 3) suggests that a complex interaction involving host, organism and environment is necessary for an epidemic to occur (Greenwood 1987; Moore 1992). A significant proportion of the global population carries the bacteria asymptomatically in the nose and throat, and never develops the invasive disease. The bacteria are transmitted from one person to another through respiratory droplets or throat secretions. Under certain circumstances, the bacteria become pathogenic, invading the naso-pharageal epithelial cells (colonization) and entering the blood stream (invasion). The case fatality rate generally ranges between 10% and 15% in the belt (Trotter & Greenwood 2007). The risk of acquiring meningococcal disease decreases with age. Disease or carriage protects against disease caused by the same serogroup (Trotter & Greenwood 2007). As carriage does not necessarily induce invasive disease (Wenzel et al. 1973), humoral immunity (Griffiss et al. 1987) and herd (i.e. population) immunity to a prevalent strain (Greenwood 1987) may be among the most important factors in the prevention of epidemics. Clone virulence or introduction of novel clones of a serogroup in a susceptible population, though population movements, contributes to trigger epidemics (Moore et al. 1989). Over the past 40 years ST-1 complex/subgroup I/II, ST-4 complex/subgroup IV, and ST-5 complex/subgroup III (shifting to ST-7 since the mid 90's) have successively caused the majority of serogroup A epidemics in the Meningitis Belt (Harrison et al. 2009). However, the introduction of novel clones during periods of susceptibility is insufficient to trigger epidemics. A clear evidence of this insufficiency is that even during major epidemic waves, meningitis is suppressed during the rainy season. During the dry season, coincident respiratory infections and climate conditions including high levels of mineral dust are thought to increase invasive disease and transmission (Moore 1992; Mueller & Gessner 2010). Other risk factors may be genetic susceptibility (Davila et al. 2010), poverty and household crowding (Moore 1992), and exposure to smoke from cooking fires (Hodgson et al. 2001).

2.4 The role of climate and dust

Environmental and climate conditions have long been highlighted as driving factors of epidemics (Waddy 1952; Lapeyssonnie 1963; Greenwood et al. 1984; Cheesbrough et al. 1995). Epidemics and seasonal upsurges in endemic disease occur during the dry season and subside at the onset of the rains (Lapeyssonnie 1963; Molesworth et al. 2002; Sultan et al. 2005). The broad spatial pattern and seasonality of meningitis suggests that certain environmental factors, such as low absolute humidity (Cheesbrough et al. 1995; Molesworth et al. 2003) and relative humidity (Dukić et al. 2012), temperature (Dukić et al. 2012) and dusty atmospheric conditions (Thomson et al. 2006; Agier et al. 2013a; Pérez García-Pando et al. 2014) play an important role. Identifying the specific climate factor that drives epidemics is challenging because many environmental variables have a prominent seasonal cycle that covaries with disease incidence. In addition, although the interactions between mucosal epithelial cells and *Neisseria Meningitidis* are well known (van Deuren et al. 2000), the effects of climate and dust on the pathogenesis and transmission of the bacteria have not been studied in vivo (Palmgren 2009).

Figure 4 displays seasonal cycle box plots of climate and dust variables by week simulated with an atmospheric dust model (Pérez et al. 2011) for 38 districts in Niger between 1986 and 2008. The seasonal cycle is controlled by the migration of the Intertropical Convergence Zone (ITCZ) that leads to the alternation during the year of two main climate regimes. In boreal winter, the Harmattan, a surface stream of dry and dusty air that is part of the African continental trade winds, sweeps from the northeast towards the Gulf of Guinea. In summer, flow from the southwest brings moisture from the oceans and coastal vegetated regions. Dry season weekly median concentrations of mineral dust range from 200 to 400 $\mu g/m^3$ with values reaching up to 500–1000 $\mu g/m^3$. The seasonal cycles of surface concentrations and column–integrated dust optical depths differ (Figure 4) because dust originating from the Sahara is transported at different altitudes during the year. From December to March, absolute humidity is at its minimum, on the order of 5 g/m^3. Temperature reaches its lowest values in December/January and steadily increases until April, while relative humidity stays below 40% during the dry season with values decreasing as temperature increases, reaching a minimum of around 20% at the end of March. Humidity exhibits strong year-to-year variability during the transition from the dry to the wet season and year-to-year changes have been shown to be gradual (Seefeldt et al. 2012).

Absolute humidity and land-cover type were found to be the best predictors in a spatial multivariate model of the broad spatial distribution of epidemics (Molesworth et al. 2003) (Figure 5). Areas with contrasting seasons in the Sahel are more likely to have epidemics with risk decreasing towards the southern peripheral regions where the dry season is shorter and less extreme. Areas without contrasting seasons, both in the humid parts of coastal and Central Africa and in the dry Saharan region show low risk of epidemics, the latter due to low population density in desert areas.

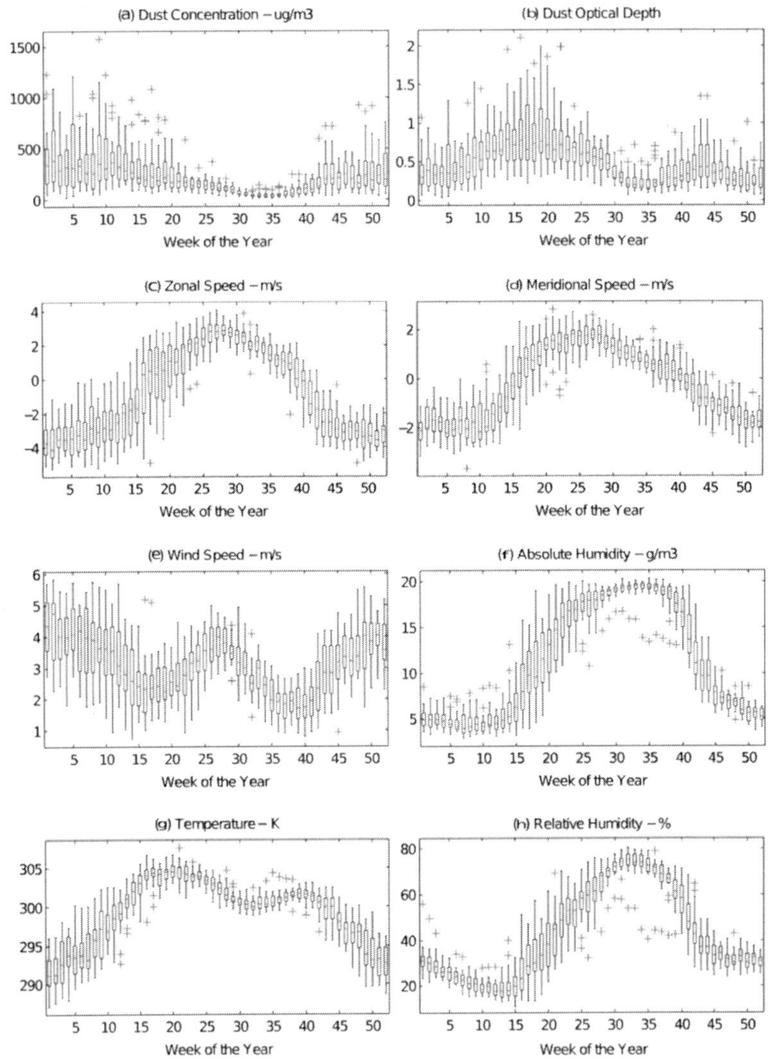

Figure 4 Seasonal cycle box plots by week for 38 districts in Niger between 1986 and 2008 for a) dust concentration, b) dust optical depth at 550 nm, c) zonal wind speed, d) meridional wind speed, e) wind speed, f) absolute humidity, g) temperature, h) relative humidity. Each box shows the range in which the central 50% of the data fall. The box borders show the first and third quartiles and the crossbar shows the median. Values (outliers) exceeding the quartile values by more than factor of 1.5, the interquartile distance, are marked as crosses. Data were derived from a long-term simulation from 1986 to 2006 with the NMMB/BSC-Dust model (Pérez et al. 2011).

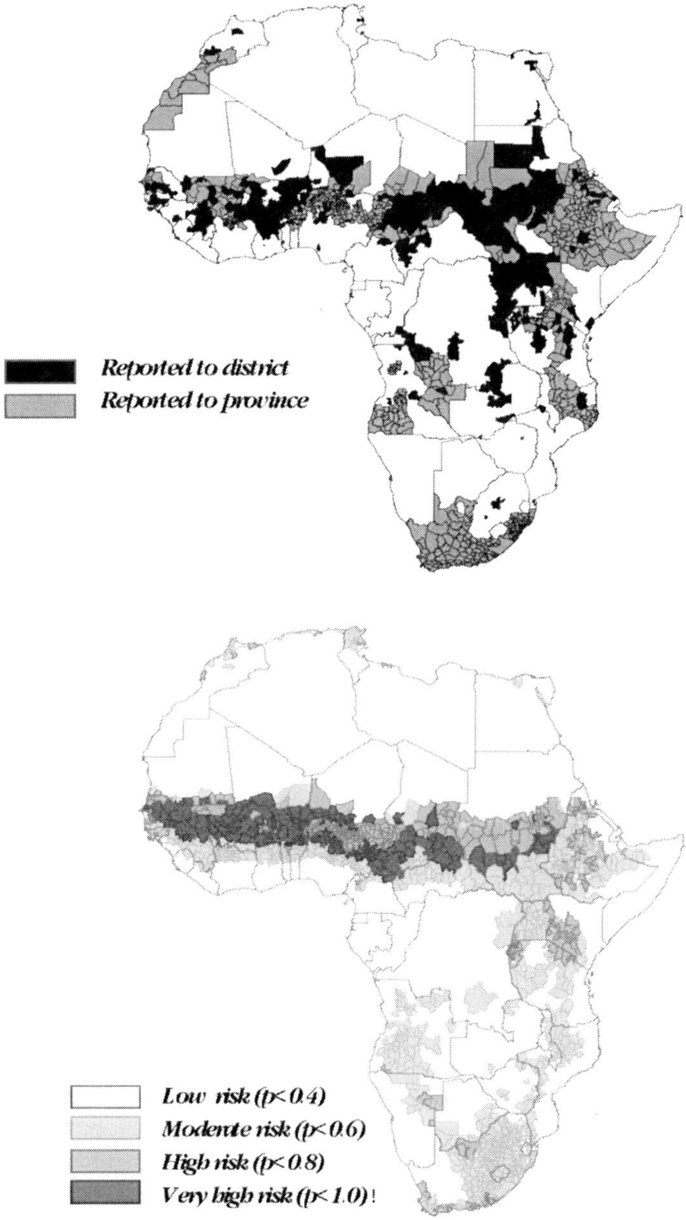

Figure 5 Distribution of districts where epidemics of meningococcal meningitis are likely to occur. Observed distribution of meningitis epidemics (1841–1999) (top panel). Predicted probability of epidemic experience based on environmental variables (absolute humidity and land-cover type) (bottom panel). Reproduced from Molesworth et al. (2003).

The abrupt shift in the seasonal cycle of the total number of cases in Mali at the sixth week of the year has been related to the dry season wind maximum (Sultan et al. 2005). In Niger, the average seasonal peak in meningitis is preceded by a week or two by a peak in dust optical depth (Martiny & Chiapello 2013). One important difficulty in uncovering drivers of infectious diseases is to identify the appropriate scale of analysis (Pascual & Dobson 2004) since local determinants of epidemics may obscure the impact of larger-scale environmental conditions. The relationship between disease and climate apparent at large spatial scales may not be appropriate to resolve local variability, which is crucial for public health interventions in the Meningitis Belt. Within multiple districts in Niger, Agier et al. (2013a) found similar time-lags between the occurrence of dust outbreaks and meningitis, which suggests that dust information may be useful in epidemiological and forecasting models. Dukić et al. (2012) applied Generalized Additive Models treating time-varying confounding processes (e.g. seasonal population migration, new strains) as a background function that varies in magnitude by year, and asked how much of the intraseasonal variability each year could be related to specific variables that vary more quickly in time. The analysis of 11 years of laboratory-confirmed meningitis cases from Navrongo, Ghana, showed that accounting for local weather improved estimates of monthly cases by up to 40%. In particular, the current maximum monthly temperature and the previous month's relative humidity and CO emissions due to fires showed the most value in anticipating meningitis cases. We note that the southern Sahel is strongly affected by biomass burning aerosols in addition to dust outbreaks during the dry season. This is consistent with the results of a survey of Navrongo-area residents, which indicted that meningitis is associated with hot and high aerosol conditions (Hayden et al. 2013).

The seasonal cycle box plot of weekly incidence for 38 districts in Niger (Figure 4) features median values below 5 weekly cases per 100,000 population and strong outliers reaching up to 50 weekly cases per 100,000 population. An ongoing debate is whether variations in climatic variables can at least in part explain year-to-year variations in seasonal incidence (amplitude) or if this influence is just related to the timing of onset, peak and end of meningitis. However, at least in some regions, the timing of onset and peak has been associated with the amplitude of epidemics (de Chabalier et al. 2000). Several studies have related anomalies in amplitude to anomalies in climate. In semi-arid, northern Benin, 14 to 34.5% of the temporal variability of the disease over 28 years was related to low absolute humidity associated with variations of the Harmattan. Rainfall anomalies in January and dust anomalies in October appeared to be the most consistent predictors of anomalies in seasonal incidence at district level in West Africa (Thomson et al. 2006), and about 25% of the year-to-year disease variance at national scale in Niger could be explained by variations in the December-averaged meridional wind (Yaka et al. 2008). Early cases, population density, wind and dust information in the early season could explain a 41% of the year-to-year spatiotemporal variability of the disease at district level in Niger (Pérez García-Pando et al. 2014).

2.5 Conceptual, mathematical and statistical models

Meningitis dynamics may be better understood and predicted with mathematical modeling. Conceptual models combining relevant factors have been proposed to guide the development of mathematical models, understand the disease and improve practical interventions (Moore 1992; Mueller & Gessner 2010). Loss of herd immunity to specific strains may have contributed to the initiation of large-scale epidemic cycles observed in the Meningitis Belt (Figure 2). The subsequent development of herd immunity due to widespread carriage of the epidemic strain during the cycle may have limited transmission ending the epidemic wave (Moore 1992). In this conceptual model, transmission is considered seasonally independent since studies haven't found a systematic variation in the carriage prevalence of meningococcal serogroups by season (Trotter & Greenwood 2007; Harrison et al. 2009). Epidemics during the dry season were then explained by a combination of climatic conditions and widespread respiratory infections decreasing mucosal protection and thus promoting invasion rather than carriage in a low herd immunity setting.

The three main hypotheses to explain the effects of climate and environment upon seasonal incidence are an increased risk of invasive disease after infection, an increased risk of transmission (including adhesion and colonization) and a combination thereof. A more recent and refined conceptual model (Mueller & Gessner 2010) describes the heterogeneous spatiotemporal dynamics of meningitis (Figure 3) distinguishing among endemic incidence during the rainy season, ubiquitous hyperendemicity during the dry season, occasional localized epidemics, and large-scale epidemic waves spanning communities or years. In this framework, the transition from endemic to ubiquitous hyperendemic conditions would be caused by an increased risk of invasion of a virulent strain due to damage of the pharyngeal mucosa by dry and dusty climate. The transition from hyperendemicity to localized epidemics would involve increased pharyngeal colonization and/or transmission through coughing and sneezing possibly caused by viral respiratory infection epidemics or other local-scale co-factors, a hypothesis that is compatible with significant increases in carriage prevalence of the specific virulent strain during epidemics (Mueller et al. 2008) and the heterogeneous and sporadic occurrence of meningococcal epidemics (Tall et al. 2012). A regional epidemic wave would result from a wider geographic spread of epidemic co-factors such as a viral epidemic or the introduction of a new virulent strain.

In both models dry and dusty conditions are considered to affect invasion rather than carriage and transmission. Yet, the contribution of climate to carriage and transmission by, for example, enhancing seasonal viral epidemics (Fuhrmann 2010) cannot be ruled out. Besides damaging the pharyngeal mucosa, some other controversial hypotheses about the role of dust include the activation of the meningococcus through the high iron content of dust particles and the impact of high dust levels on human behavior, including crowding and reduced ventilation (Thomson et al. 2009). Results from a deterministic compartmental model showed that the irregular timing of meningitis epidemics could be caused by the interaction of temporary immunity conferred by carriage together with seasonal changes in transmission rather than seasonal changes in invasion (Irving et al. 2011). Results from a parsimonious statistical model fitted to data in Nigeria supported the hypothesis of a generalized regional increase in the rate of invasive disease and suggested that the end of an epidemic was driven more by the reduction in susceptible individuals and a decrease in transmission rather than by a decline in the rate of invasive disease (Jandarov et al. 2012).

2.6 Potential tools and models to improve epidemic response

Response to meningitis epidemics is based on weekly incidence thresholds at the district level (World Health Organization 2000): the alert threshold, defined as 5 cases per 100,000 inhabitants per week for populations greater than 30,000 inhabitants and 2 cases in one week for populations of less than 30,000 inhabitants, is used to launch an investigation at the start of an epidemic, check epidemic preparedness, start a vaccination campaign if there is an epidemic in a neighbouring district and prioritize areas for vaccination campaigns in the course of an epidemic. The epidemic threshold, which is used to step up mass vaccination, is defined as 10 cases (or 15 cases depending on context) per 100,000 inhabitants for populations greater than 30,000 inhabitants and 5 cases in one week for populations of less than 30,000 inhabitants. This strategy depends on timely surveillance, and rapid response, which are difficult to achieve in less-developed countries. The strategy can therefore benefit from improved surveillance and forecasting tools.

2.7 Sub-district level surveillance

Data on confirmed cases at a sub-district spatial scale in Niger and Burkina Faso have highlighted significant intra-district heterogeneity and interannual variability (Paireau et al. 2012; Tall et al. 2012). Besides improving future mathematical modeling, surveillance and redefinition of weekly epidemic thresholds at the sub-district Health Centre level could improve reactive vaccination response. In Burkina Faso, if an epidemic threshold of 5 cases per 100,000 in one week was used, epidemics could have been spotted more precisely and in one quarter of instances at least one week earlier than with the current district level strategy (Tall et al. 2012).

2.8 Forecasting meningitis

Despite progress in surveillance and research, the incomplete understanding of meningitis epidemic patterns and the lack of quality data at the required temporal and spatial scales limit the use of mathematical modeling to forecast epidemics. EWSs for infectious diseases aim to identify whether an epidemic will occur and to predict the magnitude of incidence (World Health Organization 2004). The latter may be unattainable at precision, regardless of the quality of information, due to small-scale stochastic effects (Drake 2005). However, from a policy perspective, the exact magnitude of incidence is of less interest in comparison to whether or not the incidence crosses an action-triggering threshold.

2.9 Statistical models based on incidence

Given the complex nature of meningitis epidemics that seem to elude description by deterministic models, statistical models have been developed to forecast the probability of exceeding WHO's incidence-based decision thresholds. The motivation was to provide an estimate of the likelihood that the thresholds would be exceeded in the near future. Due to uncertainties in the factors that drive epidemics at the weekly time scale, these models were developed using an empirical approach i.e. no assumptions as to what drives changes in incidence were made, with the original forecasts based primarily on previously observed incidence but without considering climate information or other factors.

Beresniak et al. (2012) evaluated the risk that an epidemic would occur in a particular district after epidemics had been reported in other districts. Through a Bayesian network, the probabilities of districts on alert or epidemic situation influencing other districts were estimated for Niger. When considering districts reaching epidemic alert or epidemic thresholds potentially influencing other districts in 2004, 2005 and 2006, the proportion of the outbreaks observed that were successfully predicted for the period 2006–2008 was 63% when the epidemic threshold was used and 91% when the alert threshold was used.

Other statistical models for continuous data (observed incidence), and discrete data (below the WHO alert threshold, between the alert and epidemic threshold and above the epidemic threshold), have been considered (Agier et al. 2013b; Stanton et al. 2013). Using dynamic (generalised) linear modelling (Durbin & Koopman 2012; West & Harrison 1997), Stanton et al. (2013), assumed that meningitis incidence followed a general seasonal trend, with flexible timing and magnitude of peaks in incidence introduced by allowing the regression coefficients to evolve over time according to previously observed incidence (temporal dependence) in both the district itself, and in neighbouring districts (spatial dependence). To estimate the weekly incidence for each district, estimates of the probability of exceeding an arbitrary threshold could also be extracted for these models. In the discrete setting, a multinomial logistic Markov chain model was used to describe the transition between the three states (below alert, between alert and epidemic, and above epidemic) as defined by WHO thresholds (Agier et al. 2013b). Transition probabilities were dependent both on a general seasonal trend an on the states of neighbouring districts in the previous week. The probability of exceeding the alert/epidemic threshold in any given week was therefore a direct output of this model. Models were evaluated and compared using different policy-linked evaluation criteria. Firstly it was assumed that the primary goal of the exercise was to predict the exact weeks over which the epidemic threshold was exceeded (Criterion 1). Secondly (and alternatively), it was assumed that the primary goal was to predict whether or not the epidemic threshold would be exceeded within a given meningitis-year prior to its onset, regardless of the true time lag between the week in which incidence was first predicted to exceed the threshold and the week in which the observed incidence first exceeded the threshold (Criterion 2). Using the Niger weekly incidence data, the sensitivity, specificity, positive predictive value (PPV) and negative predictive value (NPV) of each of the models' predictions were calculated using both evaluation criteria. Sensitivity, also known as the true positive rate is defined as the proportion of 'positive' events that are correctly predicted. Specificity is the proportion of 'negative' events that are correctly predicted. PPV is the proportion of positive event predictions that are correct, and NPV is the proportion of negative event predictions that are correct. A key point to note when considering these measures is that both PPV and NPV are affected by the overall prevalence of the event in question, such that by fixing sensitivity and specificity, a rare event will have a lower PPV and higher NPV than a more common event. As such, a model with low PPV is more a reflection on the rarity of crossing the threshold than on the performance of the model. Therefore, to determine whether or not the models are of use to policy makers, it is of more use to consider the values of sensitivity and specificity achieved (Table 1). In comparing the models, in general the Markov chain model outperforms the dynamic models when considering evaluation Criterion 1, whereas the opposite is true when considering evaluation Criterion 2. Using a dynamic linear model and evaluation Criterion 2, 71% of epidemics are correctly predicted prior to their onset (sensitivity) whereas 58% of years during which an epidemic is predicted to occur are correct (PPV). Using a Markov chain model and evaluation Criterion 1, 71% of instances where the epidemic threshold is exceeded are correctly predicted, whereas 78% of instances where the epidemic threshold is predicted to be exceeded are correct.

Table 1 Model evaluation summaries for weekly-scale statistical models (Agier et al. 2013b ; Stanton et al. 2013) with respect to how well they predict whether or not the incidence the following week will exceed the epidemic threshold (Criterion 1), and how well they forecast an epidemic prior to its onset during a 12 month period (Criterion 2)

Model	Evaluation criterion	SENS	SPEC	PPV	NPV
Dynamic	1	0.649	0.981	0.705	0.979
Markov chain	1	0.710	0.994	0.780	0.991
Dynamic	2	0.706	0.793	0.579	0.870
Markov chain	2	0.549	0.792	0.510	0.816

2.10 Climate/weather-based models

There is a lack of spatially and temporally resolved data on carriage, herd immunity, previous vaccination type and coverage, serogroup type, clonal virulence and coincident respiratory infections that are thought to explain the epidemic pattern. Although gridded climate and dust data generated by models and/or observations contain uncertainties, their availability at the required spatial and temporal scales is an advantage for potential meningitis forecasting to the extent that they influence the evolution of epidemics.

An exploratory analysis was undertaken on extending empirical statistical models at weekly and district scales (Stanton et al. 2013) in Niger using weather variables including temperature, specific and relative humidity, wind and dust (Stanton et al. 2011) derived from long-term integrations with a regional dust model (Pérez et al. 2011) constrained by the National Centers for Environmental Prediction/National Center for Atmospheric Research (NCEP/NCAR) reanalysis (Kalnay et al. 1996). Lagged weekly and four-weekly averaged weather variables were considered for inclusion both in a linear model with a fixed seasonal trend, and a dynamic linear model with a flexible seasonal trend. More specifically, in order to disentangle the relationship between climate variables and meningitis incidence induced through a common seasonal and/or linear trend from a potential causal relationship, the climate variables included in the models were in the form of residuals obtained by subtracting both a linear trend plus a harmonic trend of order 3 from each of the climate variables under consideration. Although there was evidence of a weak relationship between residuals of weather and meningitis, particularly when taking average conditions over the preceding 4-week period, the strength of the relationship was insignificant in comparison to the strong seasonal trend already introduced by the incidence data. In addition, the inclusion of weather in these specific statistical models decreased the models' performance in terms of being able to predict when the epidemic threshold would be exceeded.

An analysis of meningitis suspected cases across the Meningitis Belt using two years of data (2007–2009) was performed Hopson et al. (2014). Meteorological variables came from the National Centers for Environmental Prediction/National Center for Atmospheric Research (NCEP/NCAR) reanalysis (Kalnay et al. 1996). The transmission of meningitis was modeled using a differential equation-based epidemiological model that has been successfully used to model Methicillin-resistant Staphylococcus aureus (Macal et al. 2012). The model distinguished among infected and susceptible population, and asymptomatic carriers, assuming a homogenous mixing of people across these groups and disease homogeneity across the belt. The number of people infected was assumed to be small compared to the overall population, with changes in district population being negligible. Both the susceptible and carriage populations were considered proportional to the overall district population, and the disease cycle was assumed to be less than two weeks. Using a linearized model and treating each district's weekly epidemiology as an independent event, cross-validated logistic regression was used to solve for the probability of crossing the alert and epidemic thresholds based only on the two years of epidemiological data. By further assuming that the proximity to neighboring districts with cases of meningitis influences the chances of having a case, it was tested if the accuracy of the predictions could be improved when the model coefficients were assumed to relate to meteorological variables. After testing over 90 meteorological variables with varying time lags, the most consistent improvement in the model's predictions came from including relative humidity first and northeasterly winds second. A relative humidity of 40 % marked an inflection point for the probability of a district exceeding the epidemic threshold (Figure 6). Given the verified correlation between meningitis cases and the average weekly relative humidity, researchers at UCAR were able to provide public health decision makers with information about the relative humidity. In addition to producing Meningitis Belt-wide humidity forecasts, the tuned differential equation-based model described earlier can be used to predict the probability of exceeding alert and epidemic thresholds one to four weeks ahead. End-of-season meningitis forecasts are produced by forecasting the date beyond which the risks of an outbreak are lower than the climatological risk for the next six weeks.

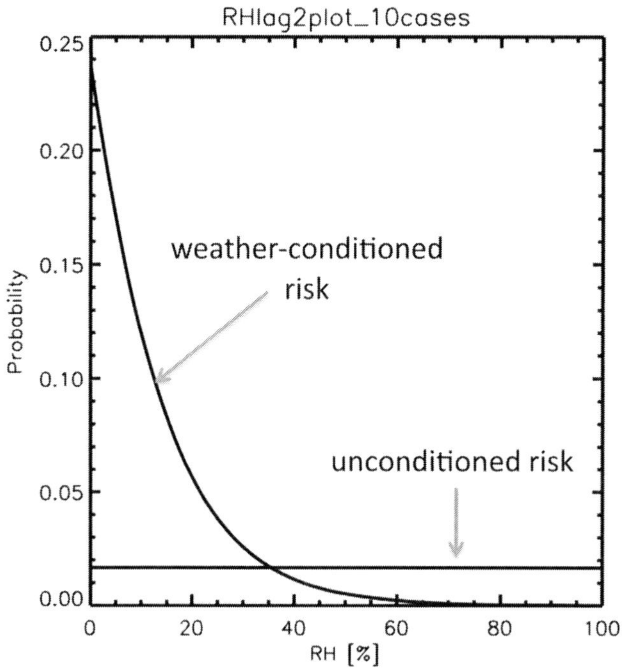

Figure 6 A log-plot of the probability of a district crossing epidemic threshold generated by a differential equation-based model of disease propagation whose coefficients depended on relative humidity and other weather variables (weather-conditioned risk) and ignored all weather variables (unconditioned risk). The line for weather-conditioned risk shows an inflection point at about 40% relative humidity, with the probability of epidemic increasing significantly in districts when the relative humidity is less than 40%. For comparison, the unconditioned risk is about 0.018, which can be understood as the probability of epidemic predicted without knowing the relative humidity. Extracted from Hopson et al. (2014).

At the seasonal scale, a study fitted a range of negative binomial generalized linear models to seasonal meningitis data (number of cases between January and May) in Niger at national and district scales based on monthly climate and dust averages, population density, and early cases prior to the onset of meningitis season in January (Pérez García-Pando et al. 2014). Early cases in December were considered as a proxy for susceptibility and/or carriage prevalence. Climate and dust data were derived from long-term integrations with a regional dust model (Pérez et al. 2011). These models were evaluated using goodness-of-fit statistics and analyzing their ability to detect whether or not a particular incidence-based threshold had been exceeded. At the national scale, the use of both early cases and November-December averaged zonal wind together provided the best fit, explaining about 57% of the year-to-year variability of the seasonal incidence (Figure 7). A national model with early cases and October-December dust concentration performed similarly. To analyze the ability of the model to detect whether or not a particular incidence-based threshold had been exceeded, if the fitted probability of the cross‑validated predicted values y_t exceeding a threshold K was greater than some value c, $0 < c < 1$, then it was predicted that $y_t > K$. At the seasonal scale, it was assumed that $K = 100$ cases per 100,000 population (de Chabalier et al. 2000). Sensitivity and specificity of this model to predict epidemics were 0.8 and 0.87, respectively, with four out of five years correctly predicted to have exceeded the threshold over a 20-year period, whereas two of the remaining fifteen years were incorrectly predicted to exceed it. Figure 7 demonstrates that during the 1994–1996 epidemic wave, years 1994 and 1996 presented anomalously high early cases while year 1995 presented an anomalously intense Harmattan during the early season. The combination of both early cases and early season climate provided a better prediction than any of the predictors alone. Whilst the model underestimated the magnitude of incidence in 1995, the general scale of incidence appeared to be well represented. At the district scale, early-season zonal wind and dust, along with the number of early cases and population density represented about 41% of the spatiotemporal variability with a cross-validated correlation of 0.55. An example of prediction for year 1995 is displayed in Figure 8, including maps of observed incidence and predicted probabilities of exceeding the threshold at district level. If equal importance is given to sensitivity and specificity, the district model achieved values of 0.68 and 0.72 respectively. Should health decision makers be concerned about wasting resources through unnecessary vaccinations, then the model could be adjusted to prioritize maximizing specificity and negative predictive value (NPV) at the expense of sensitivity, resulting in values of 0.27, 0.98, 0.77 and 0.85 for sensitivity, specificity, PPV and NPV respectively. Considering the simplicity of the model, and more importantly, the availability of the data required to fit these models, this type of predictions early in the season could have an operational value.

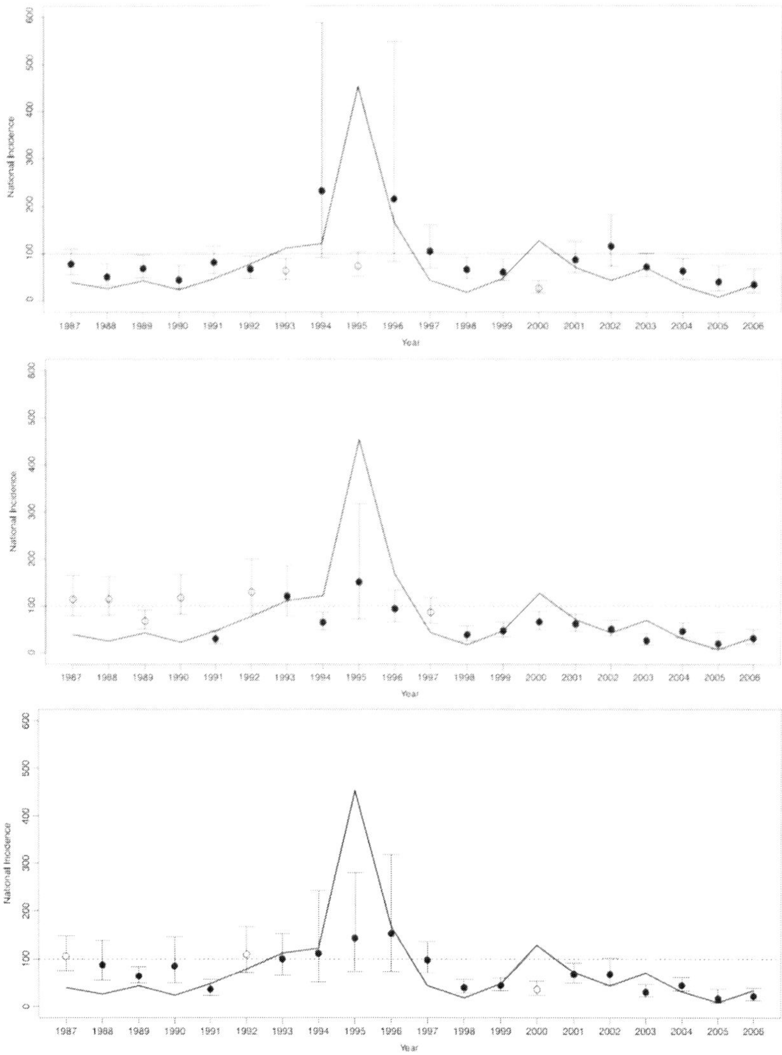

Figure 7 Observed incidence (solid black line) and cross-validated national incidence predictions (circles), with 95% confidence limits obtained by fitting a negative binomial model to the Niger national count data, using and December incidence as predictor (top panel), November-December zonal wind as predictor (middle panel), and December incidence and November-December zonal wind as predictors (bottom panel). Closed circles denote those predictions that were correctly assigned to be either above or below 100 cases per 100,000, whereas open circles are incorrect predictions. The bottom panel was extracted from Pérez García-Pando et al. (2014).

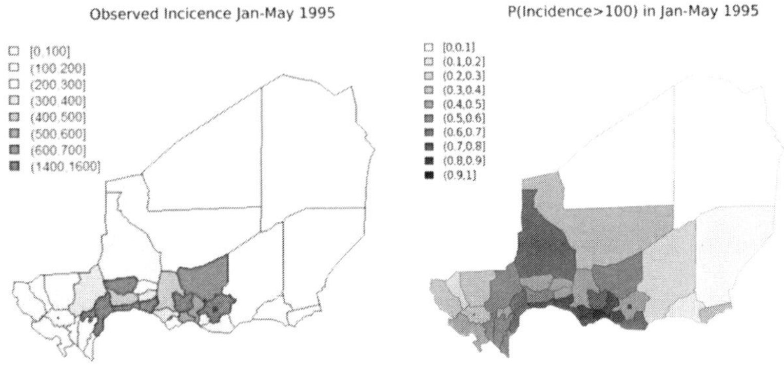

Figure 8 Maps of observed incidence for year 1995 at district level in Niger between January and May (right panel) and predicted probabilities of exceeding the seasonal epidemic threshold from a model including December incidence (national, district-level, and average of neighbors) and climate (national and district-level deviations from the national-level zonal wind and dust concentration) and population density (left panel). The modeling approach and detailed results are available in Pérez García-Pando et al. (2014).

2.11 Practice and future

Model predictions contain uncertainty, which translates into complexity for decision-makers. Another challenge is the introduction of new tools to a diverse operational community in an environment that evolves rapidly as new interventions, technologies and policy options become available. As indicated above, the reactive vaccination strategy is currently being superseded by a preventive vaccination strategy focused on the delivery of the meningitis A conjugate vaccine. Statistical prediction models need to be amended to account for the introduction of the conjugate vaccine and their value needs to be tested and validated in close collaboration with decision-makers.

In order to initiate the process whereby environmental information and predictive models might be used operationally in meningitis epidemic response the MERIT community undertook a pilot outbreak prediction exercise conducted during the 2011–2012 meningitis season in Africa. During this exercise public health professionals from several WHO and Ministry of Health (MoH) offices engaged with modelers, statisticians and climate/weather and environmental scientists to explore outputs from predictive models alongside the 'real time' evolving meningitis situation in four countries: Benin, Chad, Nigeria and Togo. The flow of information was continuous and led to the regular collection and quality control of both epidemiological and climate-related information. The observed and forecasted relative humidity and dust events in the region were also considered in order to analyze the environmental conditions favorable for epidemics. This exercise, which was facilitated by weekly teleconference calls and the sharing of data was also an opportunity to assess the effect of exchange of information between partners at the international and country levels. Learning from the 2011–2012 exercise, the 6th MERIT Technical Meeting, held in Accra, Ghana in November 2012 agreed to a similar exercise to be conducted in 2013 meningitis season (MERIT 2012), the results and conclusions of which will be published in the near future by the relevant participants.

3. CONCLUSIONS

Establishing EWSs for climate related disasters is a particular focus of the 'Global Framework for Climate Services' (Hewitt et al. 2012) and communities concerned with disaster risk reduction and climate change adaptation (Thomson 2013). Developing such EWSs for epidemic diseases (a specific type of disaster) requires the full participation of the health community and an understanding of their norms and practices. *A priori* a strong relationship between climate and environmental drivers must be shown to exist for such information to be considered useful to decision-makers. Research to date suggests that the geographic location, seasonality and year-to-year variability of meningococcal meningitis are associated in part with climatic and environmental factors. Although the determination of specific seasonal climate drivers including dust are difficult to identify, current models that incorporate environmental data, previous incidence, and/or other risk factors have shown sufficient skills to be further amended and tested for operational use. Also models are expected to improve as fine-scale surveillance and other sources of data on risk factors become available in the Meningitis Belt. The cooperation among diverse research and practice communities within MERIT provides a model for facilitating a better management of climate-sensitive diseases and provides guidance on how climate services for health might develop over time.

ACKNOWLEDGEMENTS

The authors thank Yang Liu for technical support and preparation of some figures. CPGP acknowledges the Earth Institute at Columbia University (EI), NOAA and DoE for their support. This work was also supported by the EI Cross-Cutting Initiative project: "Atmospheric aerosol impacts on health in sub-Saharan Africa", and NASA ROSES Earth sciences applications feasibility studies, and a generous grant from the Google.org Foundation. SH is a staff member of the World Health Organization. The author alone is responsible for the views expressed in this publication and they do not necessarily represent the decisions, policy or views of the World Health Organization.

REFERENCES

1. Agier L, Deroubaix A, Martiny N, Yaka P, Djibo A, Broutin H (2013a) Seasonality of meningitis in Africa and climate forcing: aerosols stand out. J R Soc Interface 10:20120814
2. Agier L, Stanton M, Soga G, Diggle PJ (2013b) A multi-state spatio-temporal Markov model for categorized incidence of meningitis in sub-Saharan Africa. Epidemiol Infect 141(8):1764–1771
3. Beresniak A, Bertherat E, Perea W, Soga G, Souley R, Duponte D, Hugonnet S (2012) A Bayesian Network Approach to the Study of Historical Epidemiological Databases: Modelling Meningitis Outbreaks in the Niger. In: Bulletin of the World Health Organization
4. Broutin H, Philippon S, de Magny GC, Courel MF, Sultan B, Guegan JF (2007) Comparative Study of Meningitis Dynamics across Nine African Countries: A Global Perspective. Inter J Health Geogr 6:29
5. Cheesbrough JS, Morse AP, Green SD (1995) Meningococcal meningitis and carriage in western Zaire- A hypoendemic zone related to climate. Epidemiol Infect 114(1):75–92
6. Davila S, Wright VJ, Khor CC, Sim KS, Binder A, Breunis WB, Inwald D, Nadel S, Betts H, Carrol ED, de Groot R, Hermans PW, Hazelzet J, Emonts M, Lim CC, Kuijpers TW, Martinon-Torres F, Salas A, Zenz W, Levin M, Hibberd ML, International Meningococcal Genetics Consortium (2010) Genome-wide association study identifies variants in the CFH region associated with host susceptibility to meningococcal disease. Nat Genet 42(9):772–776
7. de Chabalier F, Hassane A, Chippaux JP (2000) Evaluation of surveillance thresholds for prediction of meningitis epidemics using ongoing surveillance data at the district level, in Niger. Trans R Soc Trop Med Hyg 94(3):251–252
8. Djingarey MH, Noazin MP, Préziosi C, Lingani S, Tiendrebéogo K, Touré K, Viviani K (2008) A 20-year retrospective analysis of epidemic meningitis surveillance data in Burkina Faso, Mali, and Niger. In: Paper presented at the International Pathogenic Neisseria Conference. Rotterdam, The Netherlands. [http://www.meningvax.org/files/Djingarey_IPNC2008_Retrospective_P166.pdf]
9. Drake JM (2005) Fundamental limits to the precision of early warning systems for epidemics of infectious diseases. PLoS Med 2(5):e144
10. Dukić V, Hayden M, Forgor AA, Hopson T, Akweongo P, Hodgson A, Monaghan A, Wiedinmyer C, Yoksas T, Thomson MC, Trzaska S, Pandya R (2012) The role of weather in

meningitis outbreaks in Navrongo, Ghana: A Generalized Additive Modeling Approach. J Agr Biol Envir St 17(3):442–460

11. Durbin J, Koopman SJ (2012) Time Series Analysis by State Space Methods: Second Edition. Oxford University Press, Oxford
12. Fuhrmann C (2010) The Effects of Weather and Climate on the Seasonality of Influenza: What We Know and What We Need to Know. Geography Compass 4:718–730
13. Greenwood BM (1987) The epidemiology of acute bacterial meningitis in tropical Africa. In: Williams JD, Burnie J (eds) Bacterial meningitis. Academic Press, London, pp 61–91
14. Greenwood B (2013) Priorities for research on meningococcal disease and the impact of serogroup A vaccination in the African Meningitis Belt. Vaccine 31(11):1453–1457
15. Greenwood BM, Blakebrough IS, Bradley AK, Wali S, Whittle HC (1984) Meningococcal disease and season in Sub-saharan Africa. Lancet 1:1339–1342
16. Griffiss JM, Brandt BL, Jarvis GA (1987) Natural immunity to Neisseria meningitidis. In: Vedros NA (ed) Evolution of meningococcal disease. CRC Press, Boca Raton, FL, pp 99–119
17. Harrison LH, Trotter CL, Ramsay ME (2009) Global epidemiology of meningococcal disease. Vaccine 27(Suppl 2):B51–B63
18. Hayden MH, Dalaba M, Awine T, Akweongo P, Nyaaba G, Anaseba D, Pelzman J, Hodgson A, Pandya R (2013) Knowledge, Attitudes and Practices Related to Meningitis in Northern Ghana. Amer J Trop Med Hyg 89:265–270
19. Hewitt C, Mason S, Walland D (2012) The Global Framework for Climate Services. Nature Clim Change 2:831–832
20. Hodgson A, Smith T, Gagneux S, Adjuik M, Pluschke G, Mensah NK, Binka F, Genton B (2001) Risk factors for meningococcal meningitis in northern Ghana. Trans R Soc Trop Med Hyg 95(5):477–480
21. Hopson TM, Pandya R, Yoksas T, Dumont A, Hugonnet S, Cibrelus L, Lingani C, Dukic V, Hayden MH, Boehnert J, Monaghan AJ, Nakazawa T (2014) More effective meningitis vaccination campaigns using weather information over Africa. In: Fifth Conference on Environment and Health. 94th Annual Meeting of the American Meteorological Society, Atlanta, GA, [https://ams.confex. com/ams/94Annual/webprogram/Paper241720.html
22. Irving TJ, Blyuss KB, Colijn C, Trotter CL (2011) Modelling meningococcal meningitis in the African Meningitis Belt. Epidemiol Infect 140:897–905
23. Jandarov R, Haran M, Ferrari M (2012) A compartmental model for meningitis: separating transmission from climate effects on disease incidence. J Agr Biol Envir St 17(3):395–416
24. Kalnay E, Kanamitsu M, Kistler R, Collins W, Deaven D, Gandin L, Iredell M, Saha S, White G, Woolen J, Zhu Y, Chelliah M, Ebisuzaki W, Higgins W, Janowiak J, Mo KC, Ropelewski C, Wang J, Leetmaa A, Reynolds R, Jenne R, Joseph D (1996) The NCEP/NCAR 40-Year Renalysis Project. Bull Am Meteorol Soc 77:437–471
25. LaForce FM, Konde K, Viviani S, Préziosi MP (2007) The Meningitis Vaccine Project. Vaccine 3(25):A97–A100
26. Lapeyssonnie L (1963) La méningite cérébro-spinale en Afrique. Bull World Health Organ 28(Suppl 1):3–114 (in French)
27. Macal C, North M, Collier N, Dukić V, Lauderdale S, David M, Schumm P, Daum R, Evans J, Wilder J, Wegener D (2012) Modeling the Spread of Community- Associated MRSA. In: Laroque C, Himmelspach J, Pasupathy R, Rose O, Uhrmacher AM (eds) the 2012 Winter Simulation Conference. [http://www.informs-sim.org/wsc12papers/includes/ files/inv214.pdf]
28. Martiny N, Chiapello I (2013) Assessments for the impact of mineral dust on the meningitis incidence in West Africa. Atmos Environ 70:245–253
29. MERIT (2012) 6th MERIT Technical Meeting Report. In: the MERIT 6th Technical Meeting. Health and Climate Foundation, Geneva [http://merit.hc-foundation.org/documents/6thMeritWorkshop/6thMERITTechnicalMeetingReport.pdf]
30. Molesworth AM, Thomson MC, Connor SJ, Cresswell MC, Morse AP, Shears P, Hart CA, Cuevas LE (2002) Where is the Meningitis Belt? Defining an area at risk of epidemic meningitis in Africa. Trans R Soc Trop Med Hyg 96:242–249

31. Molesworth AM, Cuevas LE, Stephen JC, Andrew PM, Thomson MC (2003) Environmental risk and meningitis epidemics in Africa. Emerg Infect Dis 9(10):1287–1293
32. Moore PS (1992) Meningococcal meningitis in sub-Saharan Africa: a model for the epidemic process. Clin Infect Dis 14:515–525
33. Moore PS, Reeves MW, Schwartz B, Gellin BG, Broome CV (1989) Inter-continental spread of an epidemic group A Neisseria meningitidis strain. Lancet 2:260–263
34. Mueller JE, Gessner BD (2010) A hypothetical explanatory model for meningococcal meningitis in the African Meningitis Belt. Int J Infect Dis 14(7):e553–e559
35. Mueller JE, Yaro S, Madec Y, Somda PK, Idohou RS, Njanpop Lafourcade BM, Drabo A, Tarnagda Z, Sangaré L, Traoré Y, Fontanet A, Gessner BD (2008) The association of respiratory tract infection symptoms and air humidity with meningococcal carriage in Burkina Faso. Trop Med Int Health 13:1543–1552
36. Paireau J, Girond F, Collard JM, Mainassara HB, Jusot JF (2012) Analysing spatio-temporal clustering of meningococcal meningitis outbreaks in Niger reveals opportunities for improved disease control. PLoS Negl Trop Dis 6(3):e1577
37. Palmgren H (2009) Meningococcal disease and climate. Glob Health Action 2
38. Pascual M, Dobson A (2004) Seasonal patterns of infectious diseases. PLoS Med 2(1):e5
Pérez García-Pando C, Stanton MC, Diggle PJ, Trzaska S, Miller RL, Perlwitz JP,
39. Baldasano JM, Cuevas E, Ceccato P, Yaka P, Thomson MC (2014) Soil dust aerosols and wind as predictors of seasonal meningitis incidence in Niger. Environ Health Perspect, doi:10.1289/ehp.1306640. [http://ehp.niehs.nih.gov/1306640/]
40. Pérez C, Haustein K, Janjic Z, Jorba O, Huneeus N, Baldasano JM, Black T, Basart S, Nickovic S, Miller RL, Perlwitz JP, Schulz M, Thomson MC (2011) Atmospheric dust modeling from meso to global scales with the online NMMB/BSC-Dust model - Part 1: Model description, annual simulations and evaluation. Atmos Chem Phys 11:13001–13027
41. Roberts L (2008) An Ill Wind, Bringing Meningitis. Science 320:1710–1715
42. Seefeldt MW, Hopson TM, Warner TT (2012) A Characterization of the Variation in Relative Humidity across West Africa during the Dry Season. J Appl Meteorol Climatol 51(12):2077–2089
43. Stanton MC, Diggle PJ, Pérez C, Trzaska S, Thomson MC (2011) Exploring the relationship between meningitis incidence and climatic factors is Niger, West Africa. In: the 17th European Young Statisticians Meeting
44. Stanton MC, Agier L, Taylor BM, Diggle PJ (2013) Towards reak-time spatio-temporal prediction of meningitis incidence in sub-Saharan Africa. J R Stat Soc, in press
45. Sultan B, Labadi K, Guegan JF, Janicot S (2005) Climate drives the meningitis epidemics onset in West Africa. PLoS Med 2:e6
46. Tall H, Hugonnet S, Donnen P, Dramaix-Wilmet M, Kambou L, Drabo F, Mueller JE (2012) Definition and characterization of localised meningitis epidemics in Burkina Faso: a longitudinal retrospective study. BMC Infect Dis 12:2
47. Thomson MC (2013) Climate change and disaster risk management: challenge and opportunities. In: Palutikof J, Karoly D, Boulter S (eds) Natural Disasters and Adaptation to Climate Change. Cambridge University Press, Cambridge UK, pp 8–32
48. Thomson MC, Molesworth AM, Djingarey MH, Yameogo KR, Belanger F, Cuevas LE (2006) Potential of environmental models to predict meningitis epidemics in Africa. Trop Med Int Health 11:781–788
49. Thomson MC, Jeanne I, Djingarey M (2009) Dust and epidemic meningitis in the sahel: A public health and operational research perspective. IOP Conference Series: Earth and Environmental Science 7(1):012017
50. Thomson MC, Firth E, Jancloes M, Mihretie A, Onoda M, Nickovic S, Broutin H, Sow S, Perea W, Bertherat E, Hugonnet S (2013) A climate and health partnership to inform the prevention and control of meningococcal meningitis in sub-Saharan Africa: The MERIT initiative. In: Hurrell J, Asrar G (eds) Priorities in Climate Research, Analysis and Prediction. Springer, pp 459–484
51. Trotter CL, Greenwood BM (2007) Meningococcal carriage in the African Meningitis Belt. Lancet Infect Dis 7(12):797–803

52. van Deuren M, Brandtzaeg P, van der Meer JW (2000) Update on meningococcal disease with emphasis on pathogenesis and clinical management. Clin Microbiol Rev 13(1):144–166
53. Waddy BB (1952) Climate and respiratory infections. Lancet 2:674–677
54. Wenzel RP, Davies JA, Mitzel JR, Beam WE Jr (1973) Nonusefulness of meningococcal carriage rates [letter]. Lancet 2:205
55. West M, Harrison J (1997) Bayesian Forecasting and Dynamic Models (2nd Ed.). Springer-Verlag New York, Inc.,New York, NY, USA.
56. World Health Organization (1998) Control of epidemic meningococcal disease. WHO practical guidelines. 2nd edition. WHO/EMC/BAC/98.3. [http://www.who.int/csr/resources/publications/meningitis/WHO_EMC_BAC_98_3_EN/en/]
57. World Health Organization (2000) Detecting meningococcal meningitis epidemics in highly-endemic African countries. Wkly Epidemiol Rec 75:306–309
58. World Health Organization (2003) The use of polysaccharide trivalent acw vaccine for the control of epidemic meningococcal disease outbreaks in countries of the african Meningitis Belt. Technical. In: Report WHO/CDS/CSR/GAR. [http://www.who.int/csr/resources/publications/meningitis/WHO_CDS_CSR_GAR_2003_14/en/]
59. World Health Organization (2004) Using climate to predict infectious disease outbreaks: A review. WHO/SDE/OEH. [http://www.who.int/globalchange/ publications/ oeh0401/en/]
60. Yaka P, Sultan B, Broutin H, Janicot S, Philippon S, Fourquet N (2008) Relationships between climate and year-to-year variability in meningitis outbreaks: a case study in Burkina Faso and Niger. Int J Health Geogr 7:34

CHAPTER 13

Space environment data acquisition monitor onboard Himawari-8 for space environment monitoring on the Japanese meridian of geostationary orbit

Tsutomu Nagatsuma Kaori Sakaguchi, Yûki Kubo, Piet Belgraver, Frédéric Chastellain, Reto Muff and Takeshi Otomo

AUTHORS' CONTRIBUTIONS

TN performed the initial analysis of SEDA data and the cross-comparison between SEDA and other satellite data. KS carried out the quality check of the high-energy electron data. YK carried out the quality check of the high-energy proton data and discussed about the cause of contamination. PB, FC, and RM manufactured and tested the SEDA instrument. TO constructed the data acquisition system of Himawari-8/SEDA.

ABSTRACT

A space environment data acquisition monitor (SEDA) has been flown onboard the Japanese meteorological satellites 'Himawari-8/9' as one of the housekeeping information monitors for satellite operation. SEDA consists of a high-energy proton sensor (SEDA-p) and a high-energy electron sensor (SEDA-e). These instruments provide near-real-time information on the conditions of the space environment on the Japanese meridian. Initial cross-comparison between SEDA and high-energy particle observation by geostationary operational environmental satellite (GOES) shows that the flux level of SEDA-e is slightly smaller than that of GOES observation, and the flux level of SEDA-p observation is a half that of GOES observation.

Keywords: Space weather, High-energy electron, High-energy proton, Geostationary, orbit Cross-comparison, Space environment monitoring

1. BACKGROUND

The geostationary earth orbit (GEO) is one of the practical orbits of satellites providing our social infrastructure. More than 400 telecommunications, broadcasting, and meteorological satellites are currently operated in this orbit (e.g., http://www.satsig.net/sslist.htm Note: Military satellites in GEO are not included in this list). GEO is located at the outer edge of the outer radiation belt. Relativistic electrons in the outer radiation belt undergo dynamic changes because of the magnetospheric disturbances driven by solar wind. In addition, GEO satellites frequently experience the injection of energetic particles related to substorm activities. Variations in the charged particle environment are one of the causes of satellite anomalies (e.g., Lanzerotti et al. 1998; Baker 2000). For these reasons, the monitoring and forecasting of the space environment in GEO is one of the key issues of space weather research and operation.

The monitoring of the high-energy particle environment in the Japanese meridian of GEO started at the beginning of Japan Meteorological Agency (JMA)'s satellite program. Space environment monitors (SEM) onboard Japan's geostationary meteorological satellite (GMS) series 'Himawari' have been started with GMS-1 in 1978 (Kurino 1985). Space environment monitoring by SEM continued up to GMS-4. However, SEM was not installed onboard the satellites that followed GMS-4 (GMS-5, MTSAT-1R, MTSAT-2). Thus, there was a large gap in space environment monitoring in the Japanese meridian of GEO from the mid-1990s to the mid-2010s.

The situation changed because a space environment data acquisition monitor (SEDA) was installed in Himawari-8/9 (Bessho et al. 2016). This action has also contributed to the World Meteorological Organization (WMO)'s recommendation of installing a space environment monitor in all meteorological satellites for housekeeping information and space environment monitoring (e.g., WMO 2013). Himawari-8 and 9 were launched on October 7, 2014, and November 2, 2016, respectively. SEDA onboard Himawari-8 and 9 has been in operation since November 3, 2014, and December 20, 2016, respectively.

The National Institute of Information and Communications Technology (NICT) has a long history of operational space weather forecasting in Japan (Nagatsuma 2013). Space environment monitoring is one of the key elements in space weather forecasting. Since 1979, we have received a daily summary plot of SEM from JMA, and a real-time data acquisition system for GMS-4/SEM has been in operation since 1994 (Den et al. 1999). On the basis of this background, we are now handling and processing Himawari-8/SEDA data for space environment monitoring in the Japanese meridian of GEO. In this paper, we briefly introduce the instrumentation of SEDA, the initial observations, and the results of our initial cross-comparison with geostationary operational environmental satellite (GOES) data.

2. INSTRUMENTATION

SEDA has been flown onboard Himawari-8 and 9 to monitor the space environment around the satellite. The basic specifications of SEDA are shown in Table 1. Figure 1 shows the fully assembled SEDA. This sensor is identical to the Environmental Monitor Unit (EMU) for the European Galileo satellite series (Bartolomé et al. 2015). SEDA consists of two sensors, namely a high-energy proton sensor (SEDA-p) and a high-energy electron sensor (SEDA-e). SEDA-p and SEDA-e have eight channels each, the energies of which are shown in Table 2. The large cylinder on the right side of the SEDA assembly in Fig. 1 is SEDA-e, whereas SEDA-p comprises the eight small cylinders on the left side in Fig. 1. The intermediate-size cylinder on the left side has no use.

Table 1 Specifications of SEDA

Items	Description
Number of channels	Protons: 8 (8 individual sensor elements)
	Electrons: 8 (8 stacked plates in one element)
Energy range	Protons: 15–100 MeV
	Electrons: 0.2–4.5 MeV

Time resolution	10 s
Field of view	Protons: ±39.35°
	Electrons: ±78.3°

Figure. 1 Fully assembled SEDA

Table 2 Energies of proton and electron channels

Ch. no.	Proton energy (MeV)	Electron energy (MeV)
0	21.6	0.2
1	29.9	0.3
2	37.9	0.45
3	45.4	0.65
4	57.8	1.0
5	68.4	1.5
6	75.2	2.0
7	81.4	4.5

SEDA obtains data every 10 s based on the clock in the SEDA assembly. Because of the time difference between the clock in the main satellite system and that in SEDA, the measurement timing drifts by about one second per day. The sampling rate sometimes shifts by 9 or 11 s owing to the jitter of the clock in SEDA.

SEDA is mounted on the eastward plane on the body of Himawari-8 and 9. Owing to the gyro motion of charged particles, SEDA can measure protons whose gyro centers lie inside GEO and electrons whose gyro centers lie outside GEO (Rodriguez et al. 2010). The geographical longitude and local noon of Himawari-8 and 9 are 140.7° and 02:37 UT, respectively.

Real-time data observed by Himawari/SEDA are downlinked to JMA's Meteorological Satellite Center, where they are routinely stored on file servers. Every 10 min, we download the latest files of Himawari/SEDA data from the server to monitor the space environment at the Japanese meridian of GEO.

2.1 SEDA high-energy electron sensor (SEDA-e)

SEDA-e is not a direct-particle-counting-type sensor but a sensor for measuring the internal charging current deposited by the energetic electron flux at various shielding levels. Because the penetrating energy of electrons depends on the material thickness, the electrons which have a specific range of energy can be accumulated using a shield and a collector plate. SEDA-e consists of eight metal plates mounted on a stack for measuring eight different energy ranges from 0.2 to 4.5 MeV. There is no additional shielding between each collector plate.

The SEDA-e sensor is based on the SURF sensor and is also used as an internal charging current monitor. The SURF sensor was first demonstrated in space by the Space Technology Research Vehicle (STRV) 1D in geostationary transfer orbit in 2000 (Ryden et al. 2001). Furthermore, the SURF sensor was installed as part of the Merlin space weather monitors onboard GIOVE A, a test bed for Galileo launched in 2005. The SURF sensor has completed 8.5 years in orbit despite being originally intended to have a two-year operational life (Ryden et al. 2015). One of the advantages of this sensor is its robustness against high-energy proton contamination compared with a semiconductor-type sensor. Because of the weak current deposited from the high-energy electron flux, the signals are amplified by an electrometer with an input bias current, which is dependent on the sensor's temperature. To estimate the flux of energetic electrons, the effect of the bias current is removed using empirical equations with temperature information of the sensor.

2.2 SEDA high-energy proton sensor (SEDA-p)

SEDA-p measures the proton flux in eight energy channels from 15 to 100 MeV via individual telescopes. Each telescope has a detector diode on the top and bottom. All energy deposition events above a certain threshold (i.e., non-coincident detection) are counted by the top detector. The coincidences for particles arriving within the detector opening angle are selected by the bottom detector. A proton that is detected simultaneously by both the top and bottom detectors is counted as coincident.

2.3 Observation of high-energy particle environment by SEDA

Figures 2 and 3 show examples of high-energy particle observations by Himawari-8/SEDA on March 15–20, 2015, and October 27–31, 2015, respectively. The 5-min running mean of SEDA-e electron differential fluxes and SEDA-p proton differential fluxes in the eight channels and the provisional Dst index are plotted from the top to bottom panels in these figures.

Figure 2 shows that a strong geomagnetic storm occurred from March 17 to 20. The minimum Dst is −223 nT at 23 UT on March 17. The daily variations of the electron fluxes from Channels (Chs.) 0–5 have a peak at around 02–03 UT, which is near the local noon of Himawari-8. This signature is consistent with previous high-energy particle observations by other GEO satellites. In contrast, the electron fluxes of Chs. 6 and 7 have different phases of daily variations, which have a peak at around 14–15 UT. However, on March 19 and 20, the electron flux variations of Ch. 6 tend to correspond to those of the lower channels. Short-term variations that are related to substorm activities during the storm, and flux enhancement after the storm can be clearly seen.

Weak flux enhancements of high-energy protons are observed in Ch. 0 from March 15 to 17 and in Chs. 1–3 on Mar. 15. The proton fluxes from Chs. 4–7 show no significant variations from March 15 to 17. An apparent enhancement of the proton flux from Chs. 2 to 7 is also observed from March 18 to 20. One of the surprising characteristics of this enhancement is that there appears to be spectral inversion, with the higher channels showing larger flux as well as possibly larger flux variations. The trend of this enhancement is almost coherent for each channel and appears to be related to the variations of high-energy electron fluxes. The relationship between Ch. 7 of SEDA-p and Ch. 6 of SEDA-e is shown in Fig. 4. The flux enhancement of SEDA-p Ch. 7 corresponds well to that of SEDA-e Ch. 6.

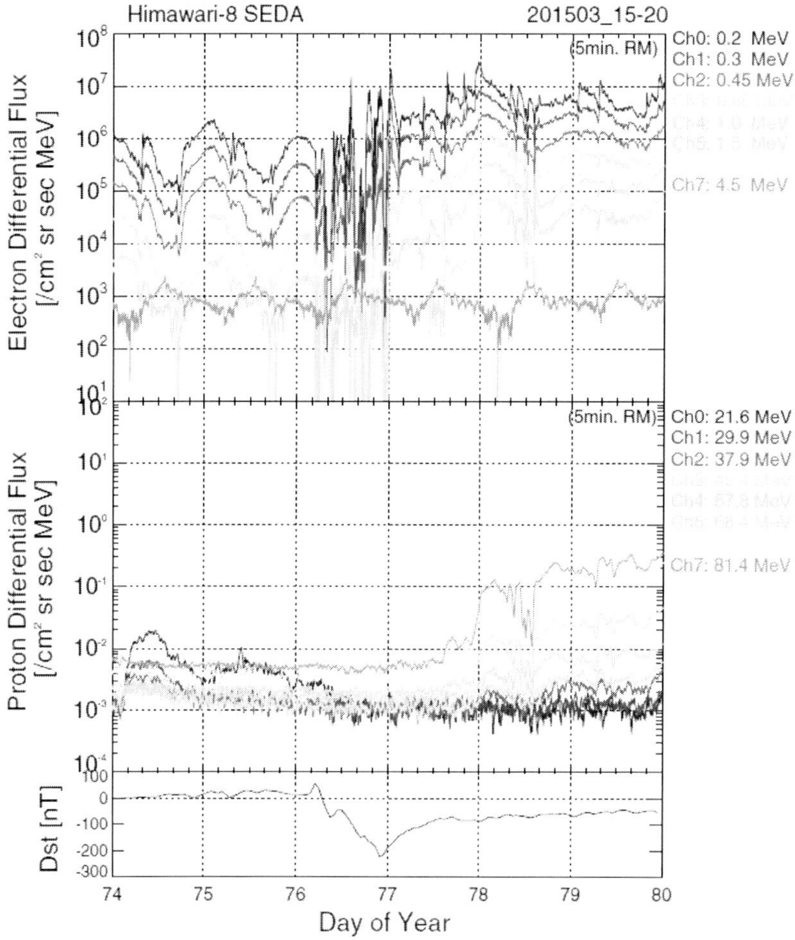

Figure. 2 Himawari-8/SEDA observations on March 15–20, 2015. The 5-min running mean of SEDA-e and SEDA-p differential fluxes in eight channels are plotted in the *top* and *second panels*, respectively. The Dst index is plotted in the *bottom panel*

An example of a minor proton event that occurred on October 29, 2015, is shown in Fig. 3. The higher proton channels show higher flux levels on October 27 and 28. In this event, the proton flux shows a two-step enhancement. Time profile of proton flux variations can be distinguished in two groups. The first rapid enhancement of the proton flux from Chs. 0–7 occurred at around 03:10 UT with energy dispersion. Higher-energy channels tend to show earlier enhancement. The second enhancement of the proton flux from Chs. 0–4 occurred at around 07:10 UT. The proton flux of the lower-energy channels shows a significant increase during the second enhancement. However, the second enhancement shows no clear energy dispersion. The reason of the different time profile of these two enhancements is not clear. It might be caused by the characteristic of response function for each channels or natural phenomena. The electron flux variation does not show any correspondence with the proton flux variation during this minor proton event.

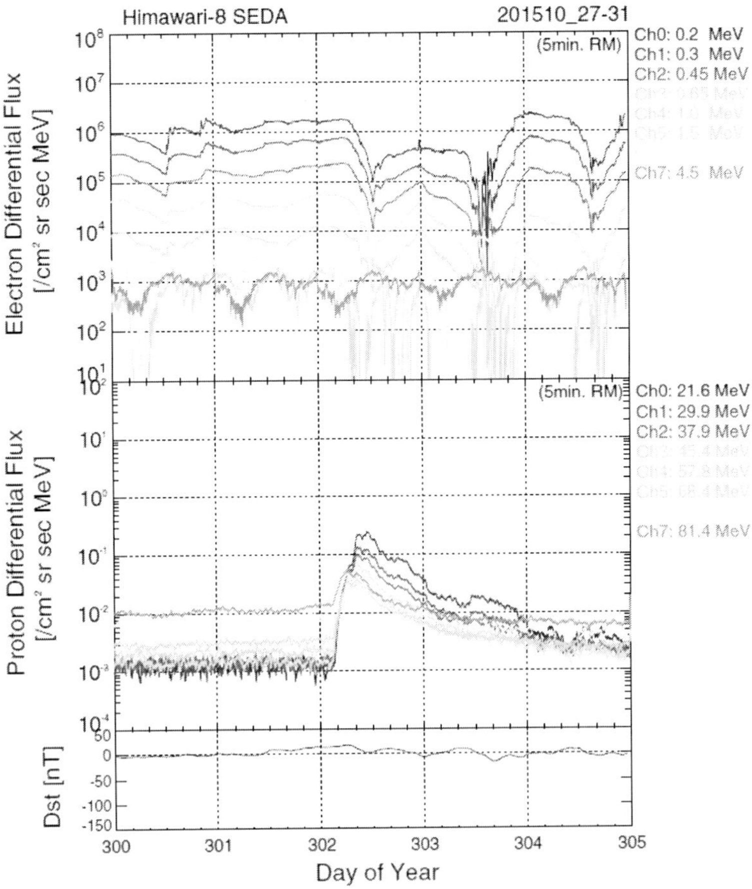

Figure. 3 Himawari-8/SEDA observations on October 27–31, 2015. The format of the plot is the same as that in Fig. 2

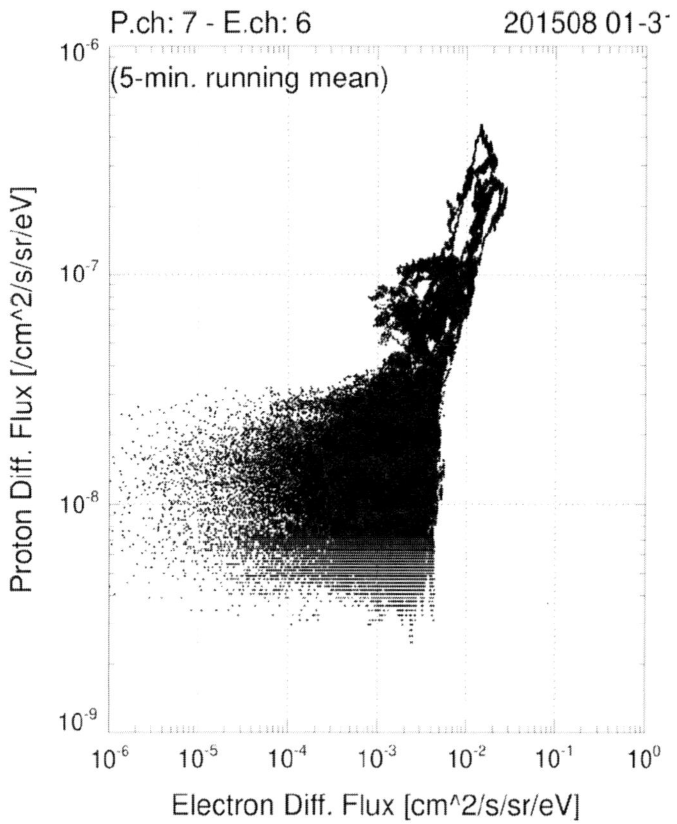

Figure. 4 Scatter plot showing relationship between proton flux (Ch. 7: 81.4 MeV) and electron flux (Ch. 6: 2.0 MeV)

2.4 Quality check of SEDA data

To examine the quality of SEDA data, it is important to compare the data with high-energy particle data observed by other GEO satellites. The differences in the *L*-value among the GEO satellites need to be taken into account for the cross-comparison of particle data, because the magnetic dipole axis is not aligned with the rotational axis of Earth. The *L*-value of each GEO satellite also changes with geomagnetic disturbances and depends on the local time. Figure 5 shows Roederer's *L* (L^*)-value of GEO estimated from the Tsyganenko 89 (T89) model under quiet conditions at 00 UT on Jan. 01, 2015 (Tsyganenko 1989), where L^* is the magnetic drift invariant proposed by Roederer (1970). In Fig. 5, the L^*-values of Himawari-8, GOES 15, and GOES 13 are 6.03, 5.94, and 6.39, respectively.

This suggests that the drift shell of GOES 15 is close to that of Himawari-8. We use high-energy particle data obtained from the Magnetospheric Electron Detector (MAGED) (Hanser 2011) and the Energetic Proton, Electron and Alpha Detector (EPEAD) (Onsager et al. 1996; Hanser 2011) onboard GOES 15 in our data analysis. The energy channels of MAGED and EPEAD are shown in Tables 3 and 4, respectively. Note that the high-energy electron channels of EPEAD measure the integral flux, whereas SEDA-e measures the differential flux of high-energy electrons.

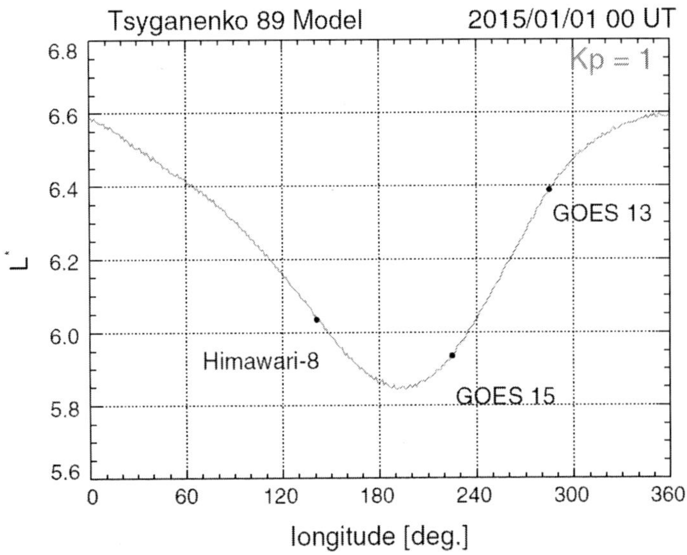

Figure. 5 L^*-value at GEO as a function of geographic longitude at 00 UT on January 01, 2015 for Kp = 1. The L^*-values of Himawari-8, GOES 13, and GOES 15 are plotted as *solid black circles*

Table 3 Energy range of GOES MAGED

Ch. no.	Energy range (keV)
1	30–50 (40)
2	50–100 (75)
3	100–200 (150)
4	200–350 (275)
5	350–500 (475)

Table 4 Energy range of GOES EPEAD

Ch. no	Proton energy (MeV)	Electron energy (MeV)
1	0.7–4	>0.8
2	4–9	>2.0
3	9–15	>4.0
4	15–40	
5	38–82	
6	84–200	
7	110–900	

2.5 Estimating differential flux from observed integral fluxes

Because EPEAD measures the integral flux of high-energy electrons, we need to estimate the differential flux from integral fluxes. Onsager et al. (2004) introduced a method of estimating the phase space density with a 90° pitch angle at the magnetic equator by assuming the exponential distribution of the phase space density and the pitch angle distribution at the equator using two channels for integral flux observation by EPEAD. We adopt the method of Onsager et al. (2004) for our data analysis. In his method, the distribution of the phase space density is expressed as the following exponential equation:

$$F(E) = f_0 \exp(-E/E0) \quad (1)$$

Equation (1) can effectively represent the energetic electron spectra observed at GEO (e.g., Cayton et al. 1989). On the basis of this assumption, the relationship between the phase space density *f*(*E*) and the observed integral flux *J*(>*E*) is written as

$$f(E) = c^2 J(>E) / \Big(\left(E_0^2 + EE_0\right) 2m_e c^2 + 2E_0^3 + 2EE_0^2 + E^2 E_0 \Big) \quad (2)$$

where c is the speed of light and m_e is the rest mass of an electron.

Using the integral flux observed in two channels (>0.8 and >2.0 MeV) from EPEAD, we can estimate f_0 and E_0. The differential flux $j(E)$ is calculated from the following equation using the estimated f_0 and E_0:

$$j(E) = p^2 f(E) = ((E/c)^2 + 2m_e c^2 E) f_0 \exp(-E/E_0), \quad (3)$$

where p is the particle momentum. Using this method, differential fluxes of 1.0, 1.5, and 2.0 MeV are estimated from EPEAD onboard GOES 15. Gannon et al. (2012) already used the same technique to estimate differential flux from GOES EPEAD data.

2.6 Cross-comparison between Himawari-8/SEDA, GOES 15/MAGED, and GOES 15/EPEAD

Figure 6 shows scatter plots revealing the relationship between Himawari-8/SEDA-e and GOES 15/MAGED (left) and that between Himawari-8/SEDA-e and GOES 15/EPEAD (right). In this analysis, we only use one-hour-averaged flux at each noon local time data for comparison. Although the geographical locations of Himawari-8 and GOES 15 on GEO are almost opposite each other, the correlation between the two particle sensors shows reasonably good agreement. The flux level of SEDA-e is slightly smaller than that of GOES observation.

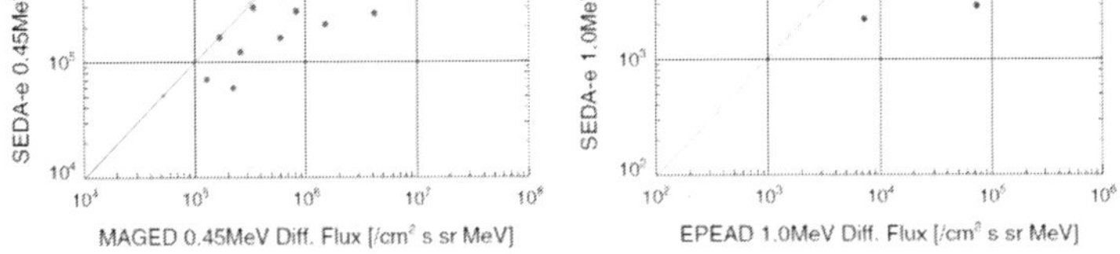

Figure. 6 A Scatter plot showing relationship between electron flux at 0.45 MeV estimated from SEDA-e and that estimated from MAGED. **B** Scatter plot showing relationship between electron flux at 1.0 MeV estimated from SEDA-e and that estimated from EPEADFigure 7 shows the combined energy spectra of electrons obtained from Himawari-8/SEDA, GOES 15/MAGED, and GOES 15/EPEAD on March 9, 2015. The energy spectra obtained from both satellites show good agreement during this period, although the flux level of Himawari-8/SEDA is slightly lower than that of GOES 15/EPEAD and MAGED.

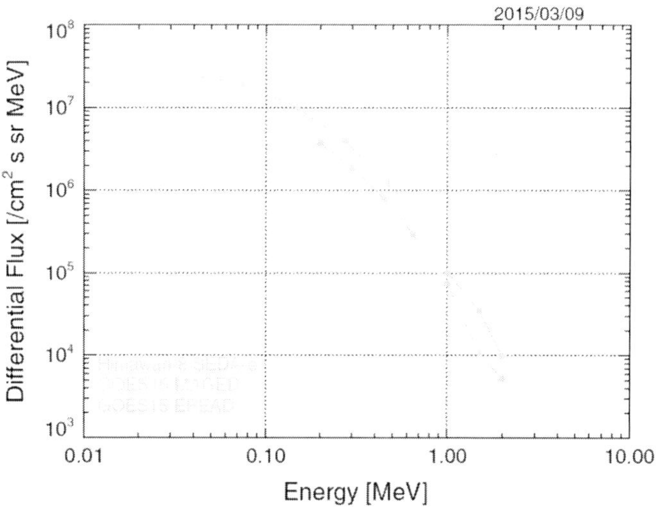

Figure. 7 Combined electron differential flux spectra from Himawari-8/SEDA (*orange*), GOES-15/MAGED (*light blue*), and GOES-15/EPEAD (*green*) on March 9, 2015

Few proton events have occurred since the Himawari-8/SEDA started operation. Thus, the proton fluxes of SEDA and EPEAD cannot be statistically examined and compared. Figure 8 shows the combined plot of proton energy spectra from Himawari-8/SEDA (red) and GOES 13/EPEAD (black) at 10:45 UT on Oct. 29, 2015. Sandberg et al. (2014) have examined the effective energies of GOES solar proton detectors using corrected interplanetary monitoring platform (IMP) 8 solar proton measurements. As shown by Rodriguez et al. (2017), agreement is greatly improved between GOES and STEREO when using the effective energies estimated by Sandberg et al. (2014). So, we also apply this effective energy of EPEAD in our analysis. Green line with filled circles shows differential flux of GOES 13/EPEAD channels with effective energies derived from Sandberg et al. (2014). Although Sandberg et al. (2014) estimated effective energies of GOES 11/EPEAD, that information can be used as effective energies of GOES 13/EPEAD with reasonable accuracy based on the inter calibration study done by Rodriguez et al. (2014). To determine the energy spectra, 5-min-averaged data with the background flux level caused by galactic cosmic rays subtracted are used. The energy spectrum of SEDA-p shows a power law distribution, except for Chs. 6 (75.2 MeV) and 7 (81.4 MeV). The flux enhancement seen in Chs. 6 and 7 appears to show contamination with high-energy electron flux, since flux of high-energy proton and that of high-energy electron has some correlation shown in Fig. 4. The energy spectrum of EPEAD shows a power-law distribution with a bump at around 30 MeV.

The proton flux of GOES 13/EPEAD with effective energy is about two times higher than that of Himawari-8/SEDA-p. However, the slope of the power law is almost the same for those energy spectra.

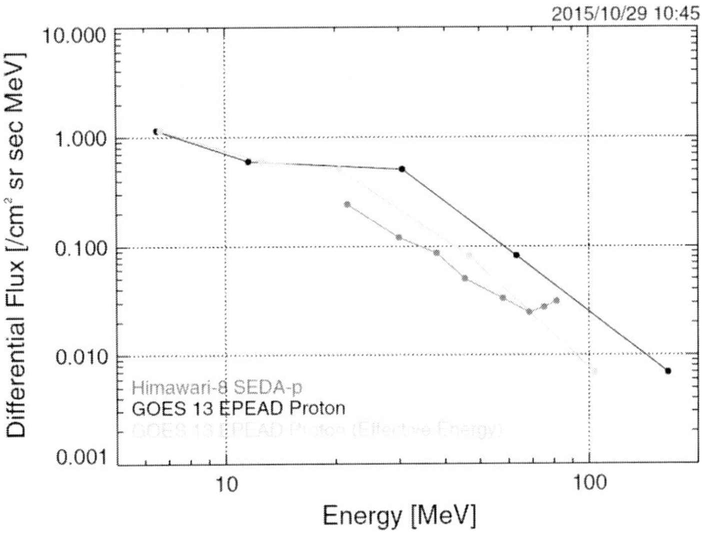

Figure. 8 Combined proton differential flux spectra from Himawari-8/SEDA (*red*), GOES-13/EPEAD (*black*) and GOES-13/EPEAD with effective energy (*green*) at 10:45 UT on Oct. 29, 2015

3. SUMMARY AND DISCUSSION

As we mentioned earlier, this instrument was originally developed as the EMU onboard the European Galileo satellite. The specifications of this instrument are optimized for observations on medium earth orbit, which is the orbit of Galileo. Thus, the sensitivity of the sensor is slightly low to cover the dynamic range of high-energy particle observations at GEO, especially for Chs. 6 and 7 of SEDA-e.

The trend of the daily variation seen in Chs. 6 and 7 of SEDA-e appears to be residuals of the bias current subtraction, because the daily variation of the temperature of SEDA is in phase (data not shown). The daily and long-term trends of the electron flux variations in Ch. 7 and their relation with the sensor temperature trend suggest that the effect of the bias current still remains. The subtraction of the bias current effect from the raw data of SEDA-e will be improved by detailed analysis of the relationship between the electron flux and the sensor temperature.

One of the advantages of SEDA-e is that high-energy proton flux contamination is negligible. Thus, high-energy electron flux variations during a proton event can be captured by SEDA-e. In contrast, contamination by high-energy electron flux is apparent in most of the SEDA-p channels. It is well known that detector diode has a sensitivity of both high-energy electrons and protons [e.g., Fig. 2 of Takagi et al. (1993)]. However, SEDA-p cannot discriminate high-energy protons and electrons on board because of simple instrument for monitoring purposes. Removing the electron flux contamination based on data processing on the ground is not easy because the high-energy electron channels (Chs. 6 and 7) can only measure the high-flux period. Another future task is to improve the data quality of SEDA-p. Problems in the removal of electron contamination from proton flux data and the improvement of bias current subtraction from electron flux data will be solved in the near future.

The cross-comparisons between differential flux of Himawari-8/SEDA-e and that of GOES 15/MAGED in Fig. 6 show good correspondence. However, each data point also shows some degree of scattering. It seems that the time difference of each differential flux at noon local time produces the difference. So, we should select the period when Himawari-8 and GOES 15 are in the same L^* at the same time for cross-calibration.

Himawari-8/SEDA data are useful for monitoring the high-energy particle environment around the Japanese meridian of GEO. Near real-time plot and archived data of SEDA are available from our web site (http://seg-web.nict.go.jp/himawari-seda/). Furthermore, combining Himawari-8/SEDA data with other satellite data in GEO (e.g., GOES 13, 15, and Kodama (JAXA's Data Relay Test Satellite)) will enable us to estimate the particle distribution in the whole area of GEO. Developing a method for cross-calibration of the particle sensor will be one of the important tasks for reconstructing the particle distributions around GEO and the inner magnetosphere.

ACKNOWLEDGEMENTS

First of all, we would like to thank the Satellite Program Division and Meteorological Satellite Center of Japan Meteorological Agency for data from SEDA onboard Himawari-8/9. MAGED and EPEAD onboard GOES are operated by NESDIS/NOAA. The data of MAGED and EPEAD were provided by NCEI/NOAA and NSSDC/NASA. A provisional Dst index was provided by World Data Center for Geomagnetism, Kyoto. To calculate Roederer's L-value, the International Radiation Belt Environment Modeling library (https://craterre.onecert.fr/prbem/irbem/description.html) was used.

Part of this work was supported by JSPS KAKENHI Grant Number JP15H05813.

COMPETING INTERESTS

The authors declare that they have no competing interests.

REFERENCES

1. Baker DN (2000) The occurrence of operational anomalies in spacecraft and their relationship to space weather. IEEE Trans Plasma Sci 28:2007–2016
2. Bartolomé JP, Maufroid X, Hernández IF, Salcedo JAL, Granados GS (2015) Chapter 2 Overview of Galileo system. In: Nurmi J, Lohan ES, Sand S, Hurskainen H (eds) GALILEO positioning tehcnology, 3rd edn. Springer, Dordrecht. doi:10.1007/978-94-007-1830-2_2
3. Bessho K, Date K, Hayashi M, Ikeda A, Imai T, Inoue H, Kumagai Y, Miyakawa T, Murata H, Ohno T, Okuyama A, Oyama R, Sasaki Y, Shimazu Y, Shimoji K, Sumida Y, Suzuki M, Taniguchi H, Tsuchiyama H, Uesawa D, Yokota H, Yoshida R (2016) An introduction to Himawari-8/9 - Japan's new-generation geostationary meteorological satellites. J Meteorol Soc Jpn. doi:10.2151/jmsj.2016-009
4. Cayton TE, Belian RD, Gary SP, Fritz TA, Baker DN (1989) Energetic electron components at geosynchronous orbit. Geophys Res Lett 16:147–150. doi:10.1029/GL016i002p00147
5. Den M, Obara T, Tomita F (1999) Development of GMS data acquisition and database systems. Adv Polar Upper Atmos Res 13:191–196
6. Gannon JL, Elkington SR, Onsager TG (2012) Uncovering the nonadiabatic response of geosynchronous electrons to geomagnetic disturbance. J Geophys Res 117:A10215. doi:10.1029/2012JA017543
7. Hanser FA (2011) EPS/HEPAD calibration and data handbook, Tech. Rep. GOESN-ENG-048D, Assurance Technol. Corp., Carlisle, Massachusetts
8. Implementation Plan for the Evolution of Global Observing Systems (EGOS-IP), WMO Integrated Global Observing System, Technical Report No. 2013-4
9. Kurino T (1985) Report on the utilization of space environment monitor (SEM) data of the GMS series satellite (in Japanese), Meteorological Satellite Center Technical Note, 12:45–55
10. Lanzerotti LJ, Breglia C, Maurer DW, Johnson GK III, Maclennan CG (1998) Studies of spacecraft charging on a geosynchronous telecommunications satellite. Adv Space Res 22:79–82. doi:10.1016/S0273-1177(97)01104-6
11. Nagatsuma T (2013) New ages of operational space weather forecast in Japan. Space Weather. doi:10.1002/swe.20050
12. Onsager TG, Grubb R, Kunches J, Matheson L, Speich D, Zwickl R, Sauer H (1996) Operational uses of the GOES energetic particle detectors. In: Proceedings of SPIE 2812, GOES-8 and Beyond, 281 (October 18, 1996). doi:10.1117/12.254075
13. Onsager TG, Chan AA, Fei Y, Elkington SR, Green JC, Singer HJ (2004) The radial gradient of relativistic electrons at geosynchronous orbit. J Geophys Res 109:A05221. doi:10.1029/2003JA010368
14. Rodriguez JV, Onsager TG, Mazur JE (2010) The east-west effect in solar proton flux measurements in geostationary orbit: a new GOES capability. Geophys Res Lett 37:L07109. doi:10.1029/2010GL042531
15. Rodriguez JV, Krosschell JC, Green JC (2014) Intercalibration of GOES 8–15 solar proton detectors. Space Weather 12:92–109. doi:10.1022/2013SW000996

16. Rodriguez JV, Sandberg I, Mewaldt RA, Daglis IA, Jiggens P (2017) Validation of the effect of cross-calibrated GOES solar proton effective energies on derived integral fluxes by comparison with STEREO observations. Space Weather 15:290–309. doi:10.1002/2016SW001533
17. Roederer JG (1970) Dynamic of geomagnetically trapped radiation, physics and chemistry in space, vol 2. Springer, Berlin
18. Ryden KA, Rodgers DJ, Morris PA, Frydland AD, Jolly HS, Dyer CS (2001) Direct measurement of internal charging currents in geostationary transfer orbit. In: Proceedings of the 6th RADECS conference, pp 44–50
19. Ryden KA, Hands ADP, Underwood CI, Rodgers DJ (2015) Internal charging measurements in medium Earth orbit using the SURF sensor: 2005–2014. IEEE Trans Plasma Sci 43:3014–3020
20. Sandberg I, Jiggens P, Heynderickx D, Daglis IA (2014) Cross calibration of NOAA GOES solar proton detectors using corrected NASA IMP-8/GME data. Geophys Res Lett 41:4435–4441. doi:10.1002/2014GL060469
21. Takagi S, Nakamura T, Kohno T, Shiono N, Makino F (1993) Observation of space radiation environment with EXOS-D. IEEE Trans Nucl Sci NS-40:1491–1497. doi:10.1109/23.273513
22. Tsyganenko NA (1989) A magnetospheric magnetic field model with a warped tail current sheet. Planet Space Sci 37:5–20

CHAPTER 14

Hydrogen mobility in transition zone silicates

Razvan Caracas[1]* and Wendy R. Panero[2]

[1]*CNRS, Ecole Normale Supérieure de Lyon, Université Claude Bernard Lyon 1, Laboratoire de Géologie de Lyon, CNRS UMR 5276, Lyon, France.*

[2]*School of Earth Sciences, Ohio State University, Columbus, OH 43210, USA.*

AUTHORS' CONTRIBUTIONS

RC and WRP proposed the topic, conceived and designed the study, analyzed the data, and constructed the manuscript. RC ran the calculations. Both authors read and approved the final manuscript.

ABSTRACT

We study the hydrogen mobility in ringwoodite and wadsleyite considering multiple charge-balanced defects, including Mg < = > 2H, Si < = > Mg + 2H, and the hydrogarnet defect, Si < = > 4H, using molecular dynamics simulations based on the density functional theory at transition zone pressures and temperatures between 1500 and 2500 K. We determine the diffusion coefficients and study in detail the mechanism of hydrogen mobility during lengthy simulations. Our results show that temperature, water concentration, and defect mechanism have a significant effect on mobility. We find that the fastest diffusion is for the Mg < = > 2H defect, while H is more mobile when incorporated as Si < = > Mg + 2H than as hydrogarnet defects. The computed diffusivities for ringwoodite are larger than for wadsleyite: at 2000 K, diffusivity is 1.13×10^{-09} m^2/s for ringwoodite compared to 0.93×10^{-09} m^2/s for wadsleyite. In general, the hydrogen atoms spend on the order of tens of picoseconds or more trapped in or around the vacancy sites with net migration between sites over timescales of tens of femtoseconds. At 2500 K, some of these hydrogen excursions take place over several angstroms, while at 2000 K, they do not always result in net diffusion. At 1500 K, most of the defects fail to make excursions from their defect sites resulting in diffusion.

Keywords: Diffusion, Transition zone, Wadsleyite Ringwoodite Transport, Electrical conductivity, Defects Hydrogen

1. INTRODUCTION

The incorporation of water in major mantle minerals affects the physical properties of these minerals, even with low concentrations of water. The introduction of water as hydrogen defects reduces seismic wave velocities (e.g., Jacobsen and Smyth 2006), enhances deformation (Kavner 2003), and increases electrical conductivity (Huang et al. 2005; Yoshino et al. 2009).

The transition zone is dominated by polymorphs of olivine, in which wadsleyite (*I2/m*) is stable from 410–520 km depth, transforming to ringwoodite (*Fd3m*), stable from 520–660 km depth. As synthesized in the laboratory, each of these polymorphs can store water in excess of 2 wt%, making them the primary focus for detecting water in the deep mantle. Wadsleyite incorporates hydrogen predominantly through Mg<=>2H defects (Smyth et al. 1997) while ringwoodite accommodates hydrous defects through multiple defect mechanisms, dominated by the same Mg<=>2H defect, but also with the hydrogarnet defect, Si<=>4H, and minor Si<=>Mg + 2H (Panero 2010; Panero et al. 2013).

The amount of water actually stored in the transition zone, however, may be significantly less than the total capacity. While an inclusion of ringwoodite in diamond suggests that localized regions of hydrated ringwoodite are possible (Pearson et al. 2014), seismic velocity and phase boundaries (Houser 2016) and electrical conductivity of the transition zone are more consistent with a mantle significantly undersaturated in water.

Electrical conductivity of the Earth's mantle varies significantly with depth and tectonic regime, including in the transition zone, in which the conductivity at the base of the transition zone is greater than the upper part of the transition zone (e.g., Kelbert et al. 2009; Shimizu et al. 2010). This variability is likely a complex function of temperature, composition (particularly H, Fe, and redox), and tectonic history. Indeed, the electrical conductivity of ringwoodite and wadsleyite vary as a function of water and iron content (e.g., Huang et al. 2005; Yoshino et al. 2009), with likely significant contributions from both proton and polaron conductivities. While disagreement remains as to the absolute values of conductivity in these materials, the electrical conductivity of hydrous ringwoodite is consistently found to be greater than hydrous wadsleyite for comparable water contents and experimental conditions but with different dependencies on absolute water content.

Direct measurements of hydrogen diffusivity in wadsleyite and ringwoodite under transition zone pressures and conditions are limited (Hae et al. 2006; Ohtani and Zhao 2009; Sun et al. 2015), in which these measurements include the complicating effects of iron that appears to enhance electrical conductivity through proton-polaron exchange mechanisms. The hydrogen diffusivity data are consistent with the electrical conductivity in which the H conductivity in ringwoodite exceeds that of wadsleyite.

We therefore calculate the diffusivity of hydrogen in ringwoodite at mantle-relevant conditions from ab initio molecular dynamics. While this is a well-established standard technique for melts and fluids (e.g., Tse 2002; Schweiger et al. 2008; Mookherjee et al. 2008), it is an unusual approach for solids due to the length of required calculations for appropriate statistics. For the latter, the large majority of the previous calculations (e.g., Bunson et al. 1999; Ammann et al. 2011) instead resolved to determine the activation energy along an assumed diffusion path, which combined with the concentration of vacancies, allowed to estimate the diffusion coefficients. Our approach is more daring: we insert hydrogen in the structures as defects in exchange for cations and obtain the self-diffusion coefficients directly from long molecular dynamics simulations; consequently, we also capture the diffusion path.

With these considerations, we investigate in detail the mobility of hydrogen in ringwoodite as a function of temperature, defect mechanism, and defect geometry. We compare our results to calculations in wadsleyite and further consider the effects of shear and strain on wadsleyite. We also examine the pathway through which hydrogen diffuses to suggest the consequences of interactions with point defects and dislocations in hydrogen diffusivity. We estimate the effect of hydrogen on the electrical conductivity of the major minerals of the Earth's transition zone through the Nernst-Einstein relation and assess the contribution of hydrogen to the measured mantle conductivity.

2. METHODS/EXPERIMENTAL

We perform first-principles molecular dynamics simulations. We allow the atoms to move in a crystal according to the Newtonian dynamics (*molecular dynamics*) under the action of interatomic forces computed within the density functional theory (DFT) (*first principles*). We use the planar-augmented wavefunction (PAW) formulation (Blochl 1994) of the DFT the VASP implementation (Kresse and Hafner 1993; Kresse and Joubert 1999) to compute the forces. The exchange-correlation energy is expressed in the Perdew-Burke-Ernzerhof form (Perdew et al. 1996) of the generalized gradient approximation. We employ a 500 eV cutoff for the kinetic energy, 700 eV for the augmented cutoff, and the Γ point to sample the reciprocal space.

We use isokinetic NVT ensembles, where the velocity of the atoms is rescaled at each time step to fit the temperature, T, at fixed volume, V, and number of atoms, N, in the simulation box. The isokinetic NVT can be seen as an extreme version of the Nose-Hoover thermostat where the oscillation period of the temperature is set to the time step. We perform the simulations at volumes corresponding to transition zone pressures.

In all our calculations, we allow for long simulation times, between 70 and 200 ps at each condition with a time step of 1 fs. A 70 ps simulation requires the equivalent of about 500 processor days of computing time.

For ringwoodite, we consider several hydrogen incorporation mechanisms (Panero 2010): $Mg^{2+} <=> 2H^+$, $Si^{4+} <=> Mg^{2+} + 2H^+$, and the hydrogarnet substitution, $Si^{4+} <=> 4H^+$. We consider two different H/O atomic ratios per simulation cell: 1/32 and 2/32 each in two different geometries: a $2 \times 1 \times 1$ supercell of the conventional cell and a $2 \times 2 \times 2$ supercell of the primitive cell, with 16 Mg_2SiO_4 formula units. The simulations are run at 1500, 2000, and 2500 K with the same initial specific volume of ringwoodite, corresponding to about 25 GPa. Temperature and the substitution mechanisms affect the actual pressure of the simulations, but in all cases, the pressure remains in an interval of about 23 to 33 GPa. Over this pressure range, we do not expect the values of the diffusivity to change significantly.

We compare the ringwoodite diffusivities to the $Mg^{2+} <=> 2H^+$ defect on the M3 site in wadsleyite, the only active defect mechanism (Tsuchiya and Tsuchiya 2009), in a simulation box of one unit cell, or eight molecular units. To accelerate the diffusion, we also study the effect of strain. In order to compensate for the smaller simulation sizes, the simulation times of wadsleyite are longer, on the order of 100 ps.

The diffusion coefficient of hydrogen, D_0, in each calculation is obtained as the slope of the mean square displacements, MSD, as a function of time:

$$D=\mathrm{MSD}/6t \qquad (1)$$

where, t is time, and 6 is a geometric factor describing motion in three dimensions. MSD is the mean square displacement defined as:

$$\mathrm{MSD}=<[r(t)-r(t_0)]^{\wedge}2> \qquad (2)$$

Where $r(t)$ is the position in three-dimensional space at time, t. MSD is computed after an initial variable thermalization type of at least 2 ps, and t-t_0 window glides over the entire remaining simulation time. Values less than $\sim 1.0 \times 10^{-10}$ m^2/s are an indication that the hydrogen atoms jump on a timescale similar to or greater than the total simulation time, and we include diffusivities for calculations only when the MSD is a linear function of time. Otherwise, we interpret that there are insufficient statistics to obtain reliable diffusion coefficients.

The diffusivity is a function temperature and can be modeled with D_0 as prefactor and the activation enthalpy H,

$$D=D_0\exp(-H/RT) \qquad (3)$$

Where R is the gas constant and T is temperature in Kelvin.

We estimate the contribution proton conductivity makes to the apparent electrical conductivity, σ, with the Nernst-Einstein relation,

$$\sigma=(F^2/R)(Dq^2c/T) \qquad (4)$$

where F is Faraday's constant, q is the charge on the ion (+1), and c is concentration.

3. RESULTS AND DISCUSSION

Typical mean square displacement plots as a function of time are shown in Fig. 1 at both 2000 and 2500 K. Other MSD plots at 2000 K show larger irregularities; in those cases only the slope in the second half of the diagram was considered, which is linear. If the MSD is unchanging with time (e.g., for Mg, Si, and O in Fig. 1), this was a sign for no diffusion during the duration of our simulation and those points were discarded.

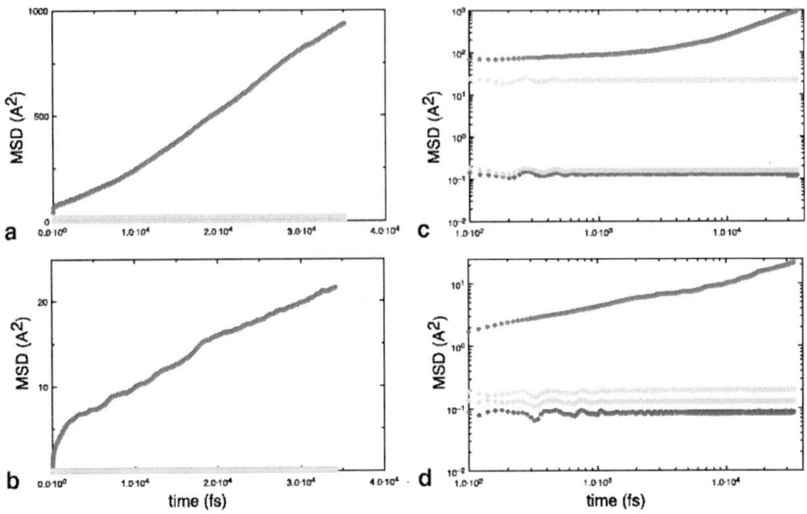

Figure 1 Examples of the mean square displacements vs time. **A, C** Mg < - > 2H $T = 2500$ K, 1.98 wt% H_2O, $2 \times 2 \times 2$ primitive supercell. **B, D** Si < - > Mg + 2H $T = 2000$K, 1.98 wt% H_2O, $2 \times 1 \times 1$ conventional supercell. The MSD for Mg, Si, O, and H are respectively in *green*, *blue*, *pink*, and *red*. The only diffusive species is H. **A** and (**B**) are linear plots, while (**C**) and (**D**) are the log-log plots

Table 1 summarizes the diffusion results of all the simulations performed for the Mg^{2+} < = > $2H^+$, Si^{4+} < = > $Mg^{2+} + 2H^+$, and Si^{4+} < = > $4H^+$ substitutions in ringwoodite and different H concentrations, and 2000 and 2500 K, and for the Mg^{2+} < = > $2H^+$ substitution in wadsleyite, at 1500 and 2000 K. There is no net diffusion in ringwoodite at 1500 K. The calculations span a range of temperatures greater than the entire transition zone. For each simulation, we also indicate the average pressure of the constant volume calculations.

Table 1 Calculations on H-bearing ringwoodite and wadsleyite

Defect	Miner al	H/O ratio	H_2O wt%	Geometry	P (GPa)	T (K)	D_H (10^{-9} m^2/s)	log(σ) (S/m)
Mg < = > 2H	Rw	1/32	0.81	2 × 1 × 1 C	23.6	2000	0.569	0.03
		1/16	1.63	2 × 2 × 2 P	25.0	2000	1.132	−1.12
		1/16	1.63	2 × 1 × 1 C	24.7	2000	a	
		1/32	0.81	2 × 2 × 2 P	25.1	2000	a	
		1/32	0.81	2 × 2 × 2 P	25.1	2000	a	

		1/32	0.81	2 × 1 × 1 C	27.1	2500	8.215	1.09
		1/16	1.63	2 × 1 × 1 C	26.8	2500	1.077	0.51
		1/32	0.81	2 × 2 × 2 P[b]	27.0	2500	8.215	1.09
		1/32	0.81	2 × 2 × 2 P[b]	27.0	2500	10.005	1.18
		1/16	1.63	2 × 2 × 2 P	26.7	2500	43.727	2.12
Si <= > Mg + 2H	Rw	1/32	0.80	2 × 1 × 1 C	22.1	2000	0.217	−0.40
		1/16	1.60	2 × 1 × 1 C	24.6	2000	0.796	0.47
		1/32	0.80	2 × 1 × 1 C	30.3	2500	2.698	0.60
		1/16	1.60	2 × 1 × 1 C	26.6	2500	5.448	1.20
Si < = > 4H	Rw	1/16	1.62	2 × 1 × 1 C	24.0	2000	0.077	−0.54
		1/8	3.27	2 × 1 × 1 C	25.2	2000	0.027	−0.70
		1/16	1.62	2 × 1 × 1 C	32.6	2500	0.263	−0.11
		1/8	3.27	2 × 1 × 1 C	27.0	2500	0.263	0.20
Mg < = > 2H and	Rw	1/16	1.62	2 × 2 × 2 P	28.0	2000	0.452	0.23
Si < = > Mg + 2H		1/16	1.62	2 × 2 × 2 P	29.8	2500	0.74	0.35
Mg < = > 2H	Wd	1/16	1.63	1 × 1 × 1	21.5	1500	0.024	−0.94
		1/16	1.63	1 × 1 × 1 5% strain	36.9	1500	0.003	−1.80
		1/16	1.63	1 × 1 × 1 7% shear	21.6	1500	0.107	−0.29
		1/16	1.63	1 × 1 × 1 11% shear	20.2	1500	0.617	0.47
		1/16	1.63	1 × 1 × 1	21.8	2000	0.926	0.52
		1/16	1.63	1 × 1 × 1	23.6	2500	88.43[c]	2.41

Summary of all the simulations performed on ringwoodite (R) and wadsleyite (W). Temperatures (T) and pressures (P) are also indicated. The 2 × 1 × 1 C and 2 × 2 × 2 P geometries represents supercell of the respectively Conventional and Primitive unit cells of ringwoodite

Rw ringwoodite structure, *Wd* wadsleyite structure

[a]There is no net diffusion observed during the duration of the simulation

[b]The two values are for two different simulations

[c]Wadsleyite at 2500 K is molten; the value is given here only as reference

Based on the computed diffusion coefficients, we fit the prefactor D_0 and the activation enthalpy, *H*, for each defect mechanism, concentration and phase according to Eq. (3). The values are listed in Table 2. They should be taken with care, as pressure increases with temperature in these isochoric calculations.

Table 2 Hydrogen diffusivity

Defect	Mineral	H/O ratio	H_2O wt%	Geometry	$D_0(m^2/s)$	H (kJ/mol)
Mg <=> 2H	R	1/16	1.63	2 × 2 × 2 P	9.73E-02	303.8
		1/32	0.81	2 × 1 × 1 C	4.16E-04	224.5
Si <=> Mg + 2H	R	1/16	1.60	2 × 1 × 1 C	1.19E-05	159.9
		1/32	0.80	2 × 1 × 1 C	6.44E-05	209.5
Si <=> 4H	R	1/8	1.62	2 × 1 × 1 C	5.05E-04	260.9
		1/16	3.27	2 × 1 × 1 C	2.26E-06	188.3
Mg <=> 2H and Si <=> Mg + 2H	R	1/16	1.62	2 × 2 × 2 P	5.32E-04	232.4
Mg <=> 2H	W	1/16	1.63	1 × 1 × 1	5.62E-05	183.2

Diffusivity fits to the computed diffusion coefficients

The hydrogen atoms typically spend several tens of picoseconds in one given cation vacancy site in between the diffusion jumps. The shape of their local trajectory depends on this site. Diffusion is realized in rapid (<100 fs) transfers between stable defect sites. During a time that is on the order of up to a tenth of a picosecond, the hydrogen atoms may undergo an excursion that can be up to several angstroms long. There are cases, especially at low temperature (1500 K) when the outcome of such an excursion is cut short and the atoms return to the initial cage. But at 2000 and 2500 K, most of these excursions result in a net displacement. This is the mechanism of the diffusion. The diffusion trajectory is a function of defect.

3.1 H migration in ringwoodite

In ringwoodite, the hydrogen atoms that are part of the Si <=> 4H defects and the Si <=> Mg + 2H defects rapidly switch between the predicted lowest energy positions on the outside face of each polyhedron, generally maintaining maximum distance between the hydrogen atoms associated with each defect. Their trajectory results in apparent crosses on the faces of the tetrahedra (Figs. 2 and 3).

Figure 2 shows a part of the trajectories of the four H atoms that substitute one Si atom for the Si <=> 4H defect. The H atoms remain around the initial vacancy. After 11 ps in the simulation one of them exits the vacancy, moves away around an O atom (Fig. 2) to be trapped in an empty adjacent octahedron, inside which it moves around describing a donut shape. Finally, the H atom undergoing the excursion falls back into a Si vacancy close to the other three hydrogen atoms but situated at a distance of one unit cell from the initial defect. During this time, it covers the equivalent net distance of one lattice parameter. The case analyzed here corresponds to one substituted Si in a $2 \times 1 \times 1$ conventional supercell. As expected, the H diffuses along the shortest path, the *y*- or *z*-axis. In Fig. 2b, c, it is clearly seen a diffusion pattern similar to a channel. An increase in temperature by 500 K leads to an increase in diffusion by more than one order of magnitude.

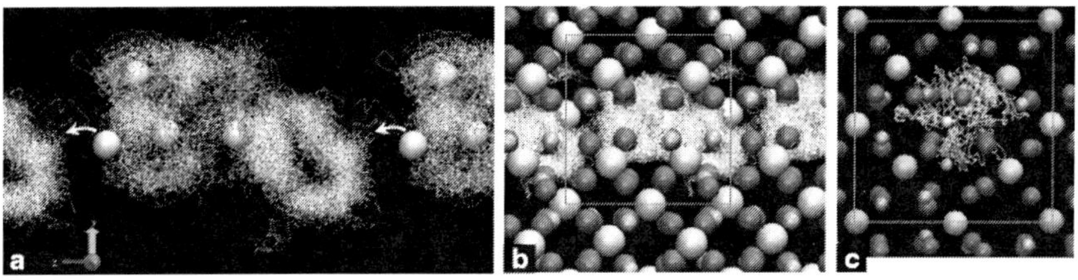

Figure 2 Diffusion path of H for the Si < = > 4H defect at 2500 K. Mg, Si, O, and H are represented respectively by *pink*, *yellow*, *red*, and *white* spheres. Sphere radii not to scale. The trajectory of the H atoms is shown with the dotted line over several picoseconds. **A** shows only the H trajectory for clarity. The diffusion path is shown in the context of the entire structure in (**B**) and (**C**). The hydrogen atoms occupy the Si vacancy. After escape, indicated by the *white arrow*, one of them described donut shapes while trapped in an empty octahedron, before falling into the neighboring Si vacancy and thus realizing a net proton transfer (**A**). The diffusion path describes channel-like patterns along the *z*-axis, as viewed in the image along *x*-axis (**B**) and *z*-axis (**C**). The unit cell is outlined with *dashed line*

In case of the Si < = > Mg + 2H substitution, an excess Mg lies on the Si vacancy site; the H atoms are on the outer parts of the SiO_6 tetrahedra. The pathway to initiate diffusion passes through an empty interoctahedral space. Not all H excursions result in net diffusion (Fig. 3) at 2000 K.

For the Mg < = > 2H defect mechanism, the hydrogen atoms migrate rapidly within a plane of the vacant Mg octahedron defined by the direction perpendicular to the edge-sharing Mg octahedra and sub-parallel to the plane defined by corner-sharing Si tetrahedra (Fig. 4) mapping out a "donut" shape about the octahedral vacancy. This plane reflects the crystallographic equivalent sites with the lowest energy positions as predicted by Panero (2010). The pair of hydrogen atoms is oriented such that their distance to the neighboring Mg octahedra is maximized, which also carry a positive charge. The actual trajectory is easily identified due to its donut shape.

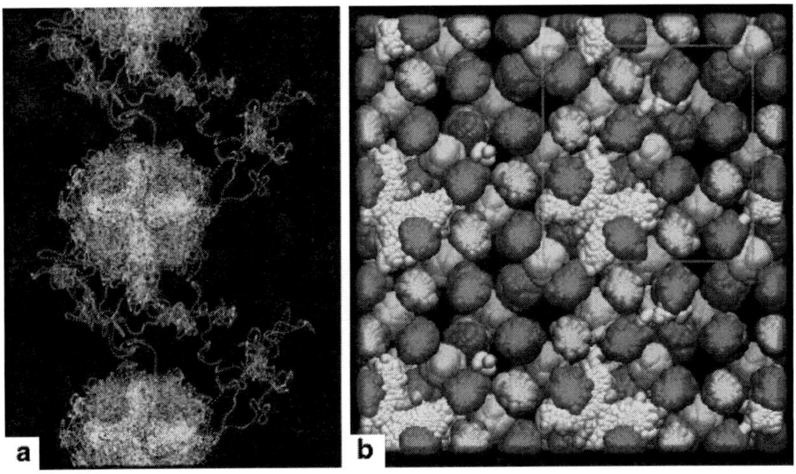

Figure 3 H diffusion for the Si ⇔ Mg + 2H defects at 2500K. They are computed in a 2 × 1× 1 convention supercell with one defect (**A**) and in a 2× 2× 2 primitive supercell containing two defects (**B**). **A** Only the trajectory of H is shown, with all the other atoms omitted for clarity. H atoms describe crosses-like patterns on the outside parts of the SiO_4 tetrahedra. H atoms undergo lengthy excursions before falling back into the vacancy site. **B** The trajectory described by all the atoms during 35 ps is represented here. *Pink*, *yellow*, *red* and *white* represent respectively the Mg, Si, O, and H atoms. Not all excursions of H result in net diffusion, as seen from the defect in the *lower left* part of the image: H falls back into the initial site after escaping into the adjacent empty octahedron for a short time. The unit cell is outlined with *dashed line*

3.2 H migration in wadsleyite

In wadsleyite, static calculations of H defects show that pairs of H in the Mg < = > 2H defect tend to align with the OH dipoles parallel to the *z*-axis (Tsuchiya and Tsuchiya 2009). In our dynamical simulations at the lowest temperatures in our simulations (1500 K), the H atoms explore the vacancy site, in *trans* positions. At 2000 K, they escape easily from their initial sites and wander around to realize a net diffusion.

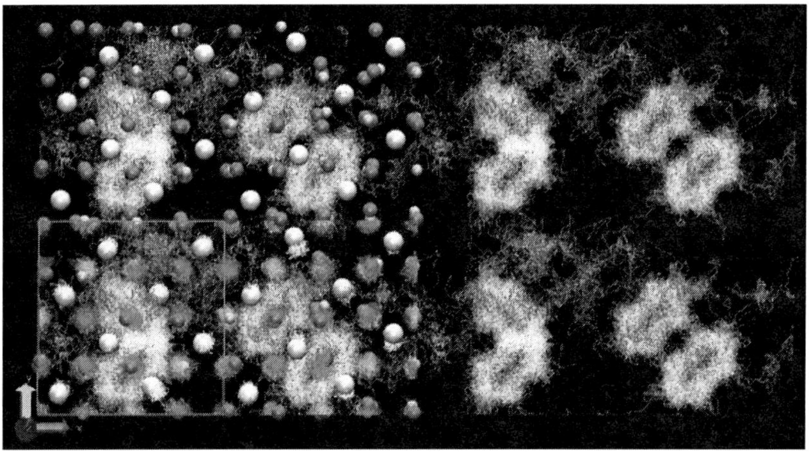

Figure 4 Trajectory of the H atoms for the Mg < = > 2H defect during about 35 ps. A structural fragment of 4 × 2 × 2 unit cells is represented; on the right half all the atoms except for H are omitted for clarity. The example shown here is for a simulation at 2500 K with two defects in a 2 × 2 × 2 primitive supercell. *Pink*, *yellow*, and *red* represent respectively the Mg, Si, and O atoms. The unit cell is outlined with *dashed line*

All Mg < = > 2H defects exchange H atoms during the course of the simulations; their diffusion paths follow not only the directions of the unit cell axis, as in the cases discussed above, but also along a diagonal.

Finally, the simulation of two adjacent defects of different types (Mg < = > 2H and Si < = > Mg + 2H) in the 2 × 2 × 2 primitive supercell shows how these defects interact. About 34.7 ps into the simulation, three hydrogen atoms are found around the Mg vacancy, and one H atom around the Si vacancy. Figure 5a shows the trajectory of the Mg atoms over about 0.8 ps and a snapshot of the position of the other atoms before the diffusion jump. It is clearly seen that the defect Mg stays inside the Si vacancy. The same snapshot with the H trajectory over 2 ps is shown in Fig. 5b: the characteristic donut and cross shapes are clearly identifiable. One hydrogen atom is shown during its jump in snapshot in Fig. 5c. The jump requires just a few tens of femtoseconds. Eventually, the H falls into the neighboring vacancy, each of which containing now two H atoms (Fig. 5d). A similar but reverse process was responsible for obtaining the three H atoms in the Mg vacancy site. Figure 4e shows a 20 ps duration of trajectories of all the atoms from the same structural fragment.

Comparable simulations on wadsleyite were performed in one unit cell, with eight formula units, so that water concentrations are comparable between the wadsleyite and ringwoodite simulations. The diffusion of hydrogen is realized in channel-like trajectories continuous along the *z*-axis and interconnected along the *y*-axis (Fig. 6). The H atoms, as in ringwoodite, diffuse by hopping between the O atoms and avoiding the Mg sites. They move around the SiO_4 tetrahedra but prefer the vacant inter-tetrahedral space.

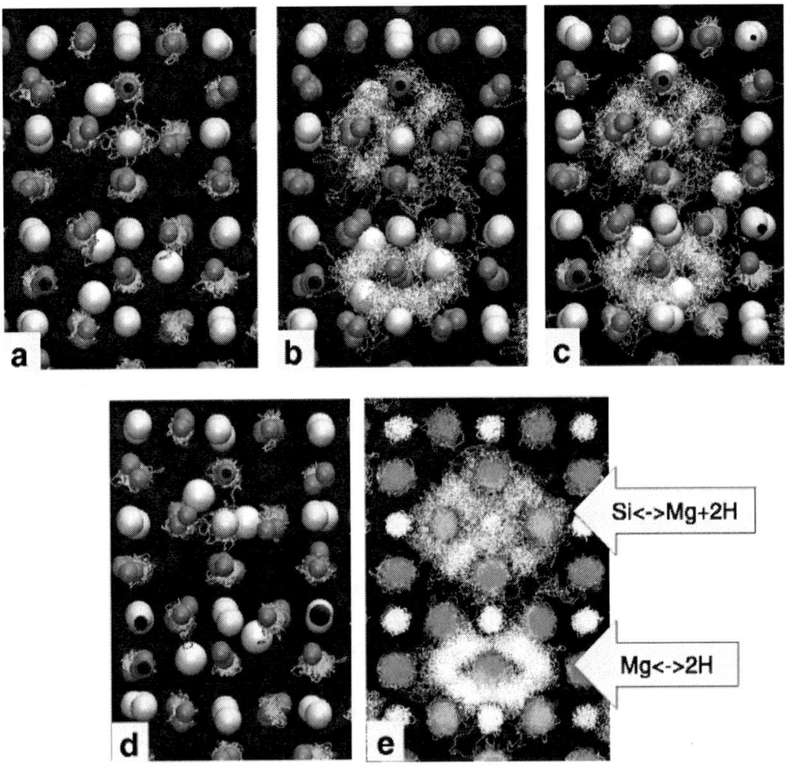

Figure 5 Snapshots of the diffusion of hydrogen in a 2 × 2× 2 primitive supercell containing two defects at 2000 K. Mg < = > 2H lies in the *lower* half of the image above and Si < = > Mg + 2H lies in the *upper* half. **A** A snapshot of the structure when 3 H atoms occupy the Mg vacancy, 1 Mg the Si vacancy, and one last H gravitates around the Si vacancy. The Mg atoms are represented with their trajectory (*pink dotted lines*) during half a picosecond. There is no diffusion of Mg. **B** The H trajectory is donut- and cross-shaped in the respective two defects as represented over 2 ps simulation time. **C** One of the three H atoms escape the Mg vacancy. The excursion lasts several tens of femtoseconds. **D** The escaped H atoms falls around the Si vacancy. **E** Trajectory of all atoms during 20 ps of simulation time. The only diffusing species is H

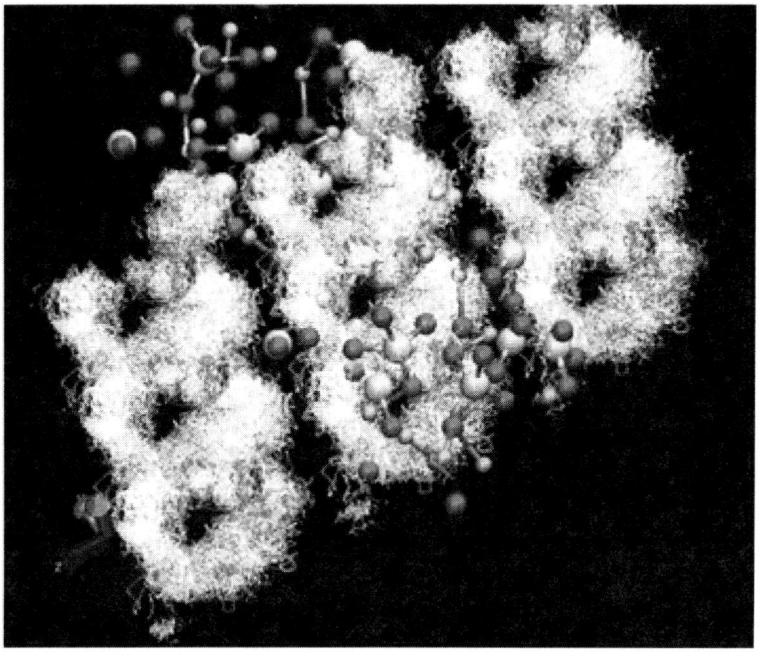

Figure 6 Diffusion of H in the wadsleyite structure at 2000 K. The structural fragment shown here represents the H trajectories during 60 picoseconds after thermalization at 2000 K spanning several unit cells. The diffusion is realized in the (010) plane, containing the shortest lattice parameters

At 2500 K and 23.6 GPa in wadsleyite (Table 1), MSD shows net displacements for all oxygen, silicon, and magnesium atoms in the structure, and therefore, we interpret that this calculation reflects a melt, consistent with the melting temperature of wadsleyite. The hydrogen diffusivity calculated in this simulation, 88.10^{-9} m^2/s, then reflects the diffusivity of H in a molten system. Our hydrogen-bearing calculations show the presence of melt of lower temperatures than both previous freezing simulations on anhydrous liquid Mg_2SiO_4 (De Koker et al. 2008) and recent static experimental data (Liebske and Frost 2012) consistent with melting-point depression for hydrated materials.

3.3 Energy considerations

There is no obvious relation between the position of the hydrogen atoms along their diffusion path and the fluctuation in total energy of the system. There are examples when the local energy maximum corresponds to migration of one H atom between sites, but in these cases, the energy profile does not resemble that of a saddle point. Furthermore, the next local energy maximum corresponds to H inside or around its new host site. This behavior reflects the complex relationship between the position of all the atoms in the lattice and the total energy of the lattice, not just the hydrogen positions. The energy fluctuations that we compute include contributions from not only the position of the diffusing atoms but also the vibrational contributions. For example, Fig. 7a shows the fluctuations in total energy computed during several picoseconds for the Si < = > 4H substitution in ringwoodite. These fluctuations are regularly observed during NVT simulations. Details in Fig. 7b are shown prior, during, and after the escape of the H atom in the Si < = > 4H substitution depicted in Fig. 2 and discussed in detail above. There is no clear relation between the position of the hydrogen atoms along the diffusion path and the total energy of the system. This might also raise questions about the validity of the activation energy approach within a partially fixed lattice in static (i.e., 0 K) or nudged-elastic band simulations.

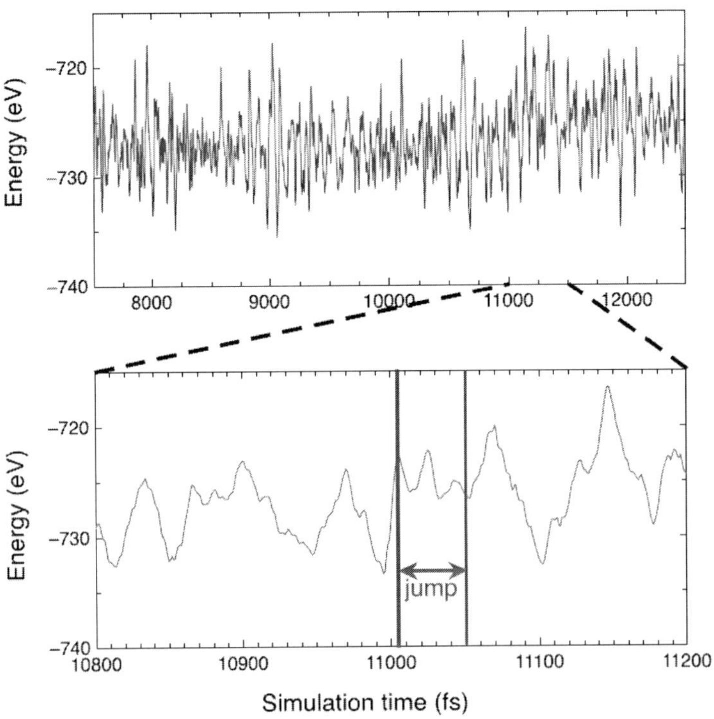

Figure 7 Energy variation during parts of the Si < = > 4H simulation at 2500 K. **A** The variations are typical for an NVT simulation. **B** The detail around the H jump from one vacancy site to the neighboring one does not show any relation between the position of H in the structure and the energy

3.4 Factors affecting diffusion rate: phase, substitution mechanism, concentration, and temperature

For ringwoodite, we find no statistically significant net migration of hydrogen at 1500 K in any of the calculations. Figure 8 shows the diffusion rates we obtained for each calculation in Table 1, with fits to Eq. (3) in Table 2. In ringwoodite, the Mg < = > 2H defect leads to the greatest H diffusivities at both 2000 and 2500 K significantly increasing with increased water concentration. We find that the apparent diffusivity is greater in the 2 × 2 × 2 primitive cell than in the 2 × 1 × 1 conventional cell, and that for constant temperature and geometry, the diffusivity increases with increasing defect density. The largest values of diffusion are greater than 10^{-9} m^2/s at 2000 K and increase to 4.4×10^{-8} at 2500 K. Such high values of the H diffusion are comparable to the ones obtained in cold hydrous $MgSiO_3$ silicate melts (Karki et al. 2010): on the order of 2×10^{-8} m^2/s at 2500 K and 0 GPa, and 3×10^{-8} m^2/s at 3000 K and 20 GPa. This suggests that either solid hydrous ringwoodite at 2500 K is superheated or that the hydrous melts at 3000 K are supercooled.

Consistently, lower hydrogen diffusivities are found in the other two defects. Hydrogen in the Si < = > Mg + 2H defect diffuse faster at each temperature than hydrogen in Si < = > 4H defects for all concentrations and supercell geometries calculated. Doubling the water content in each calculation changes the inferred hydrogen diffusivity by a factor of two but with opposite effect: it increases mobility of hydrogen associated with the Si < = > Mg + 2H defect yet decreases when associated with the Si < = > 4H defect.

The simulations with the combined Mg < = > 2H and Si < = > Mg + 2H defects yield diffusion coefficients that fall, as expected, between the equivalent concentration Mg < = > 2H calculations and the Si < = > Mg + 2H calculations. While each defect is active in ringwoodite, the Mg < = > 2H defect is dominant, particularly at concentrations below 1 wt% (Panero et al. 2013). Therefore, as the mechanism with both the greatest hydrogen diffusivity and greatest concentration, we consider the diffusivity of hydrogen in the Mg < = > 2H defect to control the diffusivity of hydrogen in ringwoodite.

The experimental hydrogen diffusion data for ringwoodite extrapolated to 2000 K falls in the middle of the range of values we obtained in our simulations (Ohtani and Zhao 2009; Sun et al. 2015). The inferred activation enthalpy in 2000–2500 K for ringwoodite in the fastest mechanism is 300 kJ/mol, significantly greater than the experimental value for iron-bearing polycrystalline and single-crystal ringwoodite, 140 and 101 kJ/mol, respectively. In contrast, the activation enthalpy in wadsleyite, 180 kJ/mol, is more similar to the polycrystalline, iron-bearing experimental measurements in wadsleyite of 123 kJ/mol. The presence of iron, additional defects, or experimental strain may explain the difference. Our computed activation enthalpies are larger than the ones obtained from static simulations (Verma and Karki 2009), by almost one order of magnitude in case of ringwoodite. Table 2 lists the diffusivity fits to our computed diffusion coefficients.

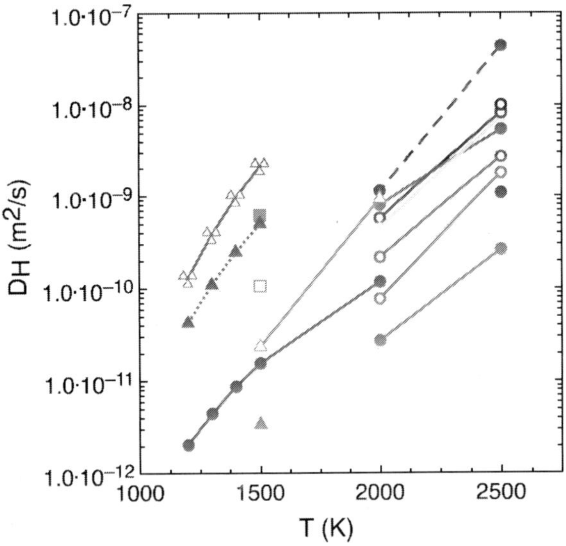

Figure. 8 Hydrogen diffusion coefficients in ringwoodite and wadsleyite as a function of temperature. Mg <=> 2H, Si <=> Mg + 2H, and Si <=> 4H defect mechanisms are respectively plotted in *blue*, *green*, and *red*. Ringwoodite results are represented with one (*open circles*) and two (*solid circles*) defects per simulation supercell in the 2 × 2 × 2 primitive (*dashed lines*) and 2 × 1 × 1 conventional (*solid lines*) supercells. The combined Mg <=> 2H and Si <=> Mg + 2H defects in the same 2 × 2 × 2 supercell is shown with *dotted circles*. Wadsleyite is in *purple*: *empty triangles* for Mg <=> 2H defect, *empty squares* with 7% shear, *solid squares* with 15% shear, *solid triangle* with 5% uniaxial strain. *Brown circles* are extrapolated experimental data from Sun et al. (2015) for ringwoodite, *brown triangles* from Hae et al. (2006) for wadsleyite. Triple empty *triangles* are for experiments in polycrystalline wadsleyite (Ohtani and Zhao 2009)

For wadsleyite however, the extrapolated experimental data (Hae et al. 2006) are greater than our theoretical results. As with the Ohtani and Zhao (2009) data, these experiments are measured in an iron-bearing polycrystalline sample, which may explain their faster diffusivities values.

At 2000 K, the H diffusion in wadsleyite for the Mg<=>2H defect is 0.93×10^{-9} m^2/s. This is 16% less than the fastest value of 1.13×10^{-9} m^2/s computed for ringwoodite for the same defect and at the same temperature, consistent to the experimental hydrogen diffusivity data (Hae et al. 2006; Ohtani and Zhao 2009) (Table 1).

To explore if the discrepancies with previously published diffusivities as discussed above are due to lattice strain, we calculated the diffusion under various systems of strains applied to the wadsleyite structure at 1500 K. We found that a 5% uniaxial strain along the *x*-axis reduces the diffusivity by one order of magnitude according to the mean square-displacement analysis. However, as the values are already below 10^{-10} m^2/s, they do not show diffusion; rather, they reflect an enhancement of the H localization. When pure sheared is applied, the values increase by about one order of magnitude for 7% shear and almost two orders for 10% shear. It is only this last case where actual net hydrogen transfers between the lattice sites is observed. While large strain values are highly unlikely to exist in a natural setting, we show that residual strains in experiments may lead to an underestimate of diffusivity due to H localization.

3.5 Electrical conductivity of the transition zone

Finally, we use the Nernst-Einstein relation to obtain the electrical conductivity from the diffusion coefficients. The values are reported on the last column of Table 1. They correspond to diffusivity inside single crystals.

The last column of Table 1 lists the computed conductivities for each simulation. For ringwoodite, the highest values computed at 2500 K are on the order of 1.1 to 2.1 log(σ) (S/m). They correspond to the Mg<=>2H defect that dominate the transport properties. Values at 2000 K are one order of magnitude lower. For wadsleyite, the computed log values of the conductivity are on the order of 0.5 log(σ) (S/m) at 2000 K.

Our values fall between the same order of magnitude down to two orders of magnitude below experimental values (Huang et al. 2005; Sun et al. 2015). This is most probably due to crystallinity: the diffusion along the grain boundaries plays a major role in the total transport. Measurements on real crystals show values order of magnitude higher as obtained for single crystal (Sun et al. 2015) than polycrystalline samples (Ohtani and Zhao 2009).

We conclude from our suite of calculations in ringwoodite that increased concentration of defects significantly enhances proton migration in the crystal, consistent with the concentration-dependent electrical conductivity (Huang et al. 2005). We find that hydrogen clearly rests in charge-neutral defect sites within the crystal for the majority of the simulation time, and therefore, it is the increased density of defects that leads to an increased conductivity through shortened hopping distance for the hydrogen. As with any diffusion-limited processes, the proton conductivity will be dominated by the fastest diffusing species, which in the case of ringwoodite, is the hydrogen atoms associated with the dominant Mg < = > 2H defect.

We find that the associated inferred conductivity of ringwoodite is greater than that of wadsleyite at realistic mantle conditions. The computed values of the conductivity and the difference between ringwoodite and wadsleyite are fully consistent with an increase in conductivity at the base of the Earth's mantle (Kelbert et al. 2009; Shimizu et al. 2010) and may help signal the presence of significant (~1 wt%) water in the transition zone.

4. CONCLUSIONS

We study the main three mechanisms of incorporating hydrogen into the ringwoodite structure via chemical substitutions: Mg < = > 2H, Si < = > Mg + 2H, and the hydrogarnet defect, Si < = > 4H. We perform first-principles molecular dynamics simulations lasting up to a couple hundred picoseconds. Three Mg < = > 2H calculations at 2000 K did not yield sufficient migration statistics to results in a reliable diffusivity in 200 ps of simulation time, which highlights that the starting conditions and the thermal history of the calculations may play a factor in the quantitative results.

We find that temperature and defect concentration increase the rate of the H diffusion. More isometric geometries, like the $2 \times 2 \times 2$ primitive supercell also show faster diffusion than more anisotropic, like the $2 \times 1 \times 1$ conventional supercell, similar to the reduction of diffusivity found in wadsleyite for 5% strain. We find that the Mg < = > 2H defect mechanism yields the greatest H diffusion rates, with the hydrogarnet substitution, Si < = > 4H, showing the slowest diffusion.

During the simulation time of several tens of picoseconds, the H atoms travel over several lattice sites. However, during their path, they spend a considerable amount of time pinned in the defect sites. The diffusion is realized via jumps between the vacancy sites; during these jumps, the hydrogen atoms move around the neighboring O atoms, such as to continuously form OH bonds. They try to avoid at all cost the Mg cations, which are also positively charged; usually they stay in the void octahedral space. There is no obvious relation in the energy fluctuations during the simulations and the positions of the hydrogen atoms in the lattice—be they in the vacancy sites or during the diffusive jumps.

The Mg < = > 2H defects are more mobile in ringwoodite than wadsleyite. This result agrees with the proton conductivity inferences of Yoshino et al. (2009) and Huang et al. (2005), as well as the observed increase in electrical conductivity with depth through the transition zone (e.g., Khan et al. 2011; Olsen 1998).

ACKNOWLEDGEMENTS

The calculations were performed on the ADA machine at the IDRIS Supercomputing Center, France, under eDARI grant x2016106368.

REFERENCES

1. Ammann MW, Brodholt JP, Dobson DP (2011) Ferrous iron diffusion in ferro-periclase across the spin transition. Earth Planet Sci Lett 302:393–402. doi:10.1016/j.epsl.2010.12.031
2. Blochl PE (1994) Projector augmented-wave method. Phys Rev B 50:17953–17979
3. Bunson PE, Di Ventra M, Pantelides ST, Schrimpf RD, Galloway KF (1999) Ab initio calculations of H+ energetics in SiO2: implications for transport. IEEE Trans Nuclear Sci 46:1568–1573
4. De Koker NP, Stixrude L, Karki BB (2008) Thermodynamics, structure, dynamics, and freezing of Mg2SiO4 liquid at high pressure. Geoch Cosmochimica Acta 72:1427–1441. doi:10.1016/j.gca.2007.12.019
5. Hae R, Ohtani E, Kubo T, Koyama T, Utada H (2006) Hydrogen diffusivity in wadsleyite and water distribution in the mantle transition zone. Earth Planet Sci Lett 243:141–148. doi:10.1016/j.epsl.2005.12.035
6. Houser C (2016) Global seismic data reveal little water in the mantle transition zone. Earth Planet Sci Lett 448:94–101. doi:10.1016/j.epsl.2016.04.018
7. Huang X, Xu Y, Karato SI (2005) Water content in the transition zone from electrical conductivity of wadsleyite and ringwoodite. Nature 434:746–749.doi:10.1038/nature03426
8. Jacobsen SD, Smyth JR (2006). Effect of water on the sound velocities of ringwoodite in the transition zone. Earth's Deep Water Cycle. In: Jacobsen SD, van der Lee S. Geophys Monogr Ser, vol. 168. pp. 131-145, AGU, Washington, DC
9. Karki BB, Bhattarai D, Mookherjee M, Stixrude L (2010) Visualization-based analysis of structural and dynamical properties of simulated hydrous silicate melt. Phys Chem Minerals 37:103–117
10. Kavner A (2003) Elasticity and strength of hydrous ringwoodite at high pressure. Earth Planet Sci Lett 214:645–654. doi:10.1016/S0012-821X(03)00402-3
11. Kelbert A, Schultz A, Egbert G (2009) Global electromagnetic induction constraints on transition-zone water content variations. Nature 460:1003–1006
12. Khan A, Kuvshinov A, Semenov A (2011) On the heterogeneous electrical conductivity structure of the Earth's mantle with implications for transition zone water content. J Geophys Res Solid Earth 116:B01103. doi:10.1029/2010JB007458
13. Kresse G, Hafner J (1993) Ab initio molecular dynamics for liquid metals. Phys Rev B 47:558
14. Kresse G, Joubert D (1999) From ultrasoft pseudopotentials to the projector augmented-wave method. Phys Rev B 59:1758

15. Liebske C, Frost DJ (2012) Melting phase relations in the MgO–MgSiO3 system between 16 and 26GPa: implications for melting in Earth's deep interior. Earth Planet Sci Lett 345–348(C):159–170. doi:10.1016/j.epsl.2012.06.038
16. Mookherjee M, Stixrude L, Karki BB (2008) Hydrous silicate melt at high pressure. Nature 452:983–986
17. Ohtani E, Zhao D (2009) The role of water in the deep upper mantle and transition zone: dehydration of stagnant slabs and its effects on the big mantle wedge. Russian Geol Geophys 50:1073–1078. doi:10.1016/j.rgg.2009.11.006
18. Olsen N (1998) The electrical conductivity of the mantle beneath Europe derived from C-responses from 3 to 720 hr. Geophys J Int 133:298–308
19. Panero WR (2010) First principles determination of the structure and elasticity of hydrous ringwoodite. J Geophys Res 115:B03203. doi:10.1029/2008JB006282
20. Panero WR, Smyth JR, Pigott JS, Liu Z, Frost DJ (2013) Hydrous ringwoodite to 5 K and 35 GPa: multiple hydrogen bonding sites resolved with FTIR spectroscopy. Amer Mineral 98:637–642. doi:10.2138/am.2013.3978
21. Pearson DG, Brenker FE, Nestola F, McNeill J, Nasdala L, Hutchison MT et al (2014) Hydrous mantle transition zone indicated by ringwoodite included within diamond. Nature 507:221–224. doi:10.1038/nature13080
22. Perdew JP, Burke K, Ernzerhof M (1996) Generalized gradient approximation made simple. Phys Rev Lett 77:3865–3868
23. Schweiger E, Sharma M, Gygi F, Galli G (2008) Melting of ice under pressure. Proc Nat Acad Sci 105:14779–14783
24. Shimizu H, Utada H, Baba K, Koyama T, Obayashi M, Fukao Y (2010) Threedimensional imaging of electrical conductivity in the mantle transition zone beneath the North Pacific Ocean by a semi-global induction study. Phys Earth Planet Inter 183:252–269
25. Smyth JR, Kawamoto T, Jacobsen SD, Swope RJ, Hervig RL, Holloway JR (1997) Crystal structure of monoclinic hydrous wadsleyite [β-(Mg, Fe)2SiO4]. Amer Mineral 82:270–275
26. Sun W, Yoshino T, Sakamoto N (2015) Hydrogen self-diffusivity in single crystal ringwoodite: implications for water content and distribution in the mantle transition zone. Geophy Res Lett 42:06448. doi:10.1002/2015GL064486
27. Tse JS (2002) Ab initio molecular dynamics with density functional theory. Ann Rev Phys Chem 62:249–290. doi:10.1146/annurev.physchem.53.090401.105737
28. Tsuchiya J, Tsuchiya T (2009) First principles investigation of the structural and elastic properties of hydrous wadsleyite under pressure. J Geophys Res 114: B02206. doi:10.1029/2008JB005841.
29. Verma, AK, Karki, BB (2009) First-principles simulations of native point defects and ionic diffusion in high-pressure polymorphs of silica. Phys Rev B 79:214115.
30. Yoshino T, Matsuzaki T, Shatskiy A, Katsura T (2009) Earth and planetary science letters. Earth Planet Sci Lett 288:291–300. doi:10.1016/j.epsl.2009.09.032.

INDEX